# Process Development Division



**Core Programming Topic at the 2012 AIChE Annual Meeting**

## Pittsburgh, Pennsylvania, USA
## 28 October - 2 November 2012

**ISBN: 978-1-62276-738-0**

**Printed from e-media with permission by:**

Curran Associates, Inc.
57 Morehouse Lane
Red Hook, NY 12571



Printed by Curran Associates, Inc. (2013)

For permission requests, please contact AIChE
at the address below.

AIChE
3 Park Avenue
New York, NY 10016-5991

Phone: (203) 702-7660
Fax: (203) 775-5177

www.aiche.org

**Additional copies of this publication are available from:**

Curran Associates, Inc.
57 Morehouse Lane
Red Hook, NY 12571 USA
Phone: 845-758-0400
Fax: 845-758-2634
Email: curran@proceedings.com
Web: www.proceedings.com

# TABLE OF CONTENTS

# (43a) Looking to the Future in Continuous Pharmaceutical Manufacturing

**Richard D. Braatz**, Chemical Engineering, Massachusetts Institute of Technology, Cambridge, MA

This presentation describes the motivations, challenges, and opportunities in continuous pharmaceutical manufacturing. Some of the motivations are particularly intriguing, such as semicontinuous and continuous-flow unit operations that have the potential to have order-of-magnitude higher controllability and/or are much easier to scale up than batch and semibatch processes. Theoretical, computational, and experimental results are provided as evidence of the potential opportunities of continuous-flow pharmaceutical manufacturing operation. One of the conclusions of reviewing such unit operations is that they can look and operate very different from the batch and semibatch unit operations currently used in the pharmaceutical industry.

Example unit operations are drawn from the published literature, which include both journal papers and patents from various academic and industrial research groups. The presentation will also review and comment on the application of Quality-by-Design methodologies to pharmaceutical manufacturing. Of particular interest is the generation of design spaces, and the pros and cons of different approaches. In particular, it is shown that many of the methods for the construction of design spaces that have been proposed in the literature are useless from a control or operations point of view.

# (43b) New Challenges and Opportunities for Pharmaceutical Manufacturing Science

**Vincent Vilker**, Office of Testing and Research, FDA, Chevy Chase, MD and Mansoor A. Khan, Cder/Ops/Otr/Dpqr, Food and Drug Administratin, Silver Spring, MD

New Challenges and Opportunities for Pharmaceutical Manufacturing Science

Vincent L Vilker and Mansoor A Khan, CDER/FDA

The increasing sophistication of modern drug products is leading to unprecedented challenges and opportunities for pharmaceutical manufacturing. At the same time, scientific and technological advancements in analytical instrumentation, coupled with implementation of the principles embedded in quality by design, enable the improved process monitoring and control that lead to better and more consistent product quality. ICH Q8(R2), the first of three International Conference on Harmonization (ICH) Guidance for Industry that spell out the concepts of Quality by Design, states that critical formulation attributes and process parameters should be identified through an assessment of the extent to which their variation can have impact on the quality of the drug product. In this presentation, examples will be given where modern instrumentation and manufacturing practices are required to maintain the critical formulation attributes of several drug products that have unique physicochemical and/or biological properties. Examples may include properties such as very low solubility, polymorphic solid states, poorly defined macromolecular size distributions, or where there is a tendency to become unstable in the formulated product.

# () Recent Advances in the Molecular Modeling Foundation of Systems-Based Pharmaceutics

**Constantinos C. Pantelides**, Department of Chemical Engineering, Imperial College London,Center for Process Systems Engineering, London, United Kingdom

There are potentially significant benefits from the adoption of a systems-based approach to pharmaceutical science and engineering based on detailed mathematical first-principles models. However, the success of such an approach is predicated on the accurate description of the thermodynamic, kinetic and mechanical properties of the materials involved.

In this presentation, we review some recent advances in the description of the solid and fluid phases arising from work by the Molecular Systems Engineering group at Imperial College London. In particular, we review the state of the art for the *ab initio* prediction of crystal structure and polymorphism, and the prediction of vapour/liquid and liquid/solid phase equilibria using the SAFT-gamma group-contribution based equation of state. Significant progress has been achieved in all these areas over the last few years, but there are also significant challenges that still need to be addressed.

# (43d) Experience with Regulatory Issues for Biomedical Devices and Vision for the Use of Advanced Control In Pharmaceutical Operations

**Francis J. Doyle III**, Chemical Engineering, UCSB, Santa Barbara, CA, Eyal Dassau, UCSB, Santa Barbara, CA and Howard Zisser, Sansum Diabetes Research Institute, Santa Barbara, CA

The landscape for clinical testing of biomedical devices has evolved rapidly over the last 5-10 years. In particular, the use of advanced control algorithms, such as model predictive control, are now being reviewed by the FDA as components of closed-loop devices. In our work with the Artificial Pancreas (AP), we have developed master files and other supporting materials to support IDE (Investigational Device Exemption) studies. In this talk, I will share some of the challenges in developing algorithms for regulatory review, including the role of simulation studies, communication architectures, algorithmic design, and IDE studies. I will extrapolate from these experiences to project the opportunities for advanced control in pharmaceutical operations.

## Chemical Engineering & Biopharmaceutics: Linking the Process with the Ultimate Quality Attribute

Ravi M. Shanker

The discovery and development of heterocyclic, small molecule new drug entities is a highly sophisticated, multi-disciplinary, scientifically advanced, collaborative, highly competitive, extremely expensive and multi-year endeavor. The heavily R&D dependent pharmaceutical industry utilizes cutting edge scientific and technological tools and techniques encompassing the broad disciplines of chemistry, biology and medicine. To provide significantly improved therapeutic treatments for various acute and chronic diseases and to be able to meet the requirements and expectations of the regulatory agencies, the patients and the society is a constant challenge for the innovative pharmaceutical industry. The challenges forge deeper dives in science among the wide array of disciplines that constitute the pharmaceutical industry and consequently have provided tremendous opportunities for the advancement of computational and predictive tools within each of these disciplines and sub-disciplines. A key segment of R&D and industrial scale manufacture is pharmaceutical sciences. It is an amalgamation of applied disciplines of chemical engineering, physical and physical organic chemistry, synthetic organic chemistry, process chemistry, analytical chemistry, polymer chemistry, material sciences, crystallization science, drug product design and manufacture and evaluation of performance of drug product in the patients. Pharmaceutical sciences have also embraced the development and use of predictive computational tools among each of its areas of specialization and disciplines to ensure the highest quality of drug product being delivered to the patient. The multi-disciplinary components of pharmaceutical sciences are operationally integrated to provide the highest quality of drug product to meet the regulatory requirements in this highly regulated industry. However, there is a lack of integration of such predictive tools across disciplines to enable the assessment of local perturbations in quality during any segment of work flow on the ultimate quality and performance of the drug product. The ultimate quality attribute of a drug product is it's performance in patients, however, linking that performance to the upstream attributes of drug substance and drug product via engineering control during manufacture with emphasis on the principles of chemical engineering continues to be unrealized and unfulfilled possibility. The vision to provide such integrated approach of existing computational and predictive tools within disciplines and sub-disciplines in pharmaceutical sciences and to provide opportunities for further development of advanced integrated tools as a holistic system is the underlying concept for the creation of Systems Based Pharmaceutics.

# (46a) Automating Chemical Kinetics Development and Analysis with the Virtual Kinetics Laboratory

Alex Kalos, Process Optimization, Engineering Sciences, The Dow Chemical Company, Freeport, TX

The Virtual Kinetics Laboratory (VKL) is a software environment that automates the development and analysis of reaction kinetic models for the purpose of materials discovery. VKL automates the process of transforming reaction mechanisms into systems of differential and algebraic equations, solving them efficiently, estimating the model parameters based on experimental data, discriminating competing models and statistically analyzing the results. Researchers may postulate ad-hoc chemical reaction mechanisms, or interface to Reaction Mechanism Generators (RMG) in order to evaluate competing multiple reaction pathways. In addition, users can choose from a variety of pre-defined reactor configurations (batch, PFR, CSTR or their combinations), or use program facilities to manually specify other more specialized configurations. Advantages over traditional kinetic analysis packages include: (1) an integrated set of tools for automating model building, (2) ability to automatically sift through and evaluate multiple candidate mechanisms, (3) customized automated data analysis of high-throughput systems such as micro-reactors, and (4) ease of deployment via a user-oriented easy-to-use, web-based interface implemented in webMathematica.

# (46b) Electric Power Generation with Sulfur-Based Thermochemical Energy Storage for Concentrated Solar Power

**Robert T. Buckingham**, General Atomics, San Diego, CA

A preliminary design of a Concentrated Solar Plant (CSP) integrated with sulfur-based Thermochemical Energy Storage (TES) for baseload electric power generation has been established. Solar heat is stored in elemental sulfur via solar decomposition of H2SO4 and disproportionation of SO2 into S and H2SO4. The system is capable both diurnal and seasonal storage. Thermodynamic modeling shows that high system pressures, high H2O:SO2 ratios, and low system temperatures drive the SO2 → S conversion. The requirements for high system pressures and high H2O:SO2 ratios have been validated by laboratory experiments. Electric power is generated by a combined-cycle power plant utilizing a sulfur-burning gas turbine and a steam turbine plant. The steam plant is integrated into the process for converting SO2 to SO3, and subsequently to concentrated sulfuric acid, as is done in sulfuric acid manufacturing. This CSP process is unusual in that the peak temperature of the heat used to generate electric power is higher than the peak solar temperature required to collect and store solar thermal energy. This is because sulfur combustion is conducted at temperatures near 1200°C, while sulfuric acid decomposition typically occurs at or below 900°C. A process flowsheet of this TES cycle will be presented, with thermodynamic analysis of efficiency and exergy considerations.

## Effective Global Process Development Team Interactions - From the Lab to the Plant and All Things Between

Paper No. 263095

*John E. Pace (SABIC, Parkersburg, WV), David B. Hall (General Electric, Niskayuna, NY), Jan van den Heijkant (SABIC, Cartagena, Spain), Jose Maria Patricio (SABIC, Cartagena, Spain), Sivakumar S (General Electric, Bangalore, India)*

We are all increasingly aware of some of the fundamental effects of the increased globalization of commerce during the past twenty years. From companies moving their manufacturing facilities to off-shore locations in pursuit of lower labor costs, to the globalization of the banking system and financial markets; significant changes have occurred in how business is conducted. As with profound and unforeseeable changes that often result when new business processes are introduced in a business setting, there have been many changes in business practices that have emerged from these globalization trends, many of which may not have been anticipated. An important change that has occurred within the chemical process industry is the way in which new process technologies are developed and commercialized. Global competition requires that current processes be continuously improved and that new processes be developed as needed. This enables the cost-effective production of existing products as well as the introduction of new products, deliverables which are necessary to improve global competitive positioning.

Prior to the globalization of the last twenty years, a “global” company usually had a manufacturing and/or a sales presence in several countries. The process development activities of those companies were, more often than not, regionally concentrated within a single country and primarily located in central research and development facilities. Companies with R&D facilities in multiple global locations tended to focus those centers on servicing their local markets rather than interacting globally. Within each center, essential communications and coordination of activities happened face to face. Most of the developers involved in the projects were usually located at the same site where the initial laboratory work and the subsequent pilot plant work were conducted, thereby allowing all of the developers the opportunity to participate in or witness the experimental work being done first hand. All of the developers involved could have a detailed understanding of the work being done, a situation that enhanced their collaboration on improvements and created a pool of technical resources for supporting the eventual commercial design and start-up.

During the past twenty years, globalization and other economic factors within the CPI have resulted in several changes in the way processes are being developed. There are many cases where large centralized research and development centers were downsized or eliminated completely in cost saving efforts. In order to offset the loss of centralized development resources, companies have employed a number of lower cost options to fill the void. Many companies moved a portion of their development activities to lower cost foreign locations. These new locations were chosen based on having relatively lower professional labor rates and also on the presence of a good supply of highly trained chemists and engineers.

In order to reduce the overhead costs of conducting developmental programs completely internally, companies have also employed a variety of measures that take advantage of outside resources to complement their internal development organizations The need to exploit outside resources also arises from the fact that this often can expedite the development work. Outside laboratory and pilot facilities can supplement internal ones, which may be fully utilized or may not have all of the required equipment in place. In addition, some of the requisite expertise may not be available internally but may be found outside of the organization. Participation in the project by individuals with technical backgrounds and work experiences that differ from those of the internal contributors can bring fresh thinking and new insights, which are catalysts for more effective development. Realizing these potential benefits, companies can choose to contract portions of the developmental work on large projects to outside companies that specialize in development and scale-up of chemical processes. Cooperation between companies is becoming more commonplace, in the forms of industry consortiums as well as joint development programs. Companies also partner with universities in programs for investigation of fundamental concepts, laboratory development, and pilot scale studies. This trend is continuing toward the establishment of universities with the equal dual objectives of advanced technical education and development of technologies for implementation within specific industries or locations. These can allow significant additional opportunities for collaboration between industry and universities. King Abdullah University of Science and Technology (KAUST) in Saudi Arabia is an example of this type of collaboration, involving many global companies including SABIC.

One result of looking increasingly to these decentralized development resources has been the distribution of the individuals and functions involved in project execution between multiple locations, and in many cases they are spread around world. This dispersion of project personnel and resources acts to increase some of the traditional challenges in development project execution. Just as change always introduces challenges, it also generally presents new opportunities. It is important to recognize the opportunities and to take advantage of them as the challenges are dealt with. Failure to take advantage of these opportunities, as well as failure to overcome the challenges, can significantly affect the quality of the end product, the commercialized process.

As some of the challenges and opportunities involved in global process development programs are examined, experience gained during a recent SABIC project provides examples of some useful and effective practices. The SABIC project involved development of a completely new process, using many of the decentralized resources discussed above. The initial proof of concept, much of the development of the chemistry of the process, and some of the initial piloting was done at General Electric's Global Research (GR) facility in Niskayuna, New York. (The Innovative Plastics business unit of SABIC was acquired from GE in 2007). At the time the initial work was being done, GE GR was an example of a totally centralized corporate research facility. In 2000 GE opened the John F. Welch Technology Centre (WTC) in Bangalore, India, recognizing the need to take a more global approach with its corporate R&D organization, and taking advantage of lower labor rates and access to an excellent resource of highly trained technical professionals. Part of the reaction optimization and much of the pilot work for a part of the new process was done at WTC.

Several external resources were utilized in portions of the developmental program as well. A portion of the piloting was done through an outside contractor. Part of the reaction optimization and piloting was done through a joint development program with another company who would later supply one of the raw materials for the new plant. Proof of concept and laboratory studies supporting development of a novel approach for separation of a highly non-ideal mixture were contracted with a university having faculty with extensive separations expertise.

This example illustrates how development project resources can become widely dispersed; requiring special emphasis to coordinate them and to maintain the proper focus so that the project schedule is maintained and costs are controlled. The challenges to be recognized and dealt with on global process development projects are not unique to these projects, but are associated with the basic elements of all project work. The challenges related to communications, collaboration, talent, and teamwork are increased by the distribution of the work over great distances. We examine these in more detail below.

## Communication Plans that Work: Effective Practices & Resources

Good communication is vitally important in the execution of projects in general. Communications between contributors working on a common portion of a project, including involvement of key project stakeholders (i.e. manufacturing, product development, design engineering, project management), are important in order that experimental plans are developed with the input of all involved, and that the data analysis and interpretation are known and understood by everyone. This feeds into the development of the plan for the next experiments and allows for changes in the experimental path as necessary. These communications are relatively easier when all of those involved are located at the same site. When contributors and stakeholders in the work are located at multiple sites, clear and frequent communications are vitally important. If contributors are not in close communication between the time of the generation of data and the discussion of plans for the next experiments, time can be wasted and working relationships can become strained. An experimental plan that is developed and executed by a portion of the team, without adequate involvement of all of the team members, risks being flawed due to unavailability of pertinent information at the time it is being developed. Data generated and conclusions drawn can later be called into question. The work can follow paths that are not useful if there is not frequent stakeholder involvement. Time is wasted if the work has to be redone, and an “us vs. them” mentality can develop between groups of contributors at different locations. All of this is very counterproductive and can cause project delays and missed objectives. Global projects insure that the contributors and stakeholders will be widely dispersed, making it very important to plan for effective communications.

One approach used during the SABIC project was weekly conference calls. These calls were kept to an hour in length, but involved in-depth discussion of the experimental results of the previous week. In one example process developers in India (WTC), doing laboratory and pilot work supporting implementation of the technology in the planned commercial plant, discussed their findings with a group including:

a. other developers in the US (GE RC) who had done the original conceptual development work
b. developers who were working on different parts of the same process, where coordination of efforts was essential
c. manufacturing engineers providing critical input on points of practical application in the plant
d. a representative from the design engineering team to give design critical input where necessary.

These stakeholders talked weekly for several years during the development, design, construction, and start-up phases of the project. Membership evolved over time depending on the work being done and the stage of the project. The nature and purpose of the interactions also evolved along with the maturity of the project.

On the initial calls, the GE RC members of the team coached the WTC members as they were building and starting to operate their lab and pilot facilities to continue the development work needed for commercialization. Weekly discussion and critique of new experiments helped to be sure that the basics of the technology were understood by the WTC team and that the results were consistent with the earlier work at the GE RC. On later calls, after the WTC team had developed a good working understanding of the process, the GE RC members continued on the calls in more of a collaborative role. The regular participation of representatives from the manufacturing team allowed them to learn the details of the technology and also to have the opportunity to influence the content of the experimental plans to include studies which would prove to be critical for plant implementation and successful start-up. The representative from the project and design staff had the opportunity to communicate timing needs for information critical for design as well as to bring new information needs to the attention of the group. After the basic technology was fully developed and the design process was well underway, the weekly discussions focused more on laboratory and pilot work that would support the plant start-up. Start-up and shutdown strategies were tested and optimized. A wide ranging list of "what if's" was developed, forming the basis for work to gain an understanding of the best responses to a variety of upset conditions and variations in raw material quality. The longer term experimental plan was regularly reviewed during these calls to make any changes suggested by the most recent results.

One of the keys to success for these globally diverse calls is an open give and take during the discussions. Care should be taken that questions are well understood and that the discussion is clear to everyone. Active listening is helpful; paraphrasing key statements or conclusions back to presenters to be sure they were correctly understood. This additional care should be taken to be sure that misunderstandings based on language or cultural differences are avoided. Differing opinions on the interpretation of data should be either resolved on the calls or noted for further study in future experiments. All participants should feel comfortable in questioning conclusions and in contributing to the discussion, which are keys to the success of these global calls. Everyone should feel confident their opinions will be heard and logically discussed. It is very important for all of the members of the conference calls, across all of the geographical and organizational lines, to feel a unity of purpose and to feel joint ownership in the content and conclusions of the discussions. It is extremely helpful if all of the participants know each other well. This can be difficult to achieve on global projects, and is addressed separately in the next section.

On the SABIC project these routine conference calls were facilitated by a principal technologist, who set the tone for the discussions with a clear agenda for each call. The facilitator brought in additional expertise where needed to help resolve difficult technical challenges. Having one person oversee these interactions during the entire project execution helps provide continuity so that key action items regarding identified project risks are not overlooked and are adequately addressed.

## Effective Collaboration Practices: Project Strategy Meetings

While the opportunity for frequent communications through the conference calls is very important for the tactical aspects of the project work, face to face meetings are critical for effective strategic planning. When all project participants are located at the same site, or at multiple sites within close geographic proximity, face to face contact can easily be a frequent, if not daily, occurrence. When project collaborators are spread across the globe, face to face meetings need to be planned for as a part of the project schedule. It is important to take the time to review the state of the experimental work as a whole. These meetings can be opportunities to discuss gaps in the understanding and significant issues remaining to be overcome. This leads to discussion and agreement on the implications for the future direction of experimentation. With participants gathered together it is much easier to have meaningful discussions on the longer term direction needed for the experimental work. Group exercises for finding and discussing potential failure modes within the process being developed are very useful tools for prioritizing the work remaining to be done. These reviews require significant time for preparation and execution, but they are critical for maintaining the correct focus for the work.

On the SABIC project, these periodic on-site meetings were referred to as summit meetings. Typically there were two summit meetings per year with each one lasting three to five days. Most of the contributors in the process work being reviewed attended these meetings. Representatives from manufacturing, project management, and engineering design attended, for the input they could provide and for their understanding of the implications of the experimental work being reviewed. Process development was approximately half of the content of the summit meetings, the other half being product development. Product developers and representatives from marketing and product management attended the summit meetings for reviews of the development status and for planning the remaining product work that was being done in parallel with the process work. Having the process and product discussions at the same site allowed valuable interchange and coordination between the two groups.

The sites of these meetings were rotated among the locations where the participants were based. By varying the location, participants had the opportunity to see the laboratory and pilot facilities where all of the experimental work was being conducted. Familiarity with the details of the experimental set-up led to better discussions on following conference calls. In some instances, these visits allowed identification and correction of critical deficiencies in the scale-up or operating plans between the lab or pilot plant and the planned commercial facilities. Suggestions for improvements in equipment configuration or in experimental techniques often resulted from these on-site tours and discussions. Prior to adjournment the meeting participants agreed on the objectives, priorities, key deliverables and timing for the experimental work to be accomplished in the next six months at each of the development locations. Agreement between all of the key stakeholders in the project was a key to maintaining a common vision for the path forward.

Development of relationships with other participants was an additional benefit of the summit meetings. Interactions over dinner, weekend sightseeing trips, and other activities outside of the conference rooms were informal ways that the participants could get to know each other better. Enjoyment of time spent together as a group is helpful for building a spirit of camaraderie and unity of purpose. This happens naturally when co-workers are co-located, but opportunity for this informal interaction should be provided when they are widely dispersed. This aspect is often overlooked, but is very helpful in the development of global teams that work together effectively.

## Talent and Teamwork: Engaging the Right Team Early

The availability of process talent and experience can be a challenge for the successful long term operation of any plant. Opportunities can be missed and small issues can grow into significant ones if personnel with sufficient knowledge and experience with the process are not readily available for support of the plant, both in day-to-day operation as well as for longer term improvements. When the plant is located far away from those individuals who have an in-depth knowledge of the process technology, the risk is greater for evolution toward sub-optimal, and in some cases unsafe, operation. This is another example where planning for the challenges of a global operation can reap dividends long into the future.

Experienced manufacturing process engineers and process chemists are vital for the safe and productive operation of any plant. One of the best ways to insure that these members of the manufacturing team are knowledgeable in the process prior to start-up is to involve them as early as practical during the development project. These individuals can provide extremely valuable input for the process developers regarding the needs for adequate attention to projected operating and capital costs, plant operability, flexibility, quality control and process safety. Taking advantage of the manufacturing staff's insight into plant operations is valuable in selecting between various options for the unit operations in the process configuration to be piloted. The engagement with the development team allows these engineers and chemists to gain understanding and insight into the process that they otherwise might only gain after plant start-up. Their involvement in the development work, and having the opportunity to influence the work being done, increases their sense of ownership in the final result.

An operations manager and manufacturing engineers and chemists were selected mid-way during the SABIC development program. These individuals were selected from experienced manufacturing personnel working at the existing manufacturing site in Spain, where the new plant was to be built. They and some of the lead operators spent several weeks working in the pilot plants in India and in the US, gaining hands-on experience with the process technology. These same individuals were extensively involved in the basic and detailed design phases of the project, being able to call on the insight they had gained to make significant contributions to the quality of the final design. During the final stages of the development work they were crucial in providing input on the weekly calls regarding additional pilot work that was needed to optimize start-up and shutdown procedures, set specifications on key operating conditions for stable operation, and for determining optimal recovery strategies from "what-if" upset conditions. During the early operation of the plant there were several instances where having this knowledge before start-up saved time and money when significant issues were avoided and time was not required to determine the best responses while the plant was running. The travel costs associated with these trips can be significant, but making provision for it in the project budget can be easily justified by the positive effect it has on the quality of the finished plant and on its operation. The SABIC experience has been that the manufacturing personnel involved in this way felt ownership in the success of the plant and were in an excellent position to train those who would follow them when they eventually moved on to other responsibilities.

Care should be taken on global development programs to provide a good basis for long term excellence in the technical support and continuing process improvement work that is necessary after plant start-up. Development personnel supporting a process are often located at a different site than the plant. In the case of global projects this may mean the development support is in a different country and can be separated from the plant by a great distance. If care is not taken to compensate for this physical separation, the development resources can become somewhat detached from the realities of the plant operation and process issues, and can fail to recognize potential development opportunities. Plant personnel can feel isolated from their support resources. They may, out of necessity, make significant changes in the process or in targeted operating conditions, in an effort to deal with problems or opportunities they recognize. This can sometimes lead to unintended changes in the product quality or in the robustness of the process. Process technology support located at or near the plant site can help to reduce the occurrence of these issues by allowing for more frequent interaction between plant and support personnel. This allows a clear and common understanding of process issues, recognition of opportunities for improvement, and agreement on the best path to accomplish changes. This communication can certainly be accomplished without a permanent on-site presence, but at the risk of issues if discussions are not frequent and if site visits are not routine.

Resident technology support resources for the SABIC plant were recruited and trained well in advance of the plant start-up. Senior chemical engineering and chemistry students from Spanish universities were interviewed and several were selected for participation in SABIC's Master's program. Tuition for their master's degree was fully paid by SABIC in return for a commitment to SABIC after graduation. These individuals generally worked for approximately one year at the Mt. Vernon, Indiana SABIC location where much of the development work was done and many of the experienced developers were based. The project work in Mt. Vernon allowed the master's graduates to gain a solid understanding of the process technology and to develop strong working relationships with many of the technologists they would collaborate with during and after plant start-up. These individuals also traveled to other locations where development work was being done, and participated in the work at those facilities to gain additional experience. They were transferred to the site in Spain in time to contribute to preparations for start-up. Laboratory and piloting facilities were provided at the plant site to allow additional experimentation in preparation for start-up and for longer term support of the plant. These engineers and chemists worked closely with the manufacturing team during start-up and formed the nucleus for ongoing process technology support afterwards. The combination of the early identification of a strong manufacturing team and the development of a capable resident technology team laid the groundwork for long term process excellence at the plant site. Provision for the members of both teams to work closely together, beginning during the piloting stage of the development work, resulted in all members feeling a shared ownership of the results and a solid commitment to the success of the project. The working relationships developed in this way paid dividends during the design, commissioning, and start-up phases of the project, when the presence of a strong cross-functional and multidisciplinary team was most crucial.

## Conclusion

Today's globally competitive environment requires that chemical process development projects be undertaken more frequently, completed in a more timely manner, with costs controlled more effectively than at any time in the past. Companies have a number of options, internally as well as externally, to leverage resources in ways that allow them to meet the timing and cost requirements of their development programs. The dispersion of portions of the overall development and subsequent implementation programs among multiple internal as well as external resources requires renewed focus on some of the key elements of traditional program management. If done effectively, not only can this diversity of resources make it easier to meet the project timing and cost objectives, it can also help improve the quality of the finished product, the operating manufacturing plant. While there is always room for improvement in execution of these projects, the cited examples of some of the methods used by SABIC to meet the challenges of global development projects are offered in the hope they can be helpful for others.

# (46d) Developing Sustainable Chemical Processes: An Integrated Approach Involving Concurrent Catalyst, Media and/or Reactor Innovations

**Bala Subramaniam**, Department of Chemical and Petroleum Engineering, Center for Environmentally Beneficial Catalysis, University of Kansas, Lawrence, KS

The increased need for sustainable chemical manufacturing processes provides many challenges/opportunities for developing novel technologies [[1]]. Sustainable alternatives to conventional chemical technologies can be developed by either developing greener process technologies based on existing feedstocks or by replacement of fossil based feedstocks with renewable ones such as those derived from biomass. Processing renewable feedstocks to produce chemicals not only requires the development of new chemistries but it is also imperative that the new conversion technologies are sustainable to preserve the "green" potential of biomass based feedstock.

At the CEBC, an integrated approach that fully exploits the synergies between catalyst design, tunable reaction media (such as supercritical fluids, gas-expanded liquids, ionic liquids) and novel reactor concepts (integrating reaction and separation) has been employed to developed alternate technology concepts for a variety of industrially important reactions. The demonstrated advantages include process intensification at mild conditions, enhancing process safety by elimination of flammable vapor phases, and waste mitigation by reduced usage of organic solvents and suppression of side reactions that generate undesired byproducts such as $CO_2$. Specific examples of novel process concepts to be discussed include the highly selective hydroformylation of higher olefins at mild conditions employing soluble polymer-attached Rh catalysts that are easily retained in solution while the product is isolated by membrane filtration [[2]]; a novel homogeneous ethylene oxide process virtually eliminating $CO_2$ formation as a byproduct [[3]]; and a one-step spray reactor concept to overcome gas-liquid transport limitations for the inherently-safe formation of polymer-grade terephthalic acid from *p*-xylene at high yields and purity [[4]]. The significant role of *quantitative* sustainability analyses (economic and environmental impact) in guiding the development of these processes will also be discussed.

**References**

---

[1]. B. Subramaniam, *Industrial & Engineering Chemistry Research,* **49**, 10218-10229 (2010); B. Subramaniam and G. R. Akien, *Current Opinion in Chemical Engineering*. **DOI:** http://dx.doi.org/10.1016/j.coche.2012.02.005 (in press)

[2]. Fang, J., Jana, R., Tunge J. A. and Subramaniam, B. *Applied Catalysis A: General.* 393, 294 (2011).

[3]. Lee, H. –J., Ghanta, M., Busch, D. H. and Subramaniam, B. *Chem. Eng. Sci.* 65, 128 (2010); M. Ghanta, H-J Lee, D. H. Busch and B. Subramaniam, *AIChE Journal* (in press).

[4]. Zuo, X., Niu, F., Snavely, W. K., Subramaniam, B. and Busch, D. H. *Green Chem*, 12, 260 (2010).

# (46e) Polar Functionalization of Polypropylene with Suppressed Molecular Weight Reduction: Near-Ambient Temperature Free Radical Chemistry During Solid-State Shear Pulverization

**J. Mirian Diop**, Department of Chemical and Biological Engineering, Northwestern University, Evanston, IL and John M. Torkelson, Departments of Chemical and Biological Engineering, Materials Science and Engineering, Northwestern University, Evanston, IL

Solid-state shear pulverization (SSSP) is a novel processing method, which utilizes a modified twin-screw extruder that operates under near-ambient temperature conditions. During SSSP, polymers are processed in their solid state using high compressional and shear forces that cause repeated fragmentation and fusion of the material. Previous work has shown that intimate mixing in polymer blends as well as effective dispersion and exfoliation in polymer nanocomposites can be achieved with SSSP. In the present work, by taking advantage of unique chemistries associated with near-ambient temperature processing, we demonstrate our ability to achieve polar functionalized polypropylene (PP) via SSSP. Furthermore, because of the differences associated with near-ambient temperature free-radical chemistry of polypropylene and high-temperature, melt-state free radical chemistry of polypropylene, we avoid the significant molecular weight reduction of PP that accompanies commercial polar modification of PP during melt-state processing.

The conventional method of functionalizing PP with maleic anhydride (MA) is via melt processing, which results in drastic molecular weight reduction of PP and consequently the loss of material properties—relative to those of neat PP from which the functionalized PP was made. This molecular weight reduction occurs as a result of β-scission, a free-radical chemistry that is highly dependent on temperature; under the high temperature conditions utilized for melt processing (~190 °C) the degree of β-scission is high. Here, we demonstrate the utility of SSSP to achieve MA functionalization of PP while suppressing molecular weight reduction due to β-scission. We have also identified our ability to functionalize PP with carbonyl groups via SSSP; this functionalization is also achieved with suppressed molecular weight reduction. Results from Fourier transform infrared spectroscopy show that carbonyl functionality cannot be obtained using conventional high temperature processing methods (when working with the same feed material as that which was used in SSSP to obtain carbonyl functionalized PP). Our success in these systems is as a result of our ability to take advantage of the unique chemistries associated with the near-ambient temperature conditions under which SSSP is carried out. Samples of MA and carbonyl functionalized PP were reacted with a fluorescent dye, 1- pyrenemethylamine (Pyr-$MeNH_2$). Fluorescence spectra suggest that both of these samples are reactive with Pyr-$MeNH_2$ and could potentially be used interchangeably in applications where MA modified PP has current or potential utility.

# (64a) Novel Molecular Design Techniques for the Design of Systems Involving Chemical Reactions

Nishanth G. Chemmangattuvalappil[1], Vikrant Dev[2] and Mario Richard Eden[2], (1)Department of Chemical and Environmental Engineering, University of Nottingham - Malaysia, Semenyih, Malaysia, (2)Department of Chemical Engineering, Auburn University, Auburn, AL

The recent developments in the area of product and molecular design enable us to optimize the process by incorporating product design considerations during the design stage. However, the existing techniques for solving such problems are limited to non-reactive systems. In this work, we are introducing an algorithm that can be used for solving integrated process and molecular design problems of reactive systems on a property based platform. Property clustering is a well-known technique for a component independent design. Property clustering technique will be extended into identifying a set of target properties for the components in the system that provides the optimum process performance. Once the property targets have been identified, a molecular design problem can be formulated to identify the potential candidate molecules that meet the targets identified in the process design problem. The molecular design involves the identification of potential molecules possible from the specific types of reactions in the process. To design molecules, a recently introduced concept known as molecular signature descriptors have been used. The signature is a systematic coding system of atom types and the signature of a molecule can be obtained as a linear combination of its atomic signatures (Visco *et al* 2002). The molecular signatures can be tailored to track the change in molecular groups in a molecule resulting from different types of chemical reactions. This task is achieved by forming the signatures for both reactants and products and by considering them as un-connected graphs. The changes in the chemical structure can be correlated with the changes in the properties of the molecule. Therefore, the changes in the molecular structure due to reactions can be represented as a function of the property. The developed algorithm applies different qualitative structure activity/property relationships (QSAR/QSPR) to estimate properties from the molecular structure. QSAR relationships make use of different topological indices and it has been proved that a number of topological indices of molecules can be represented in terms of molecular signatures and it is possible to correlate the topological indices to the actual properties and biological activities (Faulon *et al* 2003). Here, the new algorithm utilizes molecular property operators based on signatures for solving the inverse problem of obtaining the molecular structures that satisfy the property targets estimated in the process design step. A new set of equations will be employed to ensure that the molecule meets the safety and environmental constraints as well. The principles in graph theory are incorporated to track signatures and to avoid the generation of infeasible molecular structures. This contribution will illustrate the developed methods and highlight their use through a case study.

# (64b) Reaction Mechanism Understanding Through Process Development

**Michelle Mahoney**, Frederic Buono, Martin Eastgate, Maxime Soumeillant and Michael Cassidy, Chemical Development, Bristol-Myers Squibb, New Brunswick, NJ

A chemical transformation to aromatize a key intermediate in the synthesis of a complex API was developed rapidly to meet product deliverable timelines and quality attributes. Initial scale-up of the process beyond kilogram scale resulted in reaction stalling and revealed that further mechanistic understanding was needed. Small scale studies to define operating ranges provided insight into the reaction mechanism and kinetics and led to several changes to the reaction conditions to improve robustness. These studies included sensitivity to water, solubility of reactants in reaction mixture, reagent decomposition kinetics, use of additives, and ultimately the importance of oxygen which led to the identification of a novel aromatization protocol. Key to engineering the successful scale-up of the chemical transformation was developing mechanistic understanding of the chemistry, quickly defining the important parameters through small-scale experimentation, and leveraging this insight to design new reaction conditions. This talk will describe the development of the process and implementation of the step at 300 kg scale.

# Scaled Transport Coefficient Correlations for Fluids*

**Lawrence T. Novak† and F. Merrill Galloway+**
**Presented at the 2012 Annual AIChE Meeting, Session 64, Paper 64c, Oct. 29, 2012, Pittsburg, PA*
*†Department of Chemical and Biomedical Engineering, Cleveland State University, Cleveland, Ohio 44115-2214*
*+SABIC Technology, Houston, Texas*

**Extended Abstract for 2012 Annual AIChE Meeting Proceedings**

## Introduction

Rate-based models suitable for equipment and insitu process modeling require a capability for predicting transport coefficients over a sufficient range of temperature and pressure. This paper illustrates a relatively simple novel approach for correlating and estimating pure component transport coefficients, for self-diffusion, viscosity, and thermal conductivity. The approach is based on the idea that the magnitude of transport coefficients over the entire fluid region is influenced by fluid molecular disorder, as quantified by entropy. As a result of this idea, two or three model fitting parameters are found suitable for correlating n-alkane transport coefficient data over the entire fluid region, except possibly near the melting line and critical point. This approach should be relatively easy to incorporate in equipment and oil-gas recovery modeling software. Also, it should decrease properties database size and reduce complexity of modeling software. Pure-component transport coefficients could be used in various fluid mixture transport models that contain pure component properties. For example: liquid mixture diffusion,[1] liquid mixture viscosity,[2] and liquid mixture thermal conductivity.[3]

Historically, transport properties have been correlated with temperature and pressure, and density. The difficulties of these approaches, over the entire fluid region, are illustrated in Figures 1 and 2 for viscosity data plots of methane[4] and n-butane.[5-7]

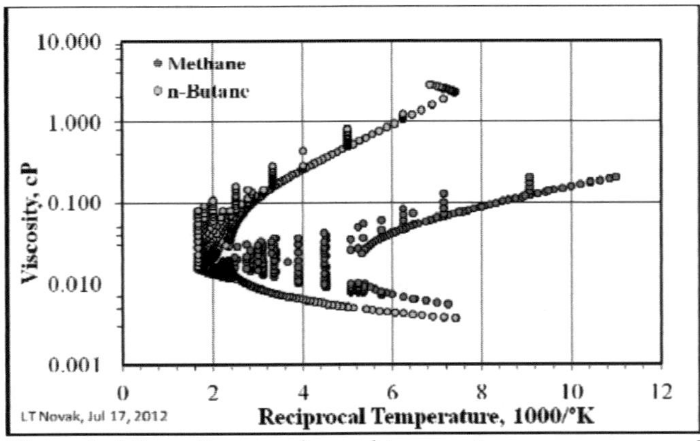


Figure 1: Viscosity vs reciprocal temperature

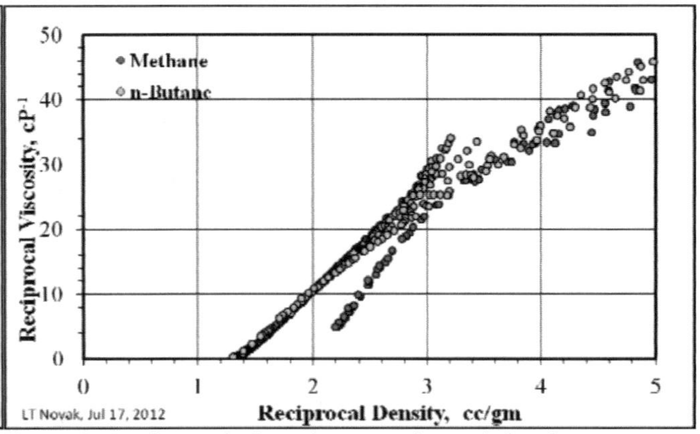


Figure 2: Reciprocal viscosity vs reciprocal density

The saturated liquid and vapor viscosity are evident in Figure 1. The elliptical region formed by the data points, at the saturation pressure, is the two phase region. Data points off the saturation curve are either in the critical region or at pressures in excess of the saturation pressure. Normally, isobaric lines would be added to Figure 1 to interpolate pressure effects. Traditional approaches in commercial simulators are to use two viscosity-temperature correlations - one for saturated liquid and another for saturated vapor. This was also the same approach used by the DIPPR project.

The above plot of reciprocals in Figure 2 is linear in the liquid region, with increasing data scatter for lower density fluids. In the early 1900's, Batshinski[8] found that many liquids produce linear plots of reciprocal viscosity versus reciprocal density. He related the x-axis intercept of reciprocal density to the Van der Waals volume parameter ("b"). This is the first instance, that this author is aware, of a relationship between viscosity and an equation of state. The work of Batshinski later led to various "free-volume" models for transport coefficients.

This paper will illustrate that the above viscosity data points in Figures 1 and 2 will all plot on a single semi-log line, over the entire fluid region when a new unifying model for transport coefficients is used. This new model results from relating Chapman-Enskog scaled viscosity to residual entropy. This relatively simple component-based modeling approach was first demonstrated for self-diffusion coefficients and viscosity.[9] Subsequently, this component-based model[9] was extended to a predictive segment-based model for viscosity.[10] We will illustrate here that this new model also describes thermal conductivity data, except for the thermal conductivity enhancement near the critical point.

In the remainder of this extended abstract, we will first illustrate our novel approach for self-diffusion coefficient, viscosity, and thermal conductivity over the entire fluid region. Extensive methane transport data will be used. Secondly, we will illustrate the component-based scaled transport coefficient model[9] for methane, propane, n-butane, and n-decane using self-diffusion coefficient and viscosity data, over the entire fluid region. Thirdly, we will demonstrate the predictability of a segment-based version of our model with n-alkanes from methane up to 2400 Mw linear polyethylene, over the entire fluid region, covering a wide range of temperature and pressure {96 °K to 650 °K and $10^{-4}$ atm to 4990 atm}.[10] We will then discuss conclusions drawn from this research and potential applications for the new unifying transport coefficient model.

**Component-Based Chapman-Enskog Scaled Transport Coefficients and Residual Entropy**

Figures 3 and 4 illustrate the Chapman-Enskog scaled transport coefficients plotted against the macroscopic variables of dimensionless density and residual entropy for the component-based model.[9] Data sources for thermal conductivity are given in references 4, 11 and 12. Although density is a function of temperature and pressure, residual entropy is a function of volume (reciprocal number density) and temperature. The use of residual entropy produces the relatively simple, two or three parameter, semi-log type curves seen below. However, this model fails to describe the thermal conductivity enhancement near the critical point. Other thermodynamic correlating variables for the x-axis, such as Helmholtz free energy, produce significant scatter.

A possible explanation for the success of residual entropy, as a transport coefficient correlation variable, is that entropy is a measure of molecular disorder. So, it is a way of quantifying molecular structure. The box in Figure 4 illustrates our simple physics-based mathematical model which is closely followed for methane viscosity over the entire fluid region.[9] There is also a consistency when using residual entropy as a correlating variable when scaling transport coefficients with Chapman-Enskog values, because Chapman-Enskog values approach the ideal gas values. This is demonstrated by asymptotes of 1.0 observed in Figures 3 and 4.

The new model, illustrated in this paper, represents a unified approach because a single model is used for the entire fluid region. This simplifies transport modeling because a single transport coefficient model eliminates the need for multiple transport coefficient models and multiple corresponding parameter data sets for saturated liquid, saturated vapor, effects of pressure, and critical phase.

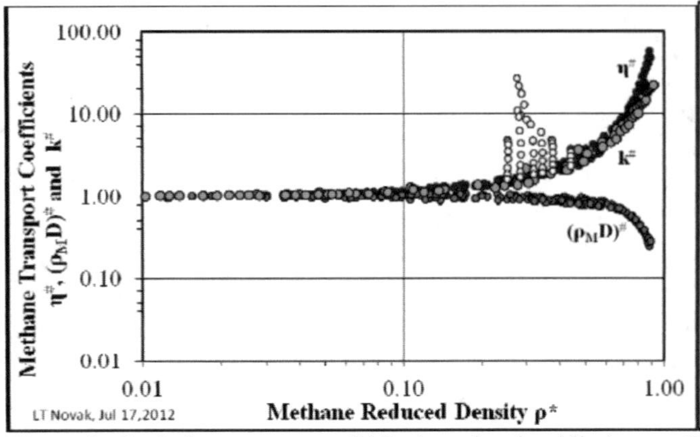


Figure 3: Scaled transport coef (dim) vs density (dim)

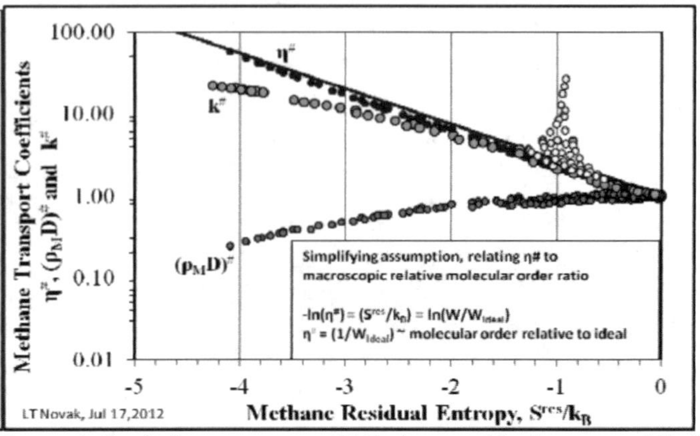


Figure 4: Scaled transport coef (dim) vs residual entropy (dim)

Using the component-based model, we next examine the effect of n-alkane chain length on viscosity and self-diffusion coefficient correlations, as shown in Figures 5 and 6. We again observe semi-log lines for scaled viscosity $(\eta^{\#})$[9] in Figure 5, with intercepts approaching one. However, the slopes increase with chain length. So, in its current form this model is not a corresponding states model.

The self-diffusion coefficient $(\rho_M D^{\#})$[9] plots in Figure 4 and 6 can be fit with a three parameter semi-log type expression, with intercepts approaching about one. A power of 1.7 to 2.0 on absolute residual entropy appears reasonable in fitting data in Figure 4 and 6.

The use of residual entropy to correlate scaled transport coefficients for <u>dense fluids (only)</u> was apparently first used by Rosenfeld in 1977.[15] He used early computer simulation data with another definition for scaled viscosity and self-diffusion for the dense fluid region ($S^{res}/k_B<-1$).[15] Novak[9] has shown the superiority of Chapman-Enskog scaling over Rosenfeld scaling, because Chapmen-Enskog scaling results in a simple mathematical expression over the entire region ($S^{res}/k_B<0$). It also has a well defined limit of 1.0 in the ideal gas case.

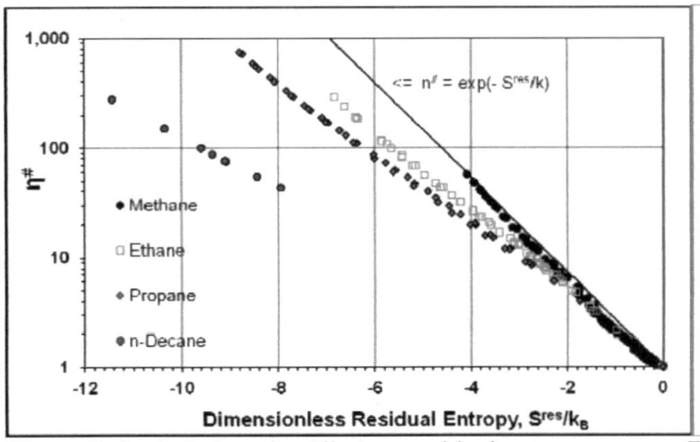


Figure 5: Scaled viscosity (dim) vs residual entropy

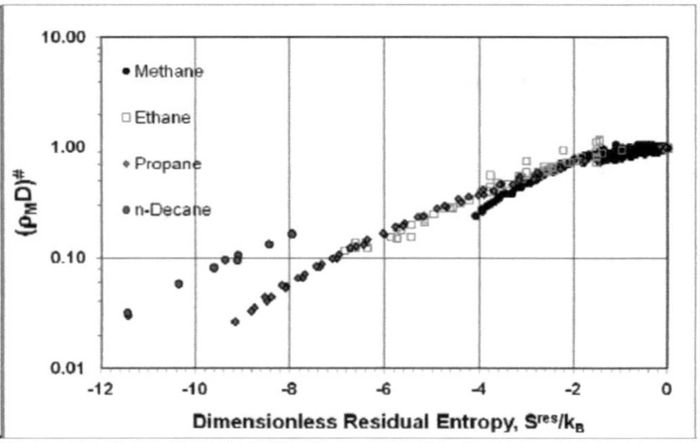


Figure 6: Scaled self-diffusion coef (dim) vs residual entropy

A limitation of the component-based approach is the need for Lennard-Jones (LJ) parameters. As molecules get large, LJ parameters are unavailable and lack a reasonable physical definition. A problem of using reference equations of state to calculate residual entropy[13] is the need for critical temperature and pressure. As molecules get large, critical properties are unavailable and may not be physically realistic. A solution to these problems is to use a segment-based approach.[2,14] Besides being more practical, it is more physically realistic and does not require Lennard-Jones parameters or critical properties.

**Segment-Based Chapman-Enskog Scaled Transport Coefficients and Residual Entropy**

The segment-based version of our model is illustrated in Figure 7 with n-alkanes from methane up to 2400 Mw linear polyethylene. Experimental data cover the entire fluid region and a wide range of temperature and pressure {96 °K to 650 °K and $10^{-4}$ atm to 4990 atm}.[10] PC-SAFT[14] was used in the calculation of segment residual entropy and PC-SAFT component segment parameters were used in the calculation of Chapman-Enskog scaling values. This assumes that PC-SAFT segment parameters represent a hypothetical size and energy parameter for describing the entity controlling transport of molecules in the dilute phase limit ($-1<S^{res}/k_B<0$). Considering the projected intercepts below, this is not the case. Two explanations may be: (1) molecular configuration in the dilute phase is different from the dense phase, with a larger size parameter, or (2) multiple segments are controlling momentum transfer in the dilute phase limit.

Fitting data for eighteen n-alkanes, ranging from methane up to 2400 Mw linear PE, produced fitting parameters that were well behaved. The semi-log slope parameters correlate with component molecular weight. And, the intercept parameters correlate with the semi-log slopes. So, a predictive equation results allowing one to calculate n-alkane viscosity over the entire fluid region, up to 2400 Mw.

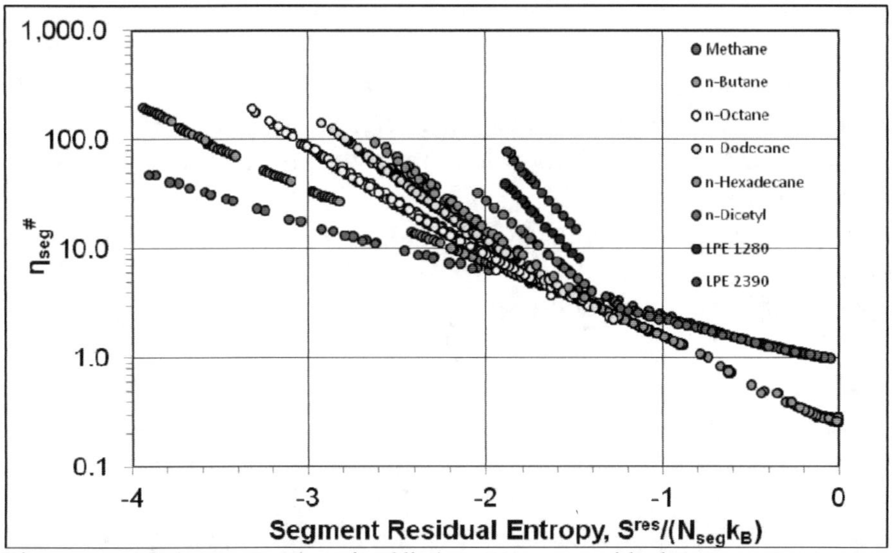


Figure 7: Scaled segment viscosity (dim) vs segment residual entropy

Figures 8 and 9 illustrate the predictability of the segment-based scaled viscosity model in the correlation mode, for the components studied. Predicted values from component correlations are plotted against actual values in Figures 8 and 9.

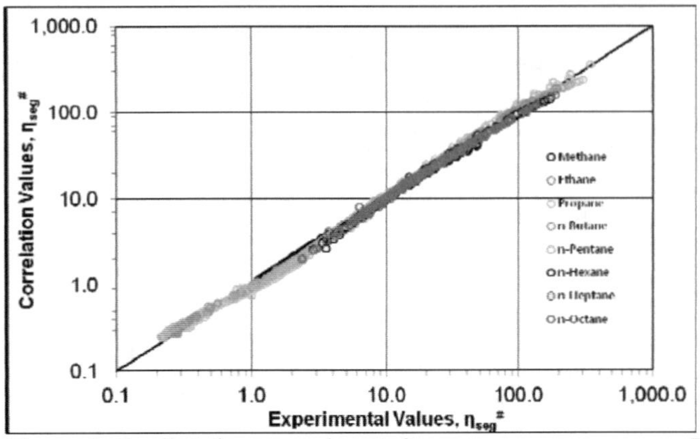


Figure 8: Predicted vs experimental

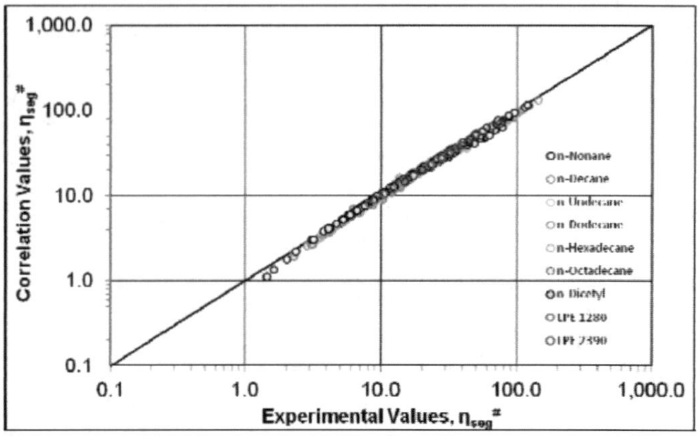


Figure 9: Predicted vs experimental



In the correlation mode, the segment-based model illustrated in Figures 7-9 was found to result in a group correlation squared ($R^2$) of -0.998 and group average absolute deviation (AAD) of 3.9%, within the range of typical viscosity measurement variation and other viscosity correlation models in the literature.[10] In the predictive mode, a group AAD of 6.7% was obtained for n-alkanes ranging from methane up to 1300 molecular weight, over the entire fluid region. The predictive mode allows one to estimate component viscosity when experimental data is limited or nonexistent.[10]

## Conclusions

The Chapman-Enskog scaling model presented here *shows promise* as:

- a unifying transport coefficient correlation model for simplifying transport modeling over the entire region, and reducing the size of corresponding model parameter databases.
- an equation of state-based transport coefficient correlation over the entire fluid region (liquid, gas, and critical fluid) for pure components.
- a relatively simple theory-based transport coefficient correlation for testing data consistency, interpolation, and extrapolation.
- most likely, an efficient transport coefficient calculation for equipment and insitu simulation models.

The segment-based Chapman-Enskog viscosity scaling model with PC-SAFT:

- is more physically realistic and more broadly applicable, for the entire fluid region and a wide range of temperature & pressure.
- does not require critical properties or Lennard-Jones parameters.
- provides correlation capability AAD's comparable to measurements and other correlation models.
- provides a predictive capability when little or no viscosity data are available.

## Applications

For industrial applications at elevated pressures, or the critical region, our model should be well suited, particularly in oil and gas recovery and supercritical separation processes. Other specific applications modeled by PC-SAFT, are listed below.[14] It is expected that the Chapman-Enskog scaled-transport vs residual entropy model described here and elsewhere[9,10], should work well for non-associating pure component molecules. Preliminary study (unpublished) suggests that the component-based model should also work well for simple like-molecule mixtures, like methane-ethane, over the entire composition range and fluid region.

Non-associating pure components with PC-SAFT:

- n-alkanes and branched alkanes, probably up to at least 2000 Mw
- cycloalkanes
- alkenes
- benzene and derivatives
- Halogenated hydrocarbons
- Ethers
- esters

Associating pure components with PC-SAFT:

- Primary alcohols and amines.
- Water
- Acetic acid

Readers interested in more historical background information on methods for correlating and estimating transport properties should consult references 1-3, 9, 10, 15-18.

**References**

(1) Bart, H-J.; Bosse, D. Prediction of Diffusion Coefficients in Liquid Systems, *Industrial & Engineering Chemistry Research*, **2006**, *45(5)*, 1822.

(2) Novak, L.T.; Chen, C-C.; Song, Y. Segment-Based Eyring-NRTL Viscosity Model for Mixtures Containing Polymers, Industrial *& Engineering Chemistry Research*, **2004**, *43*, 6231.

(3) Rowley, R.L. A Local Composition Model for Multicomponent Liquid Mixture Thermal Conductivities, *Chemical Engineering Science*, **1982**, *37*, 897.

(4) Friend, D.G.; Ely, J.F.; Ingham, H. Thermophysical Properties of Methane, *Journal of Physical Chemical Reference Data*, **1989**, *18(2)*, 583.

(5) Carmichael, L.T. and Sage, B.H. Viscosity of Hydrocarbons. N-Butane, *J. Chem. Eng. Data*, **1963**, *8(4)*, 612.

(6) Diller, D.E. and Van Poolen, L.J. Measurements of the Viscosities of Saturated and Compressed Liquid Normal Butane and Isobutane, *International Journal of Thermophysics*, **1985**, *6(1)*, 43.

(7) Kuchenmeister, C. and Vogel, E. The Viscosity of Gaseous n-Butane and its Initial Density Dependence, *International Journal of Thermophysics*, **1998**, *19( 4)*, 1085.

(8) Batshinski , Z. *Physik.Chem.*, **1913**, *84*, 643.

(9) Novak, L. Self-Diffusion Coefficient and Viscosity in Fluids, *International Journal of Chemical Reactor Engineering*, **2011**, *Vol. 9*, Article 63. Available at: http://www.bepress.com/ijcre/vol9/A63.

(10) Novak, L.T. Fluid Viscosity-Residual Entropy Correlation, *International Journal of Chemical Reactor Engineering*, **2011**, *Vol. 9*, Article 107. Available at: http://www.bepress.com/ijcre/vol9/A107'

(11) Carmichael, L.T.; Reamer, H.H.; Sage, B.H. Thermal Conductivity of Fluids. Methane, *J. Chem. Eng. Data*, **1966**, *11(1)*, 52.

(12) Sakonidou, E.P.; van den Berg, H.R.; Seldam, C.A. The thermal conductivity of methane in the critical region, *J. Chem. Phys.*, **1996**, *105(23)*, 10535.

(13) Setzmann, U.; Wagner, W. A New Equation of State and Tables of Thermodynamic Properties for Methane Covering the Range from the Melting Line to 625 K at Pressures up to 1000 MPa, *Journal of Physical Chemical Reference Data*, **1991**, *20(6)*, 1061.

(14) Gross, J.; Sadowski, G. Perturbed-Chain SAFT: An Equation of State Based on a Perturbation Theory for Chain Molecules, *Industrial & Engineering Chemistry Research*, **2001**, *40*, 1244.

(15) Rosenfeld, Y. Relation between the transport coefficients and the internal entropy of simple systems, *Physical Review A*, **1977**, *15(6)*, 2549.

(16) Reid, R.C.; Prausnitz, J.M.; Poling, B.E. *The Properties of Gases & Liquids, 4th ed*; McGraw-Hill International: New York, 1987.

(17) Assael, M.J.; Trusler, J.M.P.; Tsolakis, Th.F., *Thermophysical Properties of Fluids. An introduction to their Prediction*; Imperial College Press, 1996.

(18) Millat, J.; Dymond, J.H.; Nieto de Castro, C.A. ed. *Transport properties of fluids: their correlation, prediction and estimation*; Cambridge University Press, 1996.

# (64d) Modeling of Foam Flow From Adjacent Pour Regions in Reactive Polyurethane Foaming

Christopher J. Hershey and **Krishnamurthy Jayaraman**, Chemical Engineering & Materials Science, Michigan State University, East Lansing, MI

Molding of flexible polyurethane foam starting from a reacting mixture of polyol, water and isocyanate is a fast process where the structure is formed in a matter of minutes. The blowing reaction usually proceeds much faster than the crosslinking reaction: the reacting foam attains gel consistency well before the gel conversion of the crosslinking reaction – see Mora et al. (1991). There is considerable literature on numerical simulations of the expanding and reacting flow – see e.g. Bikard et al. (2007). However, predictions of unfilled regions near the end of fill were found to be inconsistent with observations in molding of seat cushion foam.

The object of this paper is to track the approach and merging of foam flow fronts from adjacent and separate pour regions in a rectangular mold cavity (see example in Figure). Numerical simulations were carried out by using the VOF method and the finite element software FLUENT with the help of user defined functions for the time varying density and viscosity of the foam. As a first step, foam flow patterns were obtained for the rectangular mold filling case reported by Mitani and Hamada (2003) with their specifications on density and viscosity and our results were found to be consistent with their results. Next, the flow field near the merging fronts was investigated for varying distance between pour regions. This was done at first with time varying but Newtonian viscosity for the expanding foam and then with a different viscosity function near the approaching fronts. In particular, conditions that locate an air pocket in the merging region were identified.

Bikard, J., Bruchon, J., Coupez, T., Silva, L., "Numerical simulation of 3D polyurethane expansion during manufacturing process," Coll. Surf. A 309, 49-63 (2007)

Mitani, T. and Hamada, H., "Prediction of flow patterns in the polyurethane foaming process by numerical simulation considering foam expansion," Polym. Eng. Sci. 43 (9), 1603-1612 (2003)

Mora, E., Artavia, L.D., Macosko, C.W., "Modulus development during reactive polyurethane foaming," J. Rheol., 35 (5), 921-940 (1991)

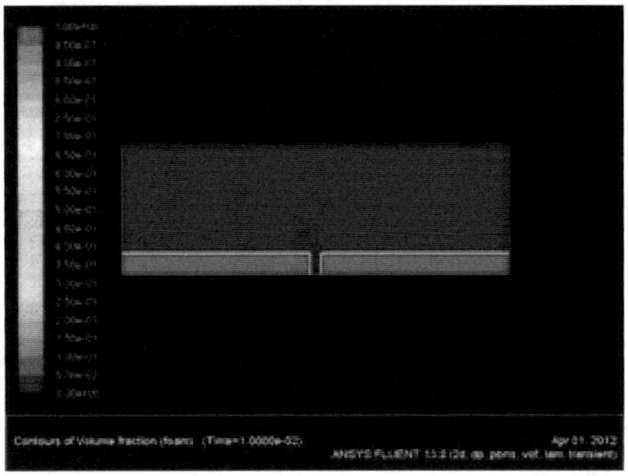

Adjacent pour regions in a rectangular mold cavity at the start of expansion

# The Role of Open Innovation in Development of Futuristic Technologies for Carbon Capture in Coal-Fired Power Plants: An Academic Perspective

Asad H. Sahir*, JoAnn S. Lighty
Department of Chemical Engineering, Institute of Clean and Secure Energy, University of Utah, Salt Lake City, UT

Lyda Bigelow,
Department of Management, David Eccles School of Business, University of Utah, Salt Lake City, UT

*Corresponding Author's Email Address: asad.sahir@utah.edu

**Abstract:**
Coal is an important fossil fuel resource for electricity generation which also contributes to significant $CO_2$ emissions. The process of capturing carbon dioxide for utilization and sequestration is an important area of research in this domain. Academia-industry collaborations are playing a significant role in the development of Oxy-fired combustion (OFC) and Chemical – looping Combustion (CLC), two technologies under consideration for burning coal in a primarily oxygenated environment to obtain a pure stream of $CO_2$.

Although OFC and CLC have different technological pathways, they have similarities in their historical development from a laboratory scale to pilot scale. Universities, in collaboration with industry, are continuing to play an important role in the development of these technologies to the pilot scale. Thus OFC and CLC could be visualized as one of the example technological platforms where 'open innovation' is being practiced in the development of carbon capture technologies. Facilitating collaboration in the pursuit of solutions to technologically complex problems can produce benefits such as decreased development time and costs. This paper will discuss the possible pathways which these technologies could undertake based on studies of power plant technologies, environment regulation technology components, and will derive insights from contemporary strategy and innovation literature. The possible challenges expected for carbon capture technologies through the open innovation model in terms of market economics, scale, and technology-enabling legislation which have the potential to frame future scenarios are also highlighted.

**Perspective:**

The development of suitable technologies for capturing carbon dioxide (i.e. obtaining a high-purity $CO_2$ stream) emitted from fossil fuel combustion has generated tremendous research interest. The Intergovernmental Panel on Climate Change (IPCC) has estimated the global temperature increase in the coming century to be between 1.8 to 4°C, which is likely attributed to the human-induced greenhouse gas (GHG) concentrations in the atmosphere (Global CCS Institute,2011). In 2010, $CO_2$ accounted for about 84% of all U.S. greenhouse gas emissions from human activities (US EPA Webpage, 2012).

44.9% of the electricity generated in the United States was derived from coal and peat in 2010 (Edison Electric Institute, 2012). The data on the U.S. energy-related carbon dioxide emissions determined on a sector basis indicates that the combustion of coal contributes to approximately 75% of the $CO_2$ emissions in the United States resulting from electric power generation (Annual Energy Outlook, 2012). In addition to the addressing the environmental impact of $CO_2$ emissions, technologies for facilitating capture of carbon dioxide merit potential consideration for generating $CO_2$ economically. The generated $CO_2$ could be utilized for injecting into depleted oil wells to recover untapped oil (McConnell, 2012). The pursuit of identifying economic and environmentally friendly technologies for the exploitation and recycling of $CO_2$ is an emerging area of active research interest (Peters et al., 2012).

Three technology routes are being currently considered for carbon dioxide capture (Markewitz et al., 2012; Riensche et al., 2012). $CO_2$ can either be captured from the flue gas (i.e. end-products of combustion) termed as 'post-combustion carbon capture', or the $CO_2$ concentration could be enhanced by burning a fuel in an oxygen-rich environment instead of air known as 'Oxy-fuel combustion'. Another class of carbon capture processes exist where $CO_2$ can be captured before the fuel is burnt called 'pre-combustion'. The 'pre-combustion' carbon capture process is achieved when the solid carbonaceous fuel (e.g. coal, petcoke) is gasified to form syngas which is subsequently converted to a mixture of $CO_2$ and $H_2$ by a shift reaction. The $CO_2$ could be recovered from the $CO_2$-$H_2$ gas mixture by a gas separation process. The objective of all the three categories of carbon capture processes ('post-combustion', 'pre-combustion' and 'oxy-fuel') is to obtain a high-purity stream of $CO_2$ which is suitable for sequestration and subsequent utilization. These processes are energy intensive and the energy requirements are derived from the power plant resulting in an efficiency loss. This efficiency loss for carbon capture processes is currently estimated to be of the order of 10% which is in addition to the efficiency penalty of 1-2% associated with pollutant removal (Riensche et al., 2012).

The subject of carbon capture utilization and sequestration is an extensive area and encompasses various technical, social, economic and ecological aspects. From the perspective of innovation, Johnsson et al. (2009) have reported results of a survey of individuals working in stakeholder organizations indicating widespread belief that technologies based on carbon capture and sequestration have the potential to gain major market entry in the next 10-20 years. Markusson et al. (2012) have developed a framework incorporating technical, economic, financial, political and societal issues associated with innovation for carbon capture and storage(CCS) technologies. In a recent review, Jiang et al. (2012) have highlighted the role of university and industry collaboration on the development of CCS technologies in the UK.

**Oxy-fuel Combustion (OFC) and Chemical-looping Combustion (CLC):**

The objective of this paper is to facilitate an understanding on the impact of open innovation on the development of two primary technological routes from the oxy-fuel category of CCS technologies – oxy-fuel combustion (OFC) and chemical-looping combustion (CLC). Open innovation has been defined by Chesbrough as "Open innovation is the use of purposive inflows and outflows of knowledge to accelerate internal innovation, and expand the markets for external use of innovation, respectively. [This paradigm] assumes that firms can and should use external ideas as well as internal ideas, and internal and external paths to market, as they look to advance their technology."(Chesbrough et al.,2006).

In OFC, the coal is burned in a composition close to pure oxygen replacing the air/coal mixture used in regular power plants. The oxygen for combustion is provided by an air separator where nitrogen is removed from the air, leaving almost pure oxygen. The coal and oxygen are subsequently sent into the boiler and ignited. The energy produced by the combustion process powers the steam turbines that generate electricity. The provision of 99.5% purity oxygen for the OFC results in an additional energy consumption of 0.29 kWh/kg $O_2$ which incurs an energy expenditure of 0.175 kWh per kWh of electrical power produced through an OFC process. Hence air separation diminishes the efficiency by almost 8% in a power plant based on OFC technology. The necessary work for compression of $CO_2$ to 110 bar for transport leads to a further loss of about 3.5%. Based on these considerations, total loss in efficiency is then estimated as 11–12% in power generation based on OFC based plant (Spliethoff, 2010). A simplified schematic of the OFC process is shown in Figure 1.

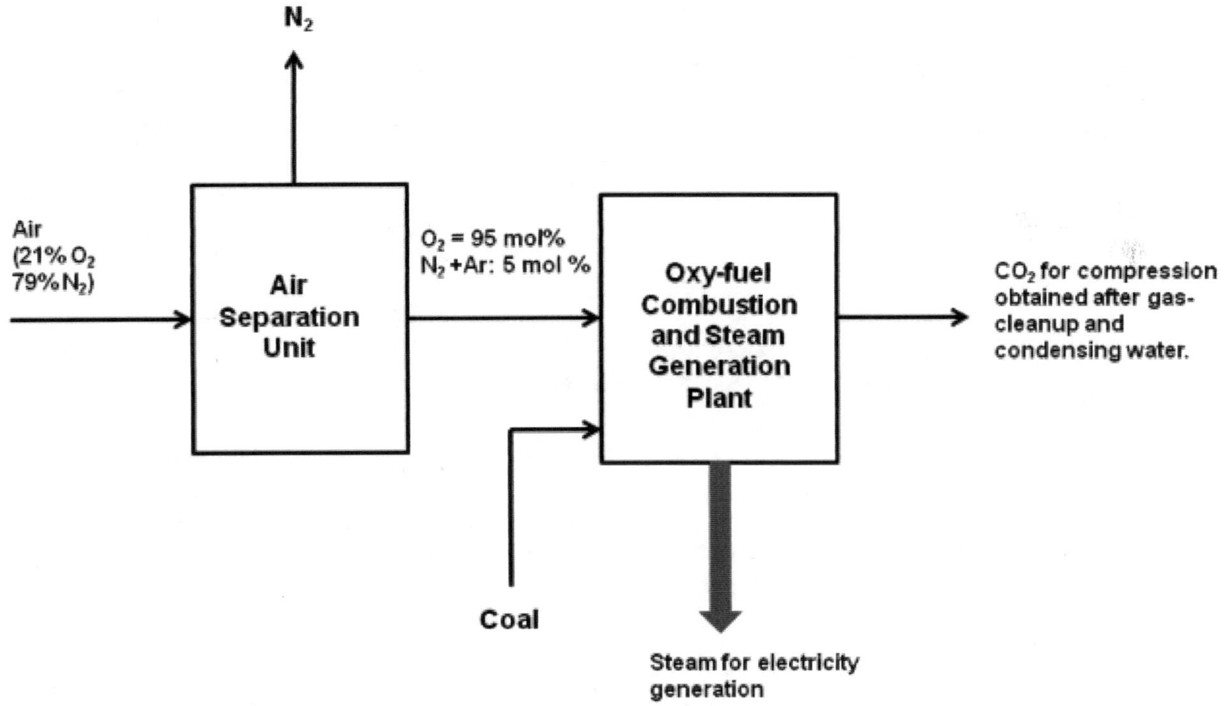


Figure 1 Simplified Schematic of an oxy-fuel combustion plant, (adapted from Riensche et al. 2011,)

To reduce the energy penalty associated with separating oxygen from air, another combustion process called Chemical-looping Combustion (CLC) is currently being investigated. CLC involves a system of two interconnected reactors with a metal oxide circulating between them. Instead of using an air separation plant, a metal oxide ($Me_xO_y$) supplies the required oxygen to burn the solid carbonaceous fuel in a reactor called the 'fuel reactor'. After losing the oxygen the reduced metal oxide ($Me_xO_{y-1}$) is regenerated by reaction with atmospheric air in another reactor called the 'air reactor'. In essence the metal oxide acts as an 'oxygen carrier' and helps in avoiding the associated energy penalty associated with an air separation unit. A loss of 4% in efficiency is in power generation has been estimated for power plants using the CLC technology (Epple, 2009). A simplified schematic of the CLC process is shown in Figure 2. Both CLC (presently in the industrial pilot-scale demonstration stage) and OFC (in the industrial pilot-scale stage) exhibit promise to cut down the cost of carbon capture significantly. The Institute of Clean and Secure Energy at the University of Utah is actively pursuing research interest in OFC and CLC technologies through its research program on clean and secure energy from coal (Sarofim et al., 2011, Wendt et al., 2012).

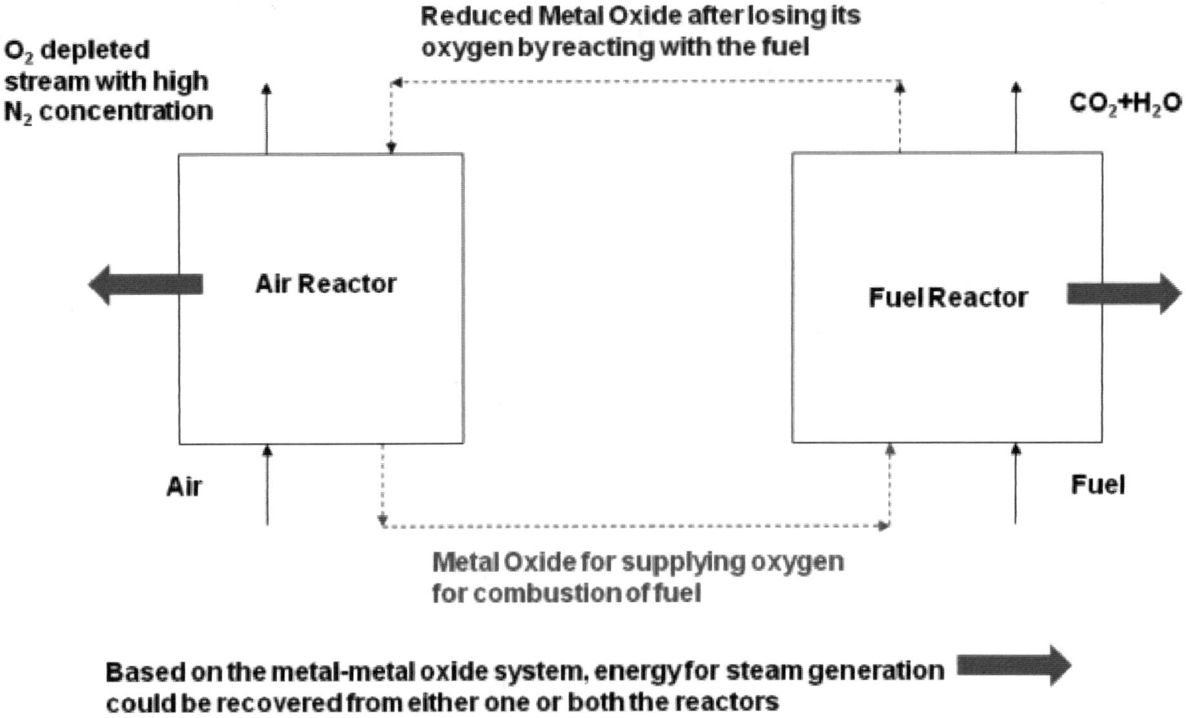


Figure 2 Simplified schematic of a chemical-looping combustion process (Adapted from Hossain and De Lasa, 2008)

According to Newell (2011), markets for energy technologies have usually emerged due to one of the following reasons. These factors also come into play when energy technologies for fossil fuels addressing carbon capture e.g. OFC and CLC are being analyzed:

1. Prices of conventional resources rise as a result of rising demand and stagnant or falling supply or production capacity.
2. Technological possibilities arise that more effectively meet energy demands
3. The government imposes new policies and regulations that affect market conditions.

OFC and CLC have the potential to revolutionize the coal-fired power plant industry. It has been difficult for recent innovations in coal-fired power plant technology (pertaining to pollutant emission reduction and novel combustion methods) to achieve significant levels of market penetration at a faster pace. The reasons attributed are the following (Newell, 2011):

1. The rate of turnover of old coal plants and the construction of new plants in developed countries has been slow.
2. The more advanced coal-fired power plants have proven to be costly for emerging economies, where new coal-fired power plants are being built.
3. The electric power industry has perceived gasification based combustion systems (an innovative technology incorporating 'pre-combustion' carbon capture), more costly and more risky. Novel technologies in the coal-fired power plant industry have faced significant difficulty in attracting private sector or utility investment which has rendered these technologies to be relatively untested.
4. Natural gas power plants have proven to be more advantageous than coal-fired power plants, as they can be built quickly, on a smaller scale, with a lower capital cost and have low levels of pollutant emissions.

The concepts associated with OFC have been investigated since the 1970's and have been incorporated in furnaces for glass, steel and cement production. The purpose then was to provide a higher flame temperature which helped in increasing energy efficiency (Chen et al., 2012). In the case of OFC and CLC the uncertainty of performance associated with first of a kind large scale plant exists. To address the challenges associated with the technology uncertainty for OFC industrial pilot scale units, power plant manufacturers, utility providers have collaborated with industrial gas suppliers who will supply the oxygen and develop gas clean up technologies for the gaseous products of combustion. Some OFC industrial pilot-scale plants in the range of 10-40 $MW_{th}$ have been developed (Wall, 2011) and are mentioned in Table 1. Larger-scale demonstration plants for OFC have also been planned. Vattenfall has supported a number of Swedish and German Universities (Chalmers, TU Dresden, TU Hamburg-Harburg, IVD-Stuttgart, Brandenburg Technical University) in operation of six oxy-fuel test-rigs to promote diversity in ideas towards OFC development (Vattenfall Webpage, 2012). The University of Newcastle is playing an important role in the development of Callide Oxy-fuel project in Australia (ABC Webpage, 2012). Thus universities are also playing a role in the development of OFC technology through industrial-academia collaborations.

Table 1: Some examples of technological alliances in oxy-fuel combustion (Adapted from Wall, 2011)

| Industrial Pilot Plant and Test Facilities | Capacity | Companies, technology providers and vendors |
|---|---|---|
| Callide A | 30 $MW_e$ | IHI, CS Energy, Air Liquide |
| Schwarze Pumpe | 30 $MW_{th}$ | Vattenfall, Alstom, Linde |
| Oxy-coalUK | 13.3 $MW_{th}$ | Doosan Babcock, Air Products |
| CIUDEN | 30 $MW_{th}$ CFB and 20 $MW_{th}$ PC | Endesa, Foster Wheeler, Air Liquide, Leni Gas and Oil |

CFB: Circulating Fluidized Bed
PC: Pulverized Coal

Chemical-looping combustion (CLC) pilot scale demonstrations in the 250 $kW_{th}$ to 3 $MW_{th}$ capacities have been constructed. Interestingly for CLC also, universities have played an important role in housing the pilot scale demonstration plants and for developing the technology in its initial phase. Notable examples in this category are Technical University of Darmstadt (1 $MW_{th}$ scale) and Ohio State University (250 $kW_{th}$). The role of the industry-academia consortia approach is quite evident from the descriptions available at project webpages and literature and is mentioned in Table 2. Technological developments in OFC have been reviewed by Toftegaard et al., 2010 and Chen et al., 2012., and various facets on development of CLC processes have been discussed by Fan, 2010 and Adanez et al., 2012.

In the authors opinion the role of universities and their impact in the development of OFC and CLC technologies is an interesting area of research which may be pursued from the perspective of 'open innovation'. The latter section of the paper is devoted to highlight contemporary literature on 'open innovation' which can offer guidance on potential challenges and opportunities from an academic perspective which lies ahead for the development of OFC and CLC development for fossil fuel power plants.

Table 2: Some examples of technological alliances in chemical-looping combustion (Adapted from Abdulally, 2012; Fan, 2012; INNOCUOUS, 2012)

| University | Capacity or Nature of the Project | Companies, technology providers and research institutions participating |
|---|---|---|
| TU Darmstadt | 1 $MW_{th}$ | TU Darmstadt, Chalmers, CSIC, SINTEF, Air Liquide, Vattenfall, Alstom |
| Ohio State | 250 $kW_{th}$ | Babcock & Wilcox (B&W), ClearSkies, CONSOL Energy, Air Products, Shell/CRI |
| Chalmers and TU Vienna(under the INNOCUOUS project) | Project for Testing Fluidized Bed Reactor design and oxygen carrier particles | Shell, CSIC,VITO, Johnson Matthey, Bertsch Energy |

**The Concept of 'Open Innovation' in the context of OFC and CLC technologies:**

Open Innovation is an innovation model where "firms commercialize external (as well as internal) ideas by deploying outside (as well as in-house) pathways to the market". The idea of 'open innovation' contrasts with the conventional idea of companies generating their own ideas in-house that they would consequently develop, manufacture, market, distribute and service themselves and is aptly termed as 'closed innovation'. The open innovation model imparts the capability to corporations to deal with 'false negatives', i.e. projects that initially lack promise but then out to be of immense value (Chesbrough, 2003). Novel carbon capture technologies like

OFC and CLC are either in the industrial pilot-scale or at the industrial pilot-scale demonstration phase where they indicate promise but significant technical challenges need to be addressed before successful deployment. Hence information in the open literature regarding OFC and CLC indicates a preference of favoring collaborations between academic institutions, research organizations and industries.

From an industry's standpoint, the open innovation model facilitates co-creation of technologies through the 'coupled process'. The 'coupled process' combines the 'outside-in process' (to gain external knowledge) with the 'inside-out process' (to bring ideas to market) to facilitate the innovation process (Enkel et al., 2009). Interestingly the development of OFC and CLC technologies for solid fuels exhibits elements of collaborations with universities or research organizations. The role of organizational dynamics between industries and universities in collaborative relationships from an open innovation perspective has been recently analyzed by Perkmann and Walsh, 2007. The concept of open innovation has the following benefits and challenges from a firm's perspective as mentioned in Table 3.

Table 3: Benefits and Challenges associated with Open Innovation from a firm's perspective (Gassmann et al., 2010 and Enkel et al., 2009):

| Benefits | Challenges |
|---|---|
| Acceleration in research and development activities through university-industry partnerships by enhancing 'absorptive capacity' and promoting 'outside-in innovation' processes. | Identification of right partners for the collaborative project |
| Provides a mechanism to reduce overcapacities, cut costs, grow through complementary assets or reduce risks | Allocation of finances and time for the open innovation project |
| Provides the alternative of 'probe and learn' to firms in addition to the traditional 'stage-gate process' | Higher coordination costs |
| Easier accessibility to diverse expertise due to globalization of R&D activities. | Management of intellectual property |

An interesting aspect which may merit consideration of researchers by studying the evolution of OFC and CLC technologies as novel carbon capture technologies for coal-fired power plants is the impact of partner heterogeneity on the collaboration as they evolve. It may be possible that cross-industry innovation may result in significant advances in these technologies in the future. These effects have recently been analyzed by Enkel and Gassmann (2010) for industries catering to the automotive, packaging, textile, sport, aircraft, chemical sectors.

**Analyzing OFC and CLC as emergent technologies from the lens of literature in technology diffusion:**

The current state of development of oxy-fuel and chemical-looping combustion technologies at either approximately 5% or 0.1% of the scale of a demonstration phase coal-fired power plant (assumed to be a 1000 $MW_{th}$ capacity) also invites interest of studying aspects pertaining to technological diffusion. Although oxy-fuel and chemical-looping are in the nascent stages of their development, the following aspects merit consideration from a financial perspective which concerns the development of energy technologies addressing the goal of CCS (Ecofin Research Foundation, 2010):

1. The importance of 'performance guarantee' for a CCS project utilizing fossil fuels is significant. The OFC or CLC technology has to be robust enough for the coal-fired power plant whose expected lifetime is at least 25-30 years. It has to ensure flexibility in different operating conditions. The first mover 'power generation company' and the 'engineering firm' has to risk its reputation in order to build the first facility on an industrial scale.

2. The companies backing an OFC or CLC project has to be a major player in the industry. Hence corporate relationships will prove to be vital in obtaining financing for the project. It also provides confidence that the major power generation company will have the engineering expertise to execute a complex large scale project with considerable technological uncertainty.

3. The importance of achieving 'grid parity' through a power generation technology based on OFC or CLC in a distributed energy generation future is also vital.

Popp (2004) studied the adoption of $NO_x$ and $SO_2$ pollution control technologies in fossil-fuel based power plants. Using patent data it was concluded that "the inventors respond to the environmental regulatory pressure in their own country, but not to foreign environmental regulations". Furthermore it was identified that for the development of $NO_X$ pollution control innovations in the United States, foreign patents served as important building blocks. It is likely that environmental regulations will play a significant role in the development of carbon capture and storage technologies. The possibility of using $CO_2$ as an oil-recovery agent and subsequent research in utilizing $CO_2$ in chemical processes mentioned in the earlier section of this paper may accelerate the development of OFC and CLC technologies. Chesbrough and Appleyard (2003) in their article comment on a fundamental change in focus from 'ownership' to the 'concept of openness' for some organizations working on innovative ideas. This shift encourages reconsideration of the concept of 'value creation' and 'value capture processes' by organizations, as has been demonstrated in the information technology industry by Linux and Wikipedia. In the same article, the 'Linus's Law' is mentioned which states, "Given enough eyeballs, all bugs are shallow" (i.e., easy to fix). It will be fascinating to observe if Linus's Law will also hold true in the development of OFC and CLC as forerunners for CCUS technologies in the future facilitated by an 'open innovation' platform.

## Conclusions:

In this paper, an attempt has been made to identify synergies between advancements in oxy-fuel combustion (OFC) and chemical-looping combustion (CLC) as carbon-capture technologies for coal combustion and contemporary management research on open innovation and technology diffusion. The evolution of these technologies is not only contingent on addressing the technical challenges associated, but also depend on adoption of environmental regulations. Hence an analysis of the literature on technological diffusion and open innovation is useful to provide guidance on the challenges associated with oxy-fuel and chemical-looping technologies in scenarios where the concept of 'open innovation' is being leveraged.

## Acknowledgements:

This material is based upon work supported by the Department of Energy under Award Number DE-NT0005015. The views and opinions of authors expressed herein do not necessarily state or reflect those of the United States Government or any agency thereof.

## References:

1. Abdulally, I. (2012), "Alstom's Chemical Looping Prototypes Program Update, Presented at 11th Annual Carbon Capture", Utilization and Sequestration Conference, Pittsburgh, PA http://www.alstom.com/Global/US/Resources/Documents/CCS12/Chemical%20Looping%20Prototype_%20Program%20Update.pdf(accessed September 13, 2012)

2. Adanez, J. et al. (2012), "Progress in Chemical-Looping Combustion and Reforming Technologies" , Progress in Energy and Combustion Science, 38 (2), 215−282.

3. Annual Energy Outlook 2012 with Projections to 2035, DOE/EIA-0383 (2012), www.eia.gov/forecasts/aeo (accessed September 13, 2012).

4. Australian Broadcasting Corporation (ABC) Webpage(2012), "Newcastle researchers involved in green power station" http://www.abc.net.au/news/2012-08-15/oxygen-fuel-project/4199112(accessed September 24, 2012)

5. Chen, L. et al. (2012), "Oxy-fuel combustion of pulverized coal: Characterization, fundamentals, stabilization and CFD modeling", Progress in Energy and Combustion Science, 38(2), 156-214.

6. Chesbrough, H.W. (2003), "The Era of Open Innovation", Sloan Management Review, 44, 3 (Spring): 35-41.

7. Chesbrough, H.W., West, J. and Vanhaverbeke, W. (2006) Open Innovation: Researching a New Paradigm. Oxford: Oxford University Press.

8. Chesbrough, H.W.,Appleyard, M.M. (2007), "Open Innovation and Strategy", California Management Review, 50(1).

9. Ecofin Research Foundation, The Climate Group and the Global CCS Institute(2010), “Carbon Capture and Storage: Mobilizing Private Sector Finance”, http://www.globalccsinstitute.com/publications/carbon-capture-and-storage-mobilising-private-sector-finance (accessed September 13, 2012)

10. Edison Electric Institute (2012), http://www.eei.org/ourissues/ElectricityGeneration/FuelDiversity/Documents/pie_fueldiversity.pdf (Accessed September 13, 2012)

11. Enkel, E. et al. (2009), “Open R&D and open innovation: exploring the phenomenon”, R&D Management, 39(4), 311-316.

12. Enkel, E., Gassmann, O. (2010) “Creative imitation: exploring the case of cross-industry innovation”, R&D Management 40(3), 256-270.

13. Epple, B.(2009), “Activities within the German Research Project “Limestone based Absorption of CO2” (LISA), Oviedo, http://www.ieaghg.org/docs/loopingpdf/16%20septiembre/C12.pdf (Accessed September 13, 2012).

14. Fan, L. S. (2010) Chemical Looping Systems for Fossil Energy Conversions; Wiley-AIChE: Hoboken, NJ, 2010.

15. Fan, L.S. (2012), “Coal Direct Chemical Looping (CDCL) Retrofit to Pulverized Coal Power Plants for In-Situ CO2 Capture”, Presented at the 2012 NETL CO2 Capture Technology Meeting, http://www.netl.doe.gov/publications/proceedings/12/co2capture/presentations/3-Wednesday/LS%20Fan-OSU-CDCL.pdf(accessed September 13, 2012)

16. Gassmann, O. et al. (2010), The future of open innovation, R&D Management, 40, 3, 213-221.

17. Global CCS Institute 2011, The global status of CCS: 2010, Canberra http://www.globalccsinstitute.com/publications/global-status-ccs-2010 (accessed September 13, 2012)

18. Hossain, M.M., de Lasa, H.I. (2008), “Chemical-looping combustion (CLC) for inherent $CO_2$ separations—a review”, Chemical Engineering Science, 63, 4433 – 4451.

19. Innovative oxygen carriers uplifting Chemical-looping Combustion, INNOCUOUS Webpage (2012), https://www.clc-innocuous.eu/ (accessed September 13, 2012)

20. Jiang, L. et al. (2012), “Current development of carbon capture and storage in the UK: a non-technical review”, International Journal of Low-Carbon Technologies, doi:10.1093/ijlct/cts008

21. Johnsson et al. (2009), "Stakeholder Attitudes on Carbon Capture and Storage - an international comparison", Energy Procedia, 1, 4819-4826.

22. Markewitz et al. (2012), "Worldwide innovations in the development of carbon capture technologies and the utilization of $CO_2$", Energy and Environmental Science, 5, 7281-7305.

23. Markusson et al. (2012), "A socio-technical framework for assessing the viability of carbon capture and storage technology", Technological Forecasting & Social Change 2012, 79, 903–918

24. McConnell, C. (2012) "Adding "Utilization" to Carbon Capture and Storage", United States Department of Energy Webpage, http://energy.gov/articles/adding-utilization-carbon-capture-and-storage , (accessed September 13, 2012).

25. Newell, R.G. (2011) "The Energy Innovation System: A Historical Perspective" In., Accelerating Energy Innovation Insights from Multiple Sectors; Henderson, R.M., Newell, R.G eds. The University of Chicago Press, Chicago and London.

26. Perkmann, M., Walsh, K. (2007), University–industry relationships and open innovation: Towards a research agenda, International Journal of Management Reviews, 9 , 4, 259–280

27. Peters et al. (2011), "Chemical Technologies for Exploiting and Recycling Carbon Dioxide into the Value Chain", ChemSusChem 4, 1216-1240.

28. Popp, D. (2004), "International Innovation and Diffusion of Air Pollution Control Technologies: The effects of $NO_x$ and $SO_2$ regulation in the U.S. Japan and Germany", NBER Working Paper Series, Working Paper 10643, http://www.nber.org/papers/w10643(accessed September 13, 2012)

29. Riensche et al. (2011), "Capture Options for Coal Power Plants", In Efficient Carbon Capture for Coal Power Plants; Stolten, D., Scherer, V. Eds;Wiley-VCH, 2011.

30. Sarofim, A.F., Lighty, J.S., Smith, P. J.; Whitty, K. J., Eyring, E.; Sahir, A., Alvarez, M., Hradisky, M., Clayton, C., Konya, G., Baraki, R., Kelly, K.(2011), Chemical Looping Combustion Reactions and Systems Task 5 Topical Report, Utah Clean Coal Program - Reporting Period October 2008 to March 2011.

31. Spliethoff, H., Power Generation from Solid Fuels, 2010, Springer.

32. Stolten, D., Scherer, V., Efficient Carbon Capture for Coal Power Plants, Wiley-VCH, 2011.

33. Toftegaard, M.B. et al. (2010) "Oxyfuel combustion of solid fuels", Progress in Energy and Combustion Science 36, (5), 581–625.

34. United States Environmental Protection Agency (US EPA) Webpage, http://www.epa.gov/climatechange/ghgemissions/gases/co2.html (accessed September 13, 2012)

35. Vattenfall Website (2012),
http://www.vattenfall.com/en/ccs/test-rigs.htm (accessed September 24, 2012)

36. Wendt, J. O. L., Eddings, E. G., Lighty, J. S., Ring, T., Smith, P. J., Thornock, J., Jia, Y., Morris, W., Pedel, J., Rezaei, D., Wang, L., Zhang, J., Kelly, K., (2012) Oxy-coal Combustion Studies Task 3 Topical Report, Utah Clean Coal Program

37. Wall, T. et al. (2011), "Demonstrations of coal-fired oxy-fuel technology for carbon capture and storage and issues with commercial deployment", International Journal of Greenhouse Gas Control, 5S, S5-S15, 2011.

# (107a) Strategies in Advancing the Pharmaceutical Pipeline - an Overview of API Pilot Plant

**Jim Ratway**, R452, Abbott, North Chicago, IL

This is a place holder that must be 122 words long entered by David Carl Attride of Eastman Chemical Corp. the Chair of the Area 12B section of the Process Development Division on Tuesday, May 1, 2012. Mary had a little lamb, little lamb, little lamb. Is that enough text to satisfy?

This is a place holder that must be 122 words long entered by David Carl Attride of Eastman Chemical Corp. the Chair of the Area 12B section of the Process Development Division on Tuesday, May 1, 2012. Mary had a little lamb, little lamb, little lamb. Is that enough text to satisfy?

This is a place holder that must be 122 words long entered by David Carl Attride of Eastman Chemical Corp. the Chair of the Area 12B section of the Process Development Division on Tuesday, May 1, 2012. Mary had a little lamb, little lamb, little lamb. Is that enough text to satisfy?

# (107b) A Pharmaceutical Engineering Process/Product Life-Cycle Approach to Drug Product Development and Manufacturing

**Jose C. Menezes**[1,2], Stephan Sacher[2], Georg Scharrer[3], Stefan Leitgeb[4], Christine Voura[5], Simon D. Fraser[5] and Johannes G. Khinast[6], (1)Institute for Biotechnology & Bioengineering, I.S.T., Lisbon, Portugal, (2)Research Center for Pharmaceutical Engineering, Graz, Austria, (3)Advanced Simulation, Research Center for Pharmaceutical Engineering, Graz, Austria, (4)TU Graz, Institute of Biotechnology and Biochemical Engineering, Research Center Pharmaceutical Engineering GmbH, Graz, Austria, (5)Research Center Pharmaceutical Engineering (RCPE), Graz, Austria, (6)Research Center Pharmaceutical Engineering GmbH, Graz, Austria

**A Pharmaceutical Engineering Process/Product Life-Cycle Approach to Drug Product Development and Manufacturing**

Jose C. Menezes, Sephan Sacher, Georg Scharrer, Stefan Leitgeb, Christine Voura, Simon Fraser, Johannes Khinast

Research Center Pharmaceutical Engineering GmbH, A-8010 Graz, Inffeldgasse 21a/II, Austria

Drug product design, development and manufacturing involve different goals and use different tools throughout product and process engineering life-cycle. At the formulation design stage a materials science-based approach should dominate. During development and especially later during manufacturing (continuous or batch) a process-centered perspective prevails. At the end of processing a drug product of required purity, safety and efficacy should be obtained. A stronger link between the different goals throughout those life-cycle stages must be established to ensure modern and competitive pharmaceutical manufacturing.

PAT monitoring tools, however, when used as such (i.e., as process state estimation and not as in-process quality controls of specific parameters) are process-centered but can be used to link process to product (viz., process parameters and attributes to product critical quality attributes). Here we discuss several case-studies involving RCPE's recent activities, highlighting the use of advanced simulation and innovative PAT tools in formulation design, process development and different operational requirements for routine manufacturing within the QbD context. A balance summary perspective of where the area is evolving into after a decade of PAT and QbD will be given.

# (107c) Leveraging HPLC Data in a QbD Approach, for Right-First Time Scale-up of API Reaction Steps

**Wilfried Hoffmann**[1], Andrew Bird[1] and Peter Clark[2], (1)Scale-up Systems, Dublin, Ireland, (2)DynoChem Inc, Yardley, PA

HPLC data that are routinely generated can be leveraged to deliver much additional value; when interpreted quantitatively, they can be used to develop mechanistic models. These models in turn can be used to probe and question the data in a way that leads to increased process understanding. There is also the opportunity to run experiments that are designed to produce impurity so that the mechanism of impurity formation can be understood. The above are core elements in developing a process that works right first time on scale-up, achieving Quality by Design.

In more detail:

Experimental data are the basis for adjusting rate parameters in kinetic models. The most powerful data sets are analytical profiles, as these data are able to show species levels lower than 1%, particularly important in the Pharmaceutical Industry. At the same time these are one of the most difficult to handle, as chromatographic detection is in the case of the most common UV-detectors linked to molecular absorbance, and this can vary by orders of magnitude.

The talk will present an approach to convert analytical data to absolute values by use of Relative Response Factors and a link to the mass balance implied in a kinetic model. The data and model may not agree and in this case either the data or the model may be wrong.

By changing one or the other, we can ultimately reach data consistent with the model's mass balance, learning much in the process about our analytical method and our reaction scheme; this consistency is a requirement to get a good match in fitting with kinetic models,. The other factor is the rate law assumption, which we also address in this talk.

In an enhancement of this approach, a method is proposed which allows the calculation of the Relative Response Factors under the conditions of overlapping mass balances, which is very often the case.

Extended Abstract for American Institute of Chemical Engineers (AIChE)
Presentation ID# 279310
Meeting: Fall Meeting 2012
Location: Pittsburgh, PA
Date: 28-Oct-2012 to 02-Nov-2012

**Title:** Continuous Distillation Operations and Application of the Wiped Film Evaporator (WFE) to Pharmaceutical Pilot Plant Processing

**Primary Author/Presenter:** Anne Mohan
Contributing Author(s): Beth Fisher, Rukyah Hennessey, Ozgur Karahan, Joe Kukura, Dave Lashen, Crystal Miranda, Glenn Spencer, Michael Ward, Ming Yue, Dimitrios Zarkadas, George Zhou

**Objective:** To provide a detailed analysis of wiped film evaporation, a type of continuous distillation technology, by utilizing a case study from a Merck pharmaceutical pilot plant. In addition, this presentation will provide a comprehensive technical explanation of the technology and its application to industrial chemical processing.

**Background:**

The utilization of continuous processing technology provides increased productivity and versatility to scaled processes. Continuous *distillation* technology, in addition to providing these general benefits, provides the opportunity to process streams at higher temperatures and lower pressure ranges than *batch* distillation operations. Wiped film evaporation (hereafter referred to as "WFE"), which is a type of continuous distillation technology, is ideal for heat sensitive and/or high-boiling products due to the short residence time (less than 60 sec) and the low vacuum capability (down to ~1 mm Hg) of the unit. In addition, the volume of batch required for WFE operation is significantly less that that required for batch distillation, making WFE appropriate for low-volume products. Wiped film evaporation (WFE) is a type of continuous distillation technology recently investigated, developed, and ultimately applied to a pharmaceutical project at the Merck Rahway site at the pilot plant scale. This technology was utilized in order to address the thermal instability and high boiling issues associated with a specific process. The implementation of wiped film evaporation to this process is an example of the appropriate application of continuous distillation technology at scale, and the benefits of such an application.

The objective of this presentation is to communicate and discuss the recent lab development and pilot plant work with wiped film evaporation technology, which is used for continuous distillation processing. The presentation will focus on the methodology used to scale-up the technology, the challenges faced in the pilot plant, how these challenges were handled, and the accomplishments of the campaign both in the lab and in the pilot plant.

**Methods:**

The 2" and 4" lab scale, and the 6" pilot plant scale WFE units were used for continuous processing technology development and for specific project-related process development during 2011 and 2012 at Merck & Co., Inc. Lab development was performed on the smaller WFE units to gain an understanding of WFE operating parameters and factors which affect the distillation operation. For 2" and 4" WFE operations, material was fed into the WFE from a Wheaton bottle or poly jug using a peristaltic pump. The feed rate was controlled under the set pressure and temperature conditions in order to obtain the desired split between the distillates and residue. These exiting streams were collected in round bottom flasks attached to the respective WFE ports, and the product concentration in both was determined by HPLC (high performance liquid chromatography) at the conclusion of the experimental runs.

The data and knowledge gained from the lab developmental work was applied to the operation of the pilot plant scale, 6", WFE unit. This larger unit was used during batch processing throughout the month of October 2011 at the Merck Rahway site. During 6" WFE operation, the batch was fed from drums to the WFE system via an in-line mass meter and diaphragm pump. Under specific temperature and pressure conditions, the batch was boiled in the column of the WFE, and the distillates and residue generated were collected into respective drums. Samples of the exiting streams were pulled at various points throughout the operations to either ascertain the split being generated, or to ensure the targeted split was achieved. In this manner, the WFE equipment was used for both technological development and process-specific operation.

The data gathered from these operations was analyzed in order to understand and characterize the heat transfer of the WFE unit and the batch performance in the WFE. Additionally, the data was used to develop a scaling procedure which can be applied to take a process from the lab to the pilot plant scale. The experiments used to accomplish these tasks explored the controllable variables of the WFE unit, such as feed rate, wiper speed, pressure, and temperature.

**Results and Conclusions:**

Through analysis of the data obtained from the operations of the WFE, the following conclusions were drawn:

1. The material of construction plays a large role in the heat transfer characterization of the WFE unit. At the lab scale, the limiting factor for heat transfer is the glass wall of the WFE body whereas at the pilot plant scale, the limiting factor for heat transfer was the solvent, or batch, itself.
2. The distillate and residue split for a given temperature and pressure condition can be controlled most tangibly by the feed rate.
3. Of the controllable factors explored, wiper speed has the least effect on the distillate and residue split during a WFE distillation.

The associated notable accomplishments of this endeavor were:

1. Increased WFE equipment knowledge regarding general operation and function, heat transfer capabilities, and overall productivity and efficiency
2. Attainment of data and knowledge regarding the calculations required to scale-up from the lab scale WFE units to the larger scale WFE units
3. Increased industrial processing knowledge in the continuous realm

This presentation will provide: (1) an in-depth overview of the wiped film evaporation operation, (2) a detailed description of the data collected and knowledge gained by all WFE units, with specific focus on the larger 6" unit, (3) a list of the advantages and disadvantages observed through the operation of all size WFE units, and (4) a report of the quantitative and qualitative results of the WFE usage for the batch processing. The technological and process-specific knowledge that was gained through the Merck WFE experiences can be used to promote the continuous platform.

The utilization of continuous processing technology provides increased productivity and versatility to scaled processes. Continuous *distillation* technology, in addition to providing these general benefits, provides the opportunity to process streams at higher temperatures and lower pressure ranges than *batch* distillation operations. Wiped film evaporation (hereafter referred to as "WFE"), which is a type of continuous distillation technology, is ideal for heat sensitive and/or high-boiling products due to the short residence time (less than 60 sec) and the low vacuum capability (down to ~1 mm Hg) of the unit. In addition, the volume of batch required for WFE operation is significantly less that that required for batch distillation, making WFE appropriate for low-volume products. Wiped film evaporation (WFE) is a type of continuous distillation technology recently investigated, developed, and ultimately applied to a pharmaceutical project at the Merck Rahway site at the pilot plant scale. This technology was utilized in order to address the thermal instability and high boiling issues associated with a specific process. The data gathered from this operation was analyzed in order to understand and characterize the heat transfer of the WFE unit and the batch performance in the WFE. Additionally, the data was used to develop a scaling procedure which can be applied to take a process from the lab to the pilot plant scale. The experiments used to accomplish these tasks explored the controllable variables of the WFE unit, such as feed rate, wiper speed, pressure, and temperature. The implementation of wiped film evaporation to this process is an example of the appropriate application of continuous distillation technology at scale, and the benefits of such an application.

The objective of this presentation is to communicate and discuss the recent lab development and pilot plant work with wiped film evaporation technology, which is used for continuous distillation processing. The presentation will focus on the methodology used to scale-up the technology, the challenges faced in the pilot plant, how these challenges were handled, and the accomplishments of the campaign both in the lab and in the pilot plant. The presentation will provide: (1) an in-depth overview of the wiped film evaporation operation, (2) a detailed description of the data collected and knowledge gained by all WFE units, with specific focus on the larger 6" unit, (3) a list of the advantages and disadvantages observed through the operation of all size WFE units, and (4) a report of the quantitative and qualitative results of the WFE usage for the batch processing. The technological and process-specific knowledge that was gained through the Merck WFE experiences can be used to promote the continuous platform.

# (107e) Optimization of Amidecoupling Reaction Conditions Using Kinetic Modeling

**Ozgur Karahan**, Chemical Process Development and Commercialization, Merck & Co. Inc., Rahway, NJ, Camille O. Anderson, Global Pharmaceutical Commercialization, Merck & Co., Inc., Rahway, NJ and Dimitrios Zarkadas, Chemical Process Development & Commercialization, Merck & Co. Inc., Rahway, NJ

An amide coupling reaction between a carboxylic acid and an amine in a solvent in the presence of EDC-HCl and an amine base was studied to optimize the reaction conditions and develop a robust process for large scale production. Results of a fractional-factorial DOE in combination with reaction understanding suggested a mechanism with multiple elementary steps, and overall apparent reaction kinetics depended on the concentration of all reactants as well as temperature. The Dynochem software was successfully utilized to model this complex reaction system and identify optimal conditions. In addition, the DOE results were used to generate a model that predicts oiling or the formation of a gummy phase in the reaction mixture, phenomena that can affect reaction performance. This information was then coupled with a cost function to determine conditions that ensure robust commercial manufacturing and minimize manufacturing costs. The amide coupling reaction with optimized conditions was successfully demonstrated on manufacturing scale.

# (107f) Scale-up of Superficial Fluidization Air Velocity for a Pharmaceutical Fluid Bed Granulation

**Brendon G. Ricart**, Pharmaceutical Commercialization Technology, Merck & Co., Lansdale, PA, Alan R. Silverman, Pharmaceutical Commercialization Technology, Merck & Co., Inc., West Point, PA, Kristin J. Ploeger, Pharmaceutical Commercialization Technology, Merck & Co, Lansdale, PA and Edward J. Smith, Merck & Co., Inc., West Point, PA

Fluid bed granulation is a common unit operation in pharmaceutical manufacturing. Although the technology is mature, the complexity of the solid-fluid system has left incomplete scale-up principles. In the current study, fluidizing air velocity scaling techniques are examined across equipment design, equipment scale, and batch size. The scale of the current study ranges from 10-20 kg pilot-scale studies to 500 kg commercial scale batches. We find that scaling superficial air velocity at the top of the fluidized bed ($v_{TOB}$) provides a more robust process scaling than at the distributor plate, which is commonly used. Additionally, a dimensional analysis is performed for general fluid bed granulations to identify 5 relevant dimensionless groups which are interpreted to describe scale-up of process moisture, fluidizing air velocity (2), bowl fill, and total spray time. We find conflicting guidance from 2 dimensionless groups for scaling fluidization air, but constant $v_{TOB}$ represents a compromise between the two. Together, these analyses led to a unique experimental design for a 500 kg scale developmental drug product campaign.

# (107g) Experimental Determination of Maximum Shear Stress in Multiphase Flow – From Micro Scale to Pilot Scale Reactors

Thomas K. Villiger, Miroslav Soos and Massimo Morbidelli, Institute for Chemical and Bioengineering, ETH Zurich, Zurich, Switzerland

Although a tremendous effort has been made to determine velocity, turbulence and shear level distribution including their maximum values in stirred tanks, resolving the small sizes of the dissipative scales applying either experimental techniques or computational approaches still remains a challenge, particularly for multiphase systems. Such flow fields occur frequently in bioreactors where mechanical stress caused by stirring or bubble burst, entrainment, coalescence and break-up has been reported to have detrimental effects on mammalian cell cultures and product quality.

In order to construct a rational process design space, we have developed an aggregate – breakup system which is able to determine the maximum shear stress present in any device with arbitrary geometry. The presented methodology is based on the measurement of the steady state aggregates size broken by the hydrodynamic stress generated by stirring, sparging or their combination. Different contractive nozzles characterized by well-defined fluid dynamic conditions are used to calibrate the aggregate size to the maximum hydrodynamic stress. This system has been applied to biotechnological equipment ranging from micro scale bioreactors (0.015 L) up to a pilot scale bioreactor (300 L). The results show that not only the size and geometry of the vessel but also the flow field has a significant impact on the highest shear stress level. A cell will therefore exhibit a profoundly different hydrodynamic stress from expansion to production which should be carefully taken into account during scale-up as well as during the entire bioprocess development.

# (119a) Effective Application of Established Treatment Technologies for Metals, Metalloids, and Oxyanions

**James Patterson**, Patterson Environmental Consultants, Chicago, IL

The underlying theories and fundamentals of application of treatment technologies for control of metals, metalloids and oxyanions are well established. However, the complexity of the wastewater matrix to be treated can introduce unanticipated responses to the applications of such technologies, including when discharge limits are unusually restrictive due for example to water quality-based effluent limitations, or in scale-up from treatability investigations to full-scale application. This Presentation will provide an overview of such theories and applications, and address pollutant speciation, wastewater matrix, and treatment technology aspects which can enhance or interfere with the effectiveness of treatability performance. Such aspects include target pollutant oxidation state and speciation, interfering (or enhancing) interactive species, wastewaters containing multiple (potentially non-compatible) target species, and other issues. Effective pretreatment can be requisite to the efficient application of the appropriate treatment technology. Pretreatment options may include extraction or chemical modification of complexing or other interfering species, electrochemical pre-oxidation or reduction of a target speci, and management of adsorptive surface area and electrical charge in applications of inherent primary, or applied secondary co-precipitation. Improved awareness and understanding of such issues support cost- and performance-effective application of treatment technology.

# (119b) Impact of Water Quality and Assessment of Adsorption Technologies for Arsenic Removal

Kashinath Banerjee, Veolia Water Solution & Tech, Moon Township, PA, Gary L. Amy, Water Desalination and Reuse Research Center, 4700 King Abdullah University of Science and Technology, Thuwal, Saudi Arabia and Tapas K. Das, School of Engineering, Saint Martin's University, Lacey, WA

Abstract

The sources of arsenic in drinking water are both natural and anthropogenic. In order to protect public health, the U.S. EPA, World Health Organization (WHO), and the European Commission have decided to lower the MCL of arsenic in drinking water to $< 10$ µg/L. This stringent arsenic standard will inevitably require many utilities to upgrade their present systems or consider new treatment options. Extensive literature and vendor surveys revealed that a large number of commercial and experimental adsorbents are available for arsenic removal. Some of the commercially available materials are matured products (e.g., activated alumina) that have been widely tested, revealing both their attributes and limitations, while others are more recent products. Results from several investigators (DeMarco *et al.*, 2003; Greenleaf and SenGupta, 2005; Jekel et al., 1999; Pena *et al.*, 2005) indicate that some of these adsorbents are very promising.

Bench-scale studies were conducted to develop a comparative assessment of several commercially available adsorbents for arsenic removal, and to assess their performance in the presence of phosphate, vanadium, and silica. The primary objective of this work was to identify the impacts of water quality, and to define adsorbent properties that will minimize the effects of water quality and/or maximize arsenic adsorption capacity. Additionally, the effect of temperature on the adsorption kinetics and capacities were investigated. Iron modified activated alumina (AA-FS50), granular ferric hydroxide (GFH), granular ferric oxide (GFO), and a titanium-based media (MetSorb G) were used as the adsorbent. In order to better understand the adsorption mechanics of the selected adsorbents, each was characterized according to its important physical/chemical properties, including particle size, bulk density, pH of zero-point charge ($pH_{zpc}$), surface area, and mineralogy. The concentrations of dissolved arsenic and vanadium were analyzed using Inductively Coupled Plasma – Mass Spectroscopy (ICP-MS); Ion Chromatography (IC) was used to analyze phosphorus. The concentrations of molybdate-reactive (monomeric) silica in the samples were measured, and polymeric silica was estimated by the difference between the total and monomeric silica.

Results revealed that pH has a significant impact on the adsorption capacities. AA-FS50 showed a moderate decrease in adsorption capacity as pH increased, and phosphate had a small impact. GFO media was only slightly affected by increased pH, but was significantly impacted by phosphate, and monomeric silica had some minor effects. GFH performance decreased as pH increased, with phosphate and monomeric silica exhibiting some impact. The adsorption performance of MetSorb G also decreased as pH increased, and phosphate and monomeric silica had some impact, although the interfering effects were less at a lower pH for this media. The two iron oxide media and MetSorb G generally exhibited comparable capacities. For iron oxide based media, rate data are best described by a pseudo-first order kinetic model at each temperature and pH condition studied. At lower pH values, arsenic (V) exhibits greater removal rates than arsenic (III). An increase in temperature increases the overall adsorption reaction rate constant values for both arsenic (V) and arsenic (III). An examination of thermodynamic parameters reveals that adsorption of arsenic (III) as well as arsenic (V) by iron oxide media is an endothermic process and is spontaneous at the specific temperatures investigated.

**References**

DeMarco, M.J., SenGupta A.K., Greenleaf J.E., 2003. Arsenic removal using a polymeric/inorganic hybrid sorbent, *Water Res.* 37, 164-176.

Greenleaf, J. E. and SenGupta, A. K., 2005, *Investigating New Arsenic Removal Technologies*, Chapter 9.3.2, p. 524-

544; In Das, T. K., (ed.), *Toward Zero Discharge: Innovative Methodology and Technologies for Process Pollution Prevention,* John Wiley and Sons, Hoboken, NJ.

Jekel, M, and Seith, R., 1999. Comparison of conventional and new techniques for the removal of arsenic in a full-scale water treatment plant. Proceedings World Water Conference, Buenos Aires, Argentina.

Pena, M.E., Korfiatis, G.P., Patel, M., Lippincott, L., Meng, X., 2005. Adsorption of As (V) and As (III) by nanocrystalline titanium dioxide. *Water Res.* 39, 2327 – 2337.

# (119c) Removal of Arsenic by a Nano-Particle Blend Hollow Fiber Membrane

Jinsong He, Civil and Env Engineering, National University of Singapore, Singapore, Singapore and **J. Paul Chen**, Department of Civil & Environmental Engineering, National University of Singapore, Singapore, Singapore

The objective of this research is to prepare a based nano-particle blend hollow fiber membrane to adsorb arsenate from groundwater. A new nano-sized metal oxide was first prepared, which was then used in fabrication of the membrane. The hydrophobicity, porosity, mechanical strength, water flux and arsenate adsorption capacity of the membrane were investigated. The additional zirconium based nano-particle increased the hydrophilicity, which resulted in a significant enhancement on the water flux of the blend membrane. The batch adsorption experiments indicated that the membrane could effectively remove arsenate in a wide pH ranging from 2 to 9. The adsorption equilibrium could be reached in 24 h, and the maximum adsorption capacity was above 120mg/g. Ionic strength did not obviously affect the adsorption. The co-anions such as $F^-$, $HCO_3^-$, $SiO_3^{2-}$and $PO_3^{2-}$ had different effects on the adsorption. In addition, the membrane could remove arsenate efficiently though the continuous filtration. The mechanism for the removal of arsenate was proposed based on the FTIR and XPS studies.

# COMPETITIVE ADSORPTION OF NICKEL, MANGANESE, CHROMIUM AND CADMIUM FROM AQUEOUS SOLUTIONS BY DIFFERENT ADSORBENTS

*Malyuba Abu-Daabes* [a,*], *Hani Abu Qdais* [b], *Hatem Alsyouri* [c]

[a] *Pharmaceutical-Chemical Engineering Department, School of Applied Medical Sciences, German-Jordanian University, Amman, Jordan*

[b] *Department of Civil Engineering, Faculty of Engineering, Jordan University of Science and Technology, Irbid, Jordan*

[c] *Department of Chemical Engineering, Faculty of Engineering and Technology, University of Jordan, Amman, Jordan*

**corresponding author:*

Telephone: +962-6-530-0666, Fax: +962-6-5300-672, Email address: malyuba.abudaabes@gju.edu.jo

**Abstract**

Landfilling is the primary solid waste disposal method in Jordan. Since Jordan is considered one of the poorest countries with water resources, it is crucial to protect its groundwater resources from any possible risk of contamination from landfill leachates. A survey on leachates from landfills of different ages and control systems in Jordan, performed by our group during the years 2009 and 2010, showed that heavy metals concentrations in leachates were high exceeding the allowable limits of both Jordan Industrial wastewater and US Environmental Protection Agency. The main heavy metals with high risk are identified to be: Cr (0.66-20.5 mg/L), Mn (0.41-10.37 mg/L), Cd (0.48-4.5 mg/L), and Ni (0.74-6.69 mg/L). In the present study, competitive adsorption of these four metals on different adsorbents including: natural untreated shrimp shells (SH), granular activated carbon (GC), acid-treated granular activated carbon (AGC) and acid-treated powder activated carbon (APC) were investigated under simulated conditions. All adsorption isotherms were generated using the bottle point method, following the ASTM standard procedure (ASTM D3860-89a). Untreated granular carbon (GC) was the best adsorbent for Ni removal from a simulated solution with 90% removal efficiency. AGC and APC both showed very high selectivity for Cr over other present metals with removal efficiency of 99%. For Mn, shrimp shells (SH) was the best adsorbent with 79% removal efficiency, while both GC and SH adsorbents have comparable efficiency for Cd removal from the simulated solution (92%). Order of metals selectivity for SH adsorbent was Cd > Mn > Ni > Cr, for GC the selectivity was Cd > Ni > Cr > Mn, for both AGC and APC the order was Cr > Cd > Ni > Mn.

**Key words:** competitive metal ions, Leachate, Nickel, Manganese, Chromium, Cadmium, activated carbon, shrimp shells, chitin.

## Introduction

Landfilling remains the predominant management method for the disposal of Municipal Solid Waste which leads to the formation of leachates. Leachates may act as a vehicle for the introduction of heavy metals to groundwater resources. Contamination of water resources by heavy metals is a great environmental problem, mainly due to their high toxicity even at very low concentrations (Jaramillo et al., 2009). Heavy metals are not biodegradable and cannot be removed by natural processes from the ecosystems. A previous study by our group on leachates collected from three landfills with different ages and control systems showed that leachates from landfills possess high contamination potential to the limited water sources in Jordan. Heavy metals, mainly Cr, Mn, Cd, and Ni, were found to exceed the allowable limits set by Jordan Industrial Estates Corporation and US Environmental Protection Agency for industrial wastewater discharge. Pollution of heavy metals may arise from different sources. Cd, Cr, Mn and Ni are mainly introduced to the environment from spent rechargeable and household batteries, electroplating, (Xara et al. 2009; Karnchanawong and Limpiteeprakan 2009) and tannery industry (Gotvajn et al., 2009).

One of the promising methods for heavy metals removal from wastewater is the use of natural and surface-modified adsorbents for selective removal of metal ions. Activated carbon is widely used in many applications. In this study the effect of modifying activated carbon surface with acidic groups on competitive metal ions removal from aquatic solution is evaluated. Shrimp shells are also used as a low cost natural adsorbent without any surface modification.

## Materials and Methods

### *1) Adsorbents*

Activated carbon was used both in granular (GC) and powder form. All carbon was boiled in deionised water, dried in an oven at 110 °C for 24 hours. Further surface modification was performed by boiling part of them with 4M $HNO_3$ solution for 1 hour with total condensation. After cooling, the solution of the acid-treated granular activated carbon (AGC) and acid-treated powder activated carbon (APC) was decanted and the solids were washed with deionised water several times until a neutral, constant pH was achieved. Shrimp shells (SH) were washed thoroughly, grounded and sieved without further modification.

### *2) Characterization of Adsorbents*

Surface area and pore size distribution of different adsorbents were measured following the Brunauer, Emmett, and Teller (BET) method (Quantachrome Instruments, Model Autosorb-1, USA).

**Table 1.** Adsorbents characterization

| Adsorbent | BET | HK cumulative pore volume | HK pore wedth |
|---|---|---|---|
| | $m^2/g$ | cc/g | nm |
| GC | 635 | 2.99E-01 | 1.398 |
| AGC | 591 | 2.79E-01 | 1.388 |
| APC | 892 | 4.22E-01 | 1.388 |
| SH | 25.6 | 9.95E-03 | 1.058 |

### *3) Adsorption isotherms*

Stock solutions of heavy metals of 100 mg/L concentration (Ni(II), Mn(II), Cr(VI)) and 50 mg/L Cd(II) were prepared separately by dissolving appropriate quantities of nickel (II) nitrate hexahydrate, cadmium nitrate in nitric acid 0.5 mol/L (Fluka, Switzerland), manganese (II) sulfate monohydrate and chromium (VI) oxide (Merck, Germany). Adsorption isotherms were generated using the bottle point method, following the ASTM standard procedure (ASTM D3860-89a). Adsorption isotherms were generated at 25 °C, using simulated solution of the main heavy metals found in the landfill leachates. The composition of the simulated solution was 100 mg Ni/L, 100 mg Cr/L, 100 mg Mn/L, and 50 mg Cd/L. The pH of the solution was adjusted to 6. Ten ml of the simulated solution was added to 15-ml bottles that contained preweighed amounts (25, 50, 100, 500 and 1000 mg) of the adsorbents. All samples were agitated for 3 days in a temperature controlled water bath to ensure that equilibrium was achieved. The samples were then centrifuged and filtered using 0.45 μm syringe filters before being analysed. The samples were analyzed for their heavy metals content using Inductive Coupled Plasma Technique (ICP, Perkin-Elmer, Optima DV2000, USA). Blank samples were used with every isotherm in order to account for any loss and adsorption on the walls of the tube.

The equilibrium adsorptive quantities of the metal ions on the adsorbents surfaces were calculated using the following relation:

$$q_e = \frac{C_0 - C_e}{M} \times V \qquad (1)$$

Where $q_e$ the amount of metal ion adsorbed per unit mass of adsorbent (mg/g), $C_0$ and $C_e$ are the original ion concentration and the equilibrium solution concentration (mg/L), respectively, $M$ is the mass of adsorbent, $V$ is the volume of solution (L).

The percent removal of the metal ion by the adsorbent was calculated using the following mass balance relation:

$$\%\, Removal = \frac{C_0 - C_e}{C_0} \times 100\% \qquad (2)$$

## Results and Discussion

### *1) Properties of the adsorbents*

The highest surface area is reported for the powdered activated carbon treated with acid (APC) (892 $m^2/g$) and the lowest one for shrimp shells (25.6 $m^2/g$), as shown in Table 1. All measured samples have microporous structure with pore sizes less than 2 nm based on Horvath-Kawazoe (HK) model for the determination of pore size distribution in microporous sorbents such as activated carbon and zeolites. Acid modification of granular activated carbon lowered its surface area from 635 $m^2/g$ to 591 $m^2/g$ for GC and AGC, respectively. Previous work showed that oxygenation of activated carbon with 4M $HNO_3$ significantly increased the concentration of acidic surface oxygen complexes on the surface (carboxylic and carbonyl groups), reduced the total number of basic groups on the surface and caused the surface to be more hydrophilic (Franz et al., 2000; Abu-Daabes and Pinto, 2004; Jaramillo et al., 2009). These acidic sites are bonded to the edges of the graphitic basal plane.

## 2) *Competitive adsorption of heavy metals*

Adsorption isotherms for all four metal ions, $Ni^{+2}$, $Cr^{+6}$, $Mn^{+2}$, and $Cd^{+2}$, are shown in Figure 1 (a-d). All isotherms don't show uniform curve which could be due to the competition between the ions for the adsorption sites as well as the repulsion and synergy between them. Similar behaviour for competitive adsorption of metal ions was observed by Covelo et al.,( 2004). In general, the mechanism of heavy metal ions adsorption on the surface of activated carbon is by electrostatic attraction to the carbon surface, ion-exchange and surface complexation (Jaramillo et al., 2009; Radovic et al., 2001). Adsorption of heavy metals on the unmodified carbon surface was suggested to be through interactions of the metal ion with the $\pi$ electrons of the carbon basal planes according to the reactions (Jaramillo et al., 2009):

$$-C\pi + H_2O \leftrightarrow -C\pi\text{-}H^+ + OH^- \quad (3)$$

$$2(-C\pi\text{-}H^+) + M^{2+} \leftrightarrow (-C\pi)_2 - M^{2+} + 2H^+ \quad (4)$$

In the presence of acidic functional groups on the surface of activated carbon the mechanism become surface complexation or ion-exchange depending on the pH. If the solution has low pH, the weakly acidic groups are more likely to be protonated and will undergo ion-exchange reactions with the metal ions, as described in reaction (5).

$$2(-COOH) + M^{2+} \leftrightarrow (-COO)_2Cu + 2H^+ \quad (5)$$

At higher pH, the acidic groups will be nonprotonated and the metal ions will bind to them by surface complexation, releasing less than one hydrogen ion by each metal ion adsorb (Jaramillo et al., 2009).

The selectivity sequences for heavy metal ions depends on the characteristics of the ions such as ionic radius ($Cd^{2+} > Ni^{+2} = Mn^{+2} > Cr^{+6}$), ionic potential or charge-to-radius ratio ($Cr^{+6} > Ni^{+2} = Mn^{+2} > Cd^{+2}$) and electronegativity (Ni > Cd > Cr > Mn) (Sposito, 1989; McBride, 1994; Covelo et al., 2004). The selectivity of each adsorbent for heavy metals under simulated conditions is shown in Figure 2 for the 500 mg adsorbents isotherms. The results show that acid treatment of both granular carbon and powder carbon (AGC and APC, respectively) changed the order of metals selectivity compared to the basic granular carbon (GC). Furthermore, acid treatment of granular carbon (AGC) increased Cr removal efficiency by 30% while it decreased for Ni, Mn and Cd by 23%, 36% and 37%, respectively. This can be explained by the much higher ionic potential of $Cr^{+6}$ compared to other ions which will lead to strong attraction to the acidic surface groups of activated carbon and less accessibility for the ions with lower ionic potential.

Shrimp shells compose of 20% chitin by mass of dry weight, 37% protein and 43% minerals (Einbu, 2007). Chitin is a natural aminopolysaccharide. The results of this study showed that untreated shrimp shells have very high selectivity for $Cd^{2+}$ and $Mn^{2+}$. This could be due to the presence of amine groups on the chitin of SH. The amine functional groups will form metal complexation. The order of metals affinity for different adsorbents under simulated conditions is summarized in Table 2.

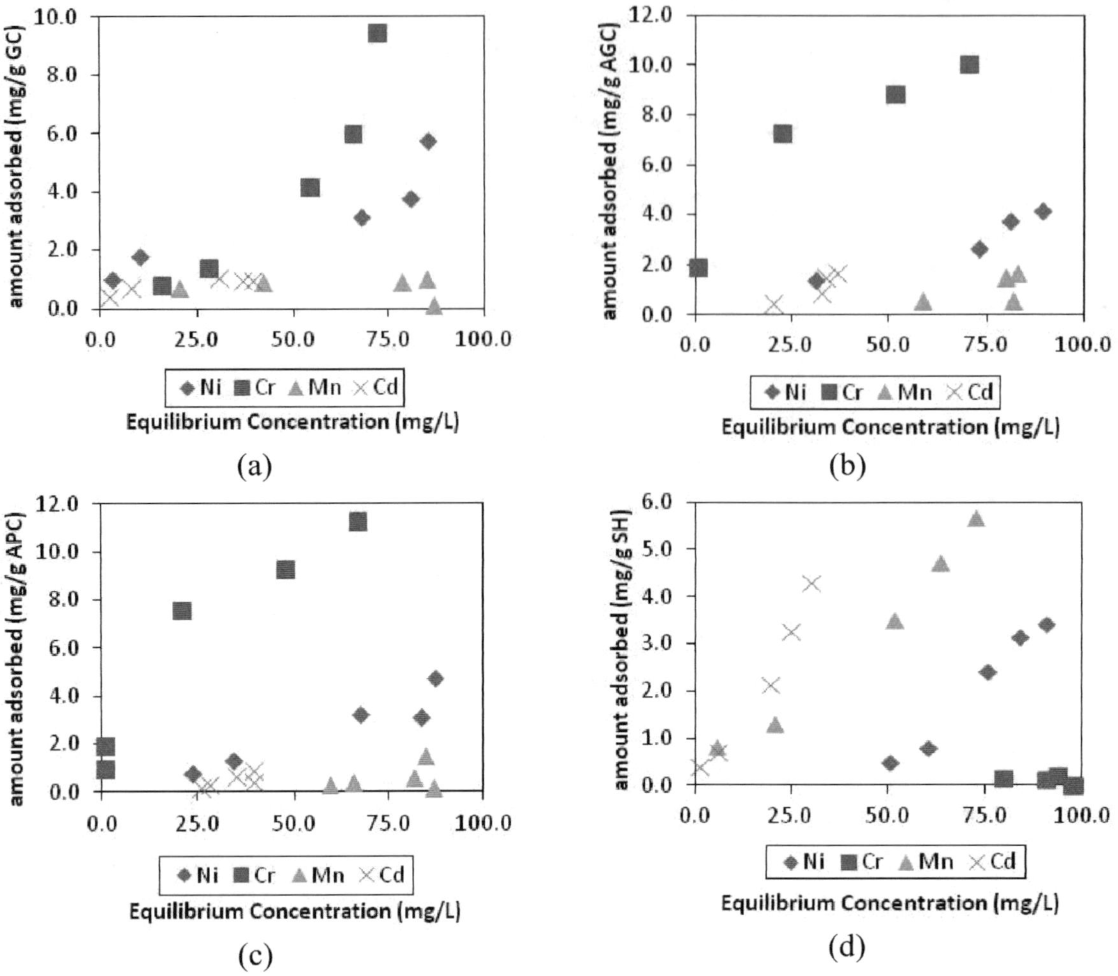


**Figure 1. Adsorption Isotherms of Metal Ions on Different Adsorbents**

**Table 2.** Order of metals selectivity for different adsorbents

| Adsorbent | Metals order |
|---|---|
| **GC** | Cd > Ni > Cr > Mn |
| **AGC** | Cr > Cd > Ni > Mn |
| **APC** | Cr > Cd > Ni > Mn |
| **SH** | Cd > Mn > Ni > Cr |

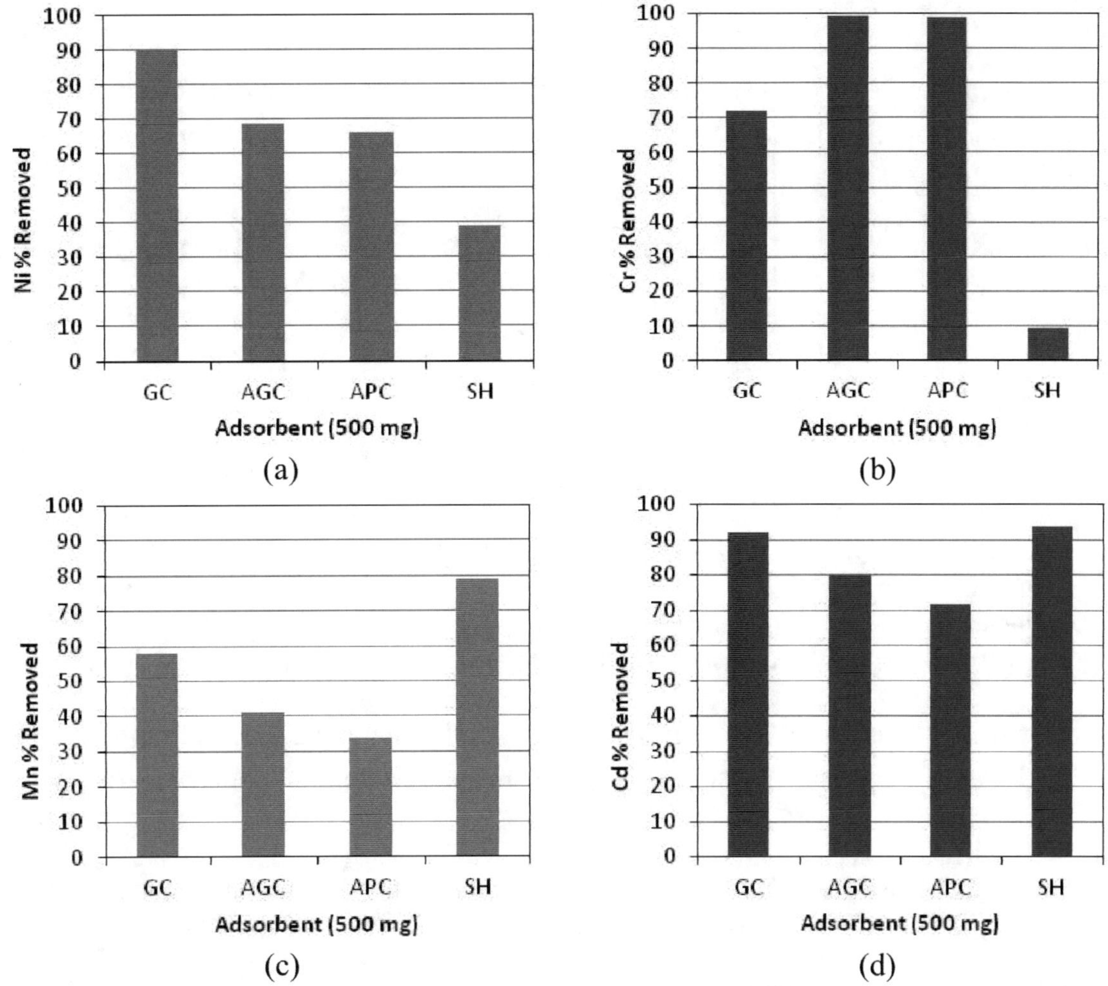


**Figure 2. Percent removal of Metal Ions on Different Adsorbents**

## Conclusion

In this study activated carbon, surface modified activated carbon and unmodified natural shrimp shells were evaluated as potential adsorbents for competitive removal of $Ni^{+2}$, $Cr^{+6}$, $Mn^{+2}$, and $Cd^{+2}$ from aqueous solutions. Untreated granular carbon was the best adsorbent for $Ni^{+2}$ removal from a simulated solution with 90% removal. AGC and APC both showed very high selectivity for $Cr^{+6}$ over other present metals with removal efficiency of 99%. This was due the high ionic potential of $Cr^{+6}$ compared to other ions. For $Mn^{+2}$, shrimp shells (SH) was the best adsorbent with 79% removal efficiency, while both GC and SH adsorbents have comparable efficiency for $Cd^{+2}$ removal from the simulated solution.


## Acknowledgments

The authors wish to thank the German-Jordanian University, the Higher Council for Science and Technology in Jordan, and the European Union presented by Support to Research and Technological Development & Innovation Initiative & Strategies in Jordan (SRTD) Project for their financial support to this project.

## References


Abu-Daabes, M., Pinto, N.G. (2004) "Effect of Surface Oxygen Complexes of Activated Carbon on Phenol Adsorption from Single and Mixed Non-Aqueous Solvents", *Separation Science and Technology*, 39, 2997-3009.

Covelo, E. F., Andrade Couce, M. L., Vega, F. A. (2004) "Competitive Adsorption and Desorption of Cadmium, Chromium, Copper, Nickel, Lead, and Zinc by Humic Umbrisols", *Communications in Soil Science and Plant Analysis*, 35, 2709-2729

Einbu, A. (2007), "Characterisation of Chitin and a Study of its Acid-Catalysed Hydrolysis", Ph.D. Thesis, Norwegian University ofScience and Technology.

Franz, M., Arafat, H.A., Pinto, N.G. (2000), "Effect of chemical surface heterogeneity on the adsorption mechanism of dissolved aromatics on activated carbon", *Carbon*, 38, 1807–1819

Gotvajn, Z., Tisler, T., Zagorc-Koncan, J. (2009), "Comparison of different treatment strategies for industrial landfill leachate". *Journal of Hazardous Materials*, 162, 1446-1456.

Jaramillo, J., Go´mez-Serrano, V., A´ lvarez P.M. (2009), "Enhanced adsorption of metal ions onto functionalized granular activated carbons prepared from cherry stones". *Journal of Hazardous Materials*, 161, 670–676

Karnchanawong, S., Limpiteeprakan, P. (2009), "Evaluation of heavy metal leaching from spent household batteries disposed in municipal solid waste". *Waste Management*, 29, 550–558.

McBride, M.B. Environmental Chemistry of Soils; Oxford University Press: New York, 1994.

Radovic, L.R., Moreno-Castilla, C., Rivera-Utrilla, J. (2001), "Carbon materials as adsorbents in aqueous solutions", in: L.R. Radovic (Ed.), Chemistry and Physics of Carbon, 27, Dekker, New York, pp. 227–405.

Sposito, G. (1989), "The Chemistry of Soils", Oxford University Press: New York,.

Strelko, V., Malik, D.J. (2002) "Characterization and metal sorptive properties of oxidized active carbon", *J. Colloid Interf. Sci.* 250 213–220

Xará, S., Delgado, J., Almeida, M., Costa, C. (2009), "Laboratory study on the leaching potential of spent alkaline batteries", *Waste Management*, 29, 2121–2131.

# (119e) Arsenic Mobilization In Alluvial Soils of Punjab, North-West India Under Flooded Conditions

**Kuldip Singh**[1], Dhanwinder Singh[2], H. S. Hundal[1] and Raj Kumar[2], (1)Soils, Punjab Agricultural University, Ludhiana, India, (2)Soil Science, Punjab Agricultural University, Ludhiana, India

Elevated arsenic contaminant in drinking water above the WHO (1996) safe limits of 10 $\mu gL^{-1}$ was reported at many locations in underground water of Punjab, North-West India. In rural areas, flooded irrigations and maintenance of standing water in paddy fields likely to create anoxic conditions in the soils over-lying above the shallow aquifers. The objective of the present investigation was to study the mechanism of arsenic mobilization in surface alluvial soils. Fifty gram of each of 18 soil samples were transferred to 200 ml polythene bottles, and 100 ml of deionized water was added to maintained flooded conditions. These soil-water systems were kept in BOD incubator at $25^0C$ for equilibration under anaerobic conditions of 24 and 240 hours period, respectively. After equilibration periods, pH, EC, soluble $HCO_3$ and $Cl^-$ were determined for each soil-water equilibrated systems. Supernatant aliquots were withdrawn using a syringe, centrifuged, filtered and acidified with a drop of nitric acid prior to analysis for Fe, Mn, $SO_4^{2-}$ and $PO_4^{3-}$ on ICAP-AES. Arsenic in the filtrates solution was determined using hydride generation assembly on ICAP-AES.

Total dissolved As in 24 hour soil-water equilibrated systems varied from 3 to 16 $\mu gL^{-1}$ with mean value of 9 $\mu gL^{-1}$ (Figure 1), whereas its concentration range in different soil-water suspensions after 240 hour of equilibration period enhanced to 33 to 1761 $\mu gL^{-1}$ with mean value of 392 $\mu gL^{-1}$. The median, 75th and 90th percentiles concentrations of As were 8, 12 and 14 $\mu gL^{-1}$ after 24 hour and eventually elevated to 190, 490 and 1013 $\mu gL^{-1}$, respectively after 240 hours of anaerobic conditions. Arsenic concentration was enhanced with decreasing redox potential. The results showed antagonistic relationship of arsenic concentration with the magnitude of redox potential status of alluvial soil-water suspensions (Figure 2a). Dissolved Fe concentration versus redox status of soil-water systems determined after two equilibration periods are illustrated in Figure 2b. As the pe decreased, the Fe concentration declined in different alluvial soil water systems.

The effect of redox potential on soluble Mn concentrations determined at different flooding periods in alluvial soil-water suspensions is shown in Figure 2c. As the pe decreased, the soluble Mn concentration enhanced in soil water systems. The soluble concentrations of Mn was ranged between 0.034 to 0.407 $mgL^{-1}$ with a mean value of 0.132 $mgL^{-1}$ after 24 hour equilibration period, but were considerably enhanced to 0.449-44.44 with mean value of 22.398 $mgL^{-1}$ after 240 hour of equilibration periods. The behavior of $SO_4^{2-}$ in different alluvial soil-water systems under different redox status is shown in Figure 2d. With decrease in pe, $SO_4^{2-}$ concentration declined in all soil water systems. Total dissolved concentration of $SO_4^{2-}$ anion in soil-water suspensions also considerably lowered to range of 0.003-25.30 $mgL^{-1}$ with mean value of 1.95 $mgL^{-1}$ after 240 hour of equilibration periods from 4.56 to 81.6 $mgL^{-1}$ with a mean value of 25 $mgL^{-1}$ after 24 hours of equilibration periods. Under reducing conditions, sulfate concentrations decreased significantly.

Pourbaix (pe-pH) diagrams for arsenic and ferromagnesian oxides (Figure 3 and 4) were constructed by to identify the chemical species for each of these elements in different soil-water suspensions. Aqueous arsenic species, $HAsO_4^{2-}$ and $H_3AsO_3^0$ along with the possible solid phase, orpiment (amorphous $As_2S_3$) as were identified in soil-water systems (Figure 3). The mono valent arsenate oxy-anion ($HAsO_4$ ) specie was predominant in all the soil-water suspensions except soils 2, 4 and 8, where arsenite ($H_3AsO_3^0$) specie was identified after 24 hour of anaerobic condition. The predominance of neutral arsenite ($H_3AsO_3^0$) specie was observed after 240 hours of reduced incubated period in all soil water systems except soils 12, 15 and 16 in which orpiment ($As_2S_3$) as solid phase was identified. The neutral arsenite ($H_3AsO_3^0$) and arsenate oxy-anions ($H_2AsO_4^-$) has been found to be predominantly occurred in the alluvial aquifers of Punjab between the pH of 7.05 to 8.25 and under reduced pe value from 0.77 to -4.49, respectively. Thus, it is conclusively evident that two chemical aqueous species controlling As chemistry and its toxicity in soils are arsenate and arsenite. Arsenate is the thermodynamically stable form under aerobic conditions. Arsenite is the predominant species under anaerobic conditions and it is a neutral species at natural pH and is more soluble, mobile and phototoxic than arsenate. The orpiment ($As_2S_3$) solid phase domain predominates near the bottom of pe-pH diagram and lying over the $H_2O$ / $H_2g$ stability line in the Pourbaix diagram (Figure 3). After 240 hours of equilibrium period, arsenite ($H_3AsO_3^0$) specie was predominant in all alluvial soil-water suspensions. The increase in arsenic concentration was also primarily due to an increase in As (III) species in all the soil-water suspensions under more reduced conditions. In some soil water systems (soil number 12, 15 and 16) apparent precipitation of orpiment was identified on pe-pH diagram. It is evident that total arsenic concentration continuously enhanced in these soil-water systems with decreased magnitude of redox potential. The simultaneously drastic decline in concentration of both Fe and $SO_4^{2-}$ in contrast to accumulation As in reduced soil-water suspensions of alluvial Holocene validated the formation of iron sulfide mineral in reduced alluvial soil-water systems of Punjab. Because of the differences in solubility of iron (DG= -117.11 Kcal mole$^{-1}$) versus arsenic (DG = -382.64 Kcal mole$^{-1}$) sulfides, precipitation of iron sulfide may remove sulfide from solution but not arsenic since the precipitation rates of iron sulfide was very rapid compared to amorphous arsenic sulfide.

The aqueous species of $Fe^{2+}$, $Mn^{2+}$ and $MnHPO_4$ and solid phase of hematite ($Fe_2O_3$) were predominant in all the alluvial soil-water equilibrated solutions after 24 hours within the pe and pH ranges of -1.75 to 0.77 and 7.25-8.25, respectively (Figure 4). After 240 hours of incubation period, reduction potential status range in soil-water suspensions declined between -4.49 to -2.74 within pH range of 7.05 to 7.94. Aqueous species of $Fe^{2+}$, $Mn^{2+}$ and $MnHPO_4$ and solid phase of hematite [$Fe_2O_3$] were predominant in soil-water suspensions of soils 1, 5, 6, 11 and 14. Pourbaix diagram (Figure 4) for iron illustrate that declined in iron was due to the predominantly precipitation of iron sulfide [$FeS_2$] and or as vivianite [$Fe_3 (PO_4)_2 8H_2O$] minerals. The reduced solid phase vivianite [$Fe_3 (PO_4)_2 8H_2O$] was found in soil 2, 3, 7 and 13, respectively. Pyrite [$FeS_2$] was found to be predominant in soil 8, 9, 10, 12, 16, 17 and 18, respectively. All the water composition falls in the stability field for aqueous $Mn^{2+}$ ion (Figure 4) and its concentrations varied from 0.45 to 44.44 mgL$^{-1}$. Stability diagram for manganese showed that $Mn^{2+}$ ion is thermodynamically more stable and predominant across a wide pe-pH range than for $Fe^{2+}$ ion. Consequently, the $Mn^{2+}$ ion remain in solution in aqueous environment in which $Fe^{2+}$ ions is oxidized to $Fe^{3+}$ and precipitated as Fe $(OH)_2.7Cl_{0.30}$ below pH 6.8 at pe > -1.8 and $Fe_2O_3$ above pH 6.8 at pe < -1.8 and thus eventually contributed to higher dissolved $Mn^{2+}$ concentration under reduced environments in soils. Iron and manganese concentration in the anoxic soil-water systems were as high as 1.634 and 44.44 mgL$^{-1}$, respectively. The largest contributor of ferrous to the equilibrated soil-water interaction appears to be an iron oxide phase (probably maghemite, Gamma-$Fe_3O_3$, an alteration of magnetite). These results showed that dissolve Fe concentration were probably controlled by hematite ($Fe_2O_3$) under alkaline oxic condition and by iron sulfide/pyrite (FeS) or in some soils by vivianite [$Fe_3(PO_4)_28H_2O$] under reduced environment, where soluble phosphate concentration was high. Under reduced condition (developed after 240 hours of flooding), soil solution Fe and $SO_4^{2-}$ are governed by newly formed precipitates of pyrite ($FeS_2$).

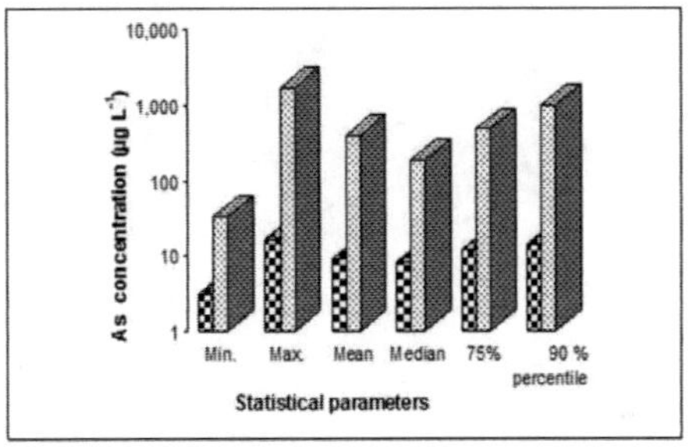


**Fig.1** As concentrations measured in 18 alluvial soil-water suspensions of Punjab after equilibration of 24 (▩) and 240( ▣) hours

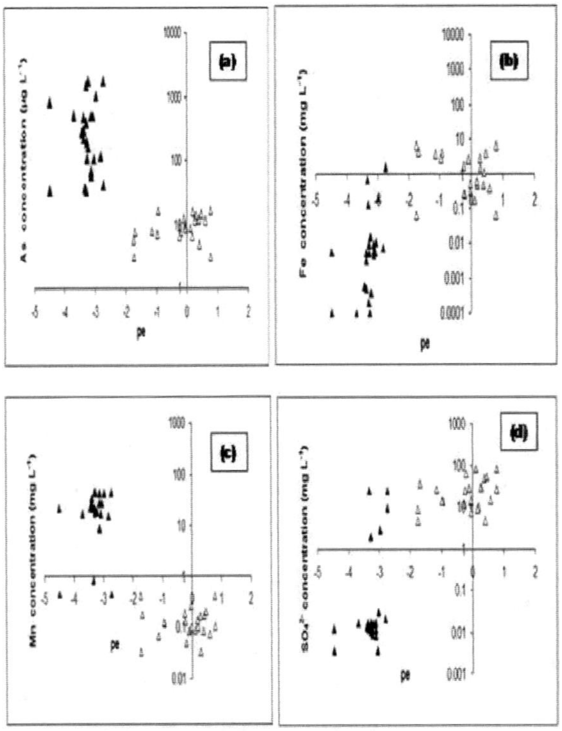


**Fig. 2** Effect of redox potential on As (a), Fe (b), Mn (c) and $SO_4^{2-}$ (d) concentrations measured in 18 alluvial soil-water suspensions after 24 (? ) and 240 (? ) hours of equilibration periods

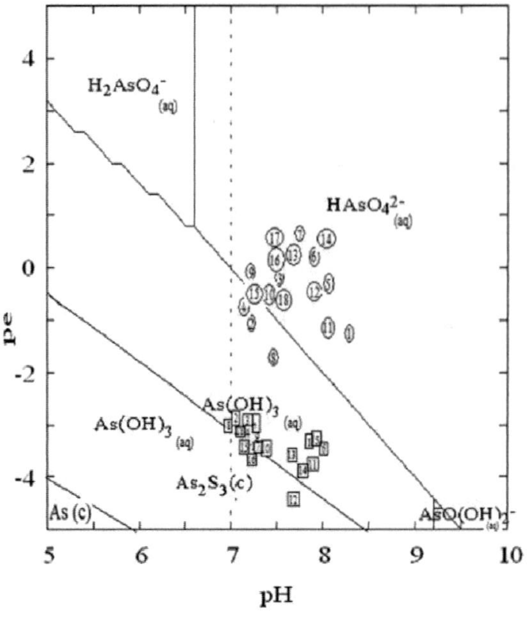


**Fig. 3** Arsenic species identified in alluvial soil-water suspensions after equilibration of 24 (O) and 240 (?) hours at 25 °C. The concentration of [$AsO_4^{3-}$] =6.70µM, [$PO_4^{3-}$]=0.37mM, [$SO_4^{2-}$]= 0.25mM,[$CO_3^{2-}$]=0.20mM, [$Cl^-$]=0.13mMand ionic strength=0.020M were fixed to construct the Pourbaix diagram

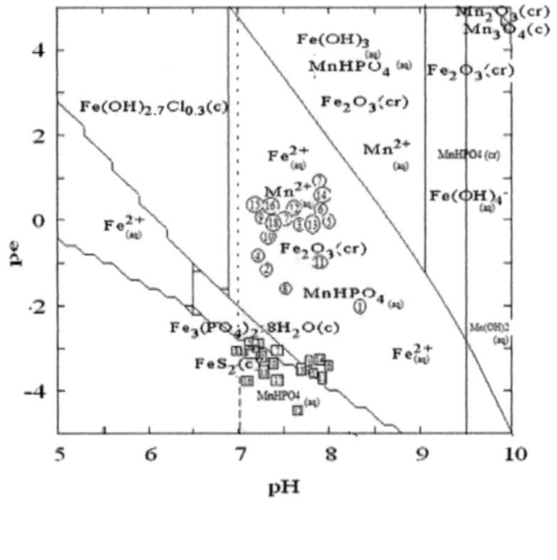


**Fig.4** Iron and manganese species identified in alluvial soil-water suspensions after equilibration of 24 (O) and 240 (?) hours at 25 °C. The concentration of [Fe=15µM], [Mn=0.27mM], =6.70uM, [PO43- ] =0.37mM, [$SO_4^{2-}$]= 0.25mM,[$CO_3^{2-}$]=0.20mM, [$Cl^-$]=0.13mM and ionic strength=0.020M were fixed to construct the Pourbaix diagram

# ADSORPTION OF LEAD FROM AQUEOUS SOLUTIONS BY COTTON FABRICS MODIFIED WITH CITRIC ACID

*Refugio Bernardo García Reyes*, Saúl H. Martínez Treviño, Alcione García González and Ricardo Gómez González*

Universidad Autónoma de Nuevo León, Facultad de Ciencias Químicas, Av. Universidad s/n, Cd. Universitaria, C.P. 66400, San Nicolás de los Garza, N. L., México. FAX: +52(81)83294000 Ext. 6282, TEL: +52(81)83294000 Ext. 6288
*E-mail: bernardogarciareyes@yahoo.com.mx

## ABSTRACT

The objective of this research was to modify cotton (i.e. cellulose) fabrics for the removal and recovery of lead (Pb(II)) from aqueous solutions. Cotton fabrics were modified with 0.6 M citric acid solutions and characterized by infrared spectroscopy (ATR-FTIR) and potentiometric titrations. Also, adsorption, desorption, and kinetic experiments in batch systems were conducted.

When modifying cotton fabrics with citric acid, density of functional groups increased up to 2.47 mmol/g in comparison with the non-modified fabric that had 0.61 mmol/g. In addition, estimated equilibrium constants for the modified fabrics can be associated in part to carboxylic groups of the citric acid. The appearance of carboxylic groups on modified fabrics was also confirmed by ATR-FTIR analyses. Maximum adsorption capacity of Pb(II) on modified fabrics, estimated with Langmuir isotherm, was 73.3 y 42.3 mg/g at pH 4 y 3, respectively, whereas non-modified cotton fabrics did not adsorb Pb(II) from water. Decrement in adsorption capacity suggests that Pb(II) and $H^+$ ions compete for the same adsorption sites. On the other hand, it was possible to desorb around 92% of Pb(II) previously adsorbed on the modified cotton with using 0.1 M HCl solutions since the chemical equilibrium is shifted. Finally, adsorption kinetics can be predicted with a pseudo-second order model.

In summary, cotton fabric treated with 0.6 M citric acid has an adsorption capacity of Pb(II) up to 3.5 times greater than the activated carbon, and the adsorbed metal can be recovered almost completely with 0.1 M HCl. These results suggest that modified cotton fabric is a highly competitive adsorbent for the treatment of metal-containing wastewater.

## 1. INTRODUCTION

Consumption of water polluted with lead (Pb(II)) is associated to human diseases such as anaemia, kidney damage, sterility, and it is suspected to be carcinogenic. For these reasons, several treatment technologies have been applied for the removal of this pollutant from water, for example: chemical precipitation, coagulation-flocculation, ion exchange, and adsorption with activated carbon, among others (Stumm y Morgan, 1996; Metcalf and Eddy, 2004). Adsorption process is one of the most commonly used technologies because it is useful as a polishing step in a water treatment plant. Nevertheless, it is important to explore the possible application of new adsorbents for removing lead from aqueous solutions. For instance, lignocelullosic materials can be used as new adsorbents because these are worldwide available and low-cost in comparison with polymeric resins and activated carbons. Moreover, it was reported that modification of lignocelullosic materials with citric acid enhanced the adsorption capacity of copper in comparison with non-treated adsorbents (Marshall *et al.*, 2007; Zhu *et al.*, 2008). In this way, the adsorption of metal cations in chemically modified lignocelullosic materials can be enhanced significantly. Nonetheless, to the best of our knowledge, at this time is not reported the density of functional groups added to the adsorbent and their equilibrium constants before and after chemical modification.

In this work, the effect of citric acid modification of cotton fabrics on lead adsorption from aqueous solutions is explored by means of ATR-FTIR analyses, potentiometric titrations, adsorption/desorption and kinetic experiments.

## 2. METHODOLOGY

**2.1 Chemical modification.** Thirty grams of cotton fabrics were added in a precipitated flask with 210 mL of 0.6 M citric acid solution. Suspension was heated at 60°C for 1 hour and, afterwards, water was evaporated at about 100°C until dryness. The remaining residues were dried at 120°C in an oven for 24 hours to carry out the esterification on cotton fabrics. Next, modified cotton fabrics were rinsed out with deionized water at 50°C and were dried at 100°C and stored in desiccators.

**2.2 Adsorbent characterization.** Non-modified and modified cotton fabrics were characterized by means of attenuated total reflection Fourier transform infrared spectroscopy (ATR-FTIR) and potentiometric titrations (Yun *et al.*, 2001; Garcia-Reyes *et al.*, 2009) to identify quantitative and qualitatively the functional groups incorporated to the adsorbent materials, respectively.

**2.3 Adsorption experiments.** Adsorption tests were conducted in batch systems at 30°C, 150 rpm and the selected pH (3 or 4) to keep lead in solution. A mass of 40 mg of cotton fabrics (non-modified and modified) were added to 40 mL of Pb(II) solutions with different concentration (5-100 mg/L). Solution pH was daily adjusted (with $HNO_3$ or NaOH) along the Pb(II) adsorption tests at the chosen value (pH 3 or 4) until the equilibrium was reached. Aliquots were taken at the beginning and the ending of adsorption experiments to determine the metal concentration by atomic absorption spectroscopy (ASS).

**2.4 Desorption trials.** Cotton fabrics, previously used in lead adsorption experiments, were regenerated with 40 mL of 0.1 N HCl solutions after being separated by filtration and rinsed with 30 mL of deionized water at pH 4. These suspensions were continuously stirred at 150 rpm during 24 hours. After that, an aliquot was taken to determine the Pb(II) concentration in solution by AAS.

**2.6 Adsorption kinetics.** To determine the time required to achieve the equilibrium and to estimate the parameters of kinetic models, adsorption rate experiments of Pb(II) on cotton fabrics were carried out in a packed column with recirculation at initial Pb(II) concentration of 0.256 mM, 30°C, initial pH of 4 and a feeding flow of 10 mL/min. Previous to the adsorption kinetic experiments, modified cotton fabrics were conditioned by stirring the adsorbent in a solution with pH 4 for 24 hours.

## 3. RESULTS AND DISCUSSION

Figure 1 shows the infrared spectra of the adsorbents used in this research. It is observed that chemical modification of cotton fabrics shown a peak about 1740 $cm^{-1}$ (absent in non-modified cotton fabrics) that can be associated to carboxylic groups of the citric acid (see Figure 1) covalently bound to hydroxyl groups of cellulose structure. It was reported that carboxylic functional groups play a major in adsorption of metal cations from aqueous solutions (Tiemann *et al.*, 1999; Garcia-Reyes *et al.*, 2009).

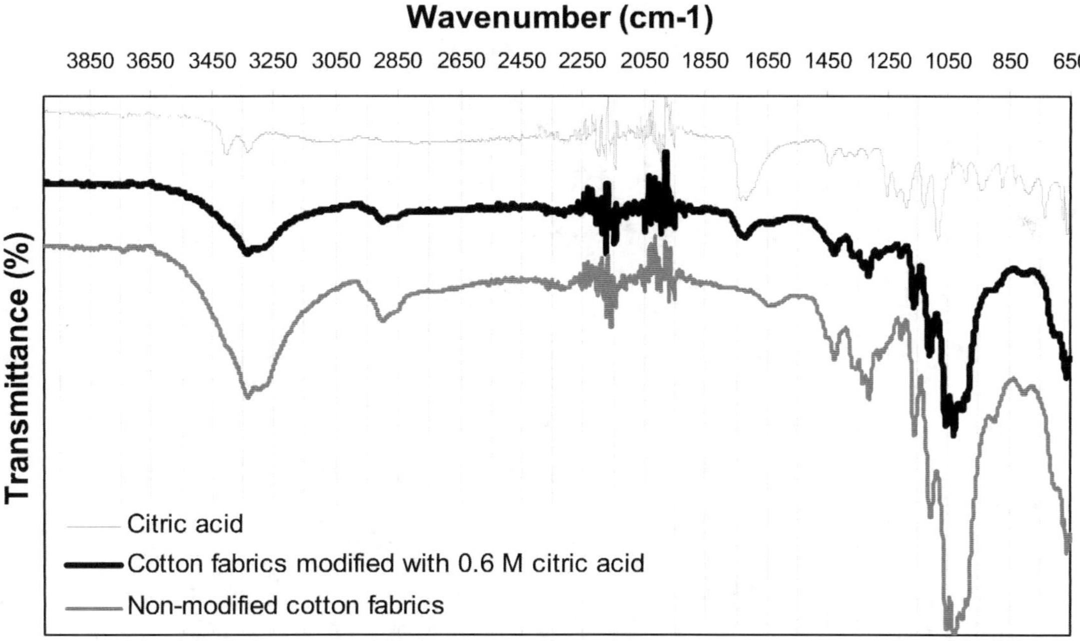


**Figure 1.** Infrared spectra (ATR-FTIR) of the studied adsorbents.

Table 1 shows the quantity of functional groups per unit mass of adsorbent ($b_j$) and their equilibrium constants ($pk_{aj}$) of non-modified and modified cotton fabrics. Non-modified cotton fabrics presented one functional group with an estimated quantity of 0.61 mmol/g and a $pk_a$ value of 8.71 probably due to the hydroxyl groups of cellulose (peaks at 3350 and 1030 $cm^{-1}$ in Figure 1) (Silverstein and Webster, 1998). In contrast, cotton fabrics modified with citric acid showed two functional groups ($b_1$=0.67 y $b_2$=1.80 mmol/g) with different $pk_a$ values (4.06 and 8.08). These equilibrium constants are close to two of three $pk_a$ values reported for citric acid (3.13, 4.8 y 6.40). It is important to mention that the functional group of modified cotton fabrics with $pk_a$=4.06 may correspond to a carboxylic group of citric acid and, therefore, these functional groups can be used as adsorption sites of positively-charged species in water such as Pb(II) species ($Pb^{2+}$, $PbNO_3^+$, $PbOH^+$, and $Pb_2OH^{3+}$) (Tiemann *et al.*, 1999; García-Reyes *et al.*, 2009) meanwhile pH of solution remains below 5.6 to avoid Pb(II) precipitation.

**Table 1.** Density of functional groups ($b_j$) and their equilibrium constants ($pk_{ai}$) of the studied adsorbents.

| Adsorbent | $b_1$ (mmol/g) | $pk_{a1}$ | $b_2$ (mmol/g) | $pk_{a2}$ |
|---|---|---|---|---|
| ***Non-modified cotton fabrics*** | n.d. | n.d. | 0.61 | 8.71 |
| ***Modified cotton fabrics*** | 0.67 | 4.06 | 1.80 | 8.08 |

n.d.: Not detected.

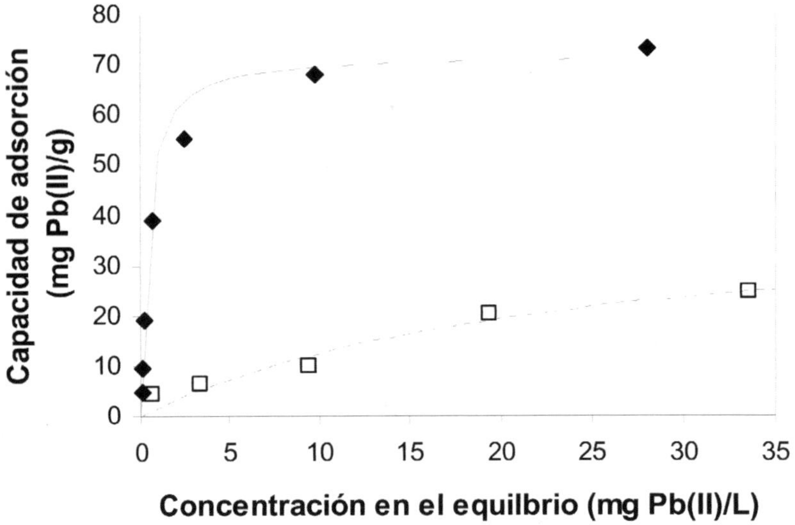


**Figure 2.** Adsorption isotherms of Pb(II) on cotton fabrics modified with 0.6 M citric acid at 30°C, pH 3 (□) y 4 (♦). Symbols represent experimental data and lines were predicted with the Langmuir isotherm.

Figure 3 shows the adsorption rate of Pb(II) on modified cotton fabrics without pH control. Kinetic experiments were conducted at initial pH 5 but final (at equilibrium) pH decreased to 3.36. This result seems to be reasonable because pH was not corrected along the kinetic experiments and Pb(II) species displaced $H^+$ ions from protonated carboxylic groups (covalently bound on the cotton fabrics). Adsorption kinetics was well predicted with the pseudo-second order model (Ho y

McKay, 1999). This result suggests that lead species interact with adsorption sites by physical forces such as electrostatic charges.

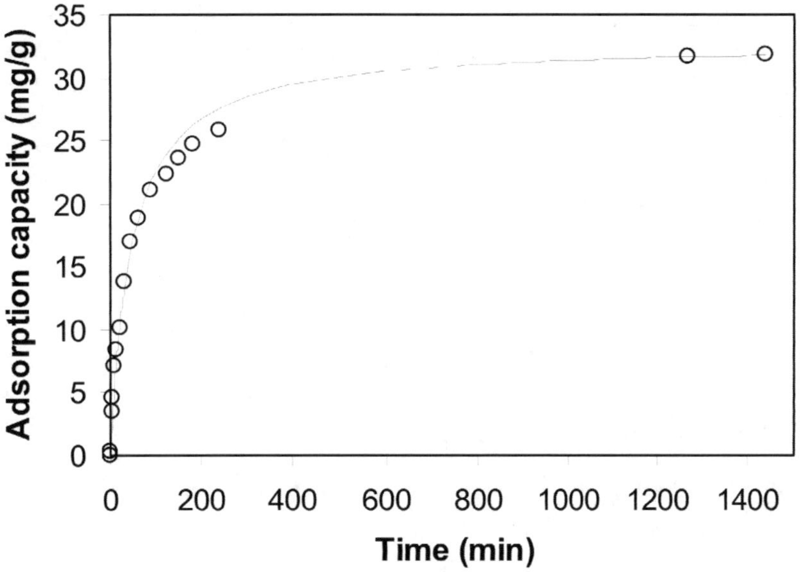


**Figure 3.** Adsorption kinetics of Pb(II) on modified cotton fabrics at 30°C, initial pH 5, initial concentration 53.18 mg/L, volumetric flow of 10 mL/min, and adsorbent dosage of 1 g/L. Experimental data are represented by symbols and line was predicted with the pseudo second order model.

## 4. CONCLUSIONS

- ✓ Chemical modification of cotton fabrics by citric acid incorporate functional groups (*i.e.* carboxyl groups) capable of binding metal cations such as Pb(II) from aqueous solutions.
- ✓ Adsorption capacity of Pb(II) on modified cotton fabrics at pH 4 was around 3.5 times greater than activated carbon.
- ✓ It is possible to recover up to 92% of the lead previously adsorbed on the modified cotton with using 0.1 M HCl solutions.
- ✓ Adsorption kinetics can be predicted by a pseudo-second order model and it suggests that lead species interact with adsorption sites by physical forces such as electrostatic charges.

## 5. ACKNOWLEDGMENTS

This research was financially supported in part by the Secretaría de Educación Pública (PROMEP/103.5/11/6627) y la Universidad Autónoma de Nuevo León (PAICYT IT577-10).

## 6. REFERENCES


Acharya, J., Sahu, J.N., Mohanty, C.R. and Meikap, B.C. (2009) Removal of lead(II) from wastewater by activated carbon developed from Tamarind wood by zinc chloride activation. Chemical Engineering Journal 149(1-3), 249-262.

Garcia-Reyes Refugio Bernardo, Rangel-Mendez Jose Rene, Alfaro-De la Torre Ma. Catalina (2009), Chromium (III) uptake by agro-waste biosorbents: chemical characterization, sorption-desorption studies, and mechanism, Journal of Hazardous Materials, 170(2-3) 845-854.

Ho, Y.S. and McKay, G. (1999) Pseudo-second order model for sorption processes. Process Biochemistry 34(5), 451-465.

Marshall, W.E., Akin, D.E., Wartelle, L.H. and Annis, P.A. (2007) Citric acid treatment of flax, cotton and blended nonwoven mats for copper ion absorption. Industrial Crops and Products 26(1), 8-13.

Metcalf and Eddy (2004), Waste Water Engineering Treatment and Reuse, 4th Ed., Ed. Mc Graw Hill, USA, 1819 + XXVIII.

Silverstein, R. M., and Webster, F. X. (1998), Spectrometric Identification of Organic Compounds. Wiley & Sons, New York.

Stumm, W. & J. J. Morgan, Aquatic Chemistry: Chemical Equilibria and Rates in Natural Waters, 3rd. ed., John Wiley & Sons, New York, 1996.

Tiemann, K. J., J. L. Gardea-Torresdey, G. Gamez, K. Dokken, S. Sias (1999), Use of X-ray absorption spectroscopy and esterification to investigate Cr(III) and Ni(II) ligands in alfalfa biomass. Environmetal Science and Technology, 33(1) 150-154.

Yun, Y. S., D. Park, J. M. Park, B. Volesky (2001), Biosorption of trivalent chromium on the brown seaweed biomass, Environmetal Science and Technology, 35(21) 4353-4358.

Zhu, B., T. X. Fan, D. Zhang (2008), Adsorption of copper ions from aqueous solution by citric acid modified soybean straw. Journal of Hazardous Materials, 153(1-2) 300-308.

# (146a) A Model-Based Approach for the Construction of Design Spaces in Quality-by-Design

Masako Kishida, University of Illinois at Urbana-Champaign, Urbana, IL and **Richard D. Braatz**, Chemical Engineering, Massachusetts Institute of Technology, Cambridge, MA

This presentation considers the characterization of the set of allowable real parameter uncertainties for the output of a nonlinear system to be located in a prespecified target range. This word "parameter" is used by FDA and in this presentation to refer to any variable held at a fixed value during a single experiment. For example, this could be a parameter in a model, an initial condition such as the initial concentrations in a semibatch chemical reactor, or an operational parameter such as the cooling rate in batch reactor.

This uncertainty analysis problem is motivated by various industrial initiatives for Quality by Design (QbD) (Juran 1992), which have the objective of designing a manufacturing process so as to ensure that the product achieves its desired quality qualifications despite perturbations in the manufacturing process. The QbD literature refers to the set of allowable parameters as the design space (Juran 1992), which in industrial practice is constructed from the data collected from a large number of time-consuming and expensive experiments (e.g., see Garcia-Munoz et al, 2010, for a description of the state-of-the-art in industrial practice and Togkalidou et al, 2001, for an example of experimental design for generating models for pharmaceutical processes). This presentation considers the alternative construction of the design space based on a first-principles, grey-box, or black-box model of the manufacturing process, as fewer experiments are required to construct a model than to directly construct a design space from experiments.

In the control literature, this design space is typically characterized in terms of a robustness margin (Safonov and Athans 1977). Numerous publications quantify the effects of a specified norm-bounded perturbation in a system model on a signal norm on the system output (Wang et al 1992; Zhou et al, 1995) or on outer bounds for state or output trajectories (Durieu et al 2001; Kishida and Braatz 2010, 2011; Kishida et al 2011; Scott and Barton, 2009). The structured singular value and quadratic stability margin quantify the amount of uncertainty allowed to maintain closed-loop stability or a signal norm characterizing output performance (Ma et al 2001, 2002; Hovd et al 2003; Zhou et al 1995; Boyd et al 1997), for linear time-invariant or time-varying systems or linear systems coupled with bounded nonlinear perturbations. This presentation considers nonlinear systems and defines allowable variations in parameters for instantaneous specified bounds on the states or outputs to be satisfied, as this characterization of output performance is used in QbD applications (Juran 1992). The overall approach applies to both open- and closed-loop systems. As the exact generation of a design space is a challenging computational problem for a large number of parameters (Braatz et al 1994), tight upper bounds are determined.

The system output is approximated by a polynomial function (Taylor series expansion) or a rational function (Pad´e approximant) followed by application of the skewed structured singular value to obtain a set of allowable real parameters that ensures that the system output will be in a target set. A target set described by upper and lower bounds on each output is considered here; the same analysis can be used for an ellipsoidal target set by slight modification of the skewed structured singular value analysis (Smith 1990). Both polynomial and rational approximants can be written in the form of a linear fractional transformation, which is used in the next step of the analysis of applying the skewed structured singular value to quantify the allowable uncertainty set. Algorithms are also discussed that generate the algebraic constructions needed by the method as well as model reductions to reduce computational cost (Russell et al 1997; Russell and Braatz 1998).

Several design spaces are constructed for models reported by pharmaceutical companies, including a crystallization operation and a nasal spray.

References:

S. Boyd, L. El Ghaoul, E. Feron, and V. Balakrishnan. Linear Matrix Inequalities in System and Control Theory.

SIAM Press, Philadelphia, PA, 1997.
R. D. Braatz, P. M. Young, J. C. Doyle, and M. Morari. IEEE T. Automat. Contr., 39:1000–1002, 1994.
C. Durieu, E.Walter, and B. Polyak. J. Optimiz. Theory App., 111:273–303, 2001.
S. Garcia-Munoz, S. Dolph, and H. W. Ward. Comput. Chem. Eng., 34:1098–1107, 2010.
M. Hovd, D. L. Ma, and R. D. Braatz. Ind. Eng. Chem. Res., 42:2183–2188, 2003.
J. M. Juran. Juran on Quality by Design: The New Steps for Planning Quality into Goods and Services. The Free Press, New York, 1992.
M. Kishida and R. D. Braatz. Opt. Contr. Appl. Met., 31:433–449, 2010.
M. Kishida and R. D. Braatz. In Proc. of IEEE CDC/ECC, pages 5671–5676, Orlando, FL, 2011.
M. Kishida, P. Rumschinski, R. Findeisen, and R. D. Braatz. Efficient polynomial-time outer bounds on state trajectories for uncertain polynomial systems using skewed structured singular values. Proc. of the Joint Symposium on Computer-Aided Control System Design and Systems with Uncertainty, Denver, Colorado, 216-221, 2011
D. L. Ma and R. D. Braatz. IEEE Trans. Control Syst. Tech., 9:766–774, 2001.
D. L. Ma, J. G. VanAntwerp, M. Hovd, and R.D. Braatz. IEE Proc. - Control Theory & Applications, 149:423–432, 2002.
E. L. Russell and R. D. Braatz. Comput. Chem. Eng., 22:913–926, 1998.
E. L. Russell, C. P. H. Power, and R. D. Braatz. Int. J. Robust Nonlin. Control, 7:113–125, 1997.
M. G. Safonov and M. Athans. IEEE TAC, 22:173–179, 1977.
J. K. Scott and P. I. Barton. Comput. Chem. Eng., 34:717–731, 2009.
R. S. R. Smith. Model Validation for Uncertain Systems. PhD thesis, California Institute of Technology, Pasadena, CA, 1990.
T. Togkalidou, R. D. Braatz, B. K. Johnson, O. Davidson, and A. Andrews. AIChE J., 47:160–168, 2001.
Y. Wang, L. Xie, and C. E. de Souza. Syst. Control Lett., 19:139–149, 1992.
K. Zhou, J. C. Doyle, and K. Glover. Robust and Optimal Control. Prentice Hall, 1995.

# (146b) Fluid Bed in a Flash

**Otto Scheibelhofer**[1], Roland Hohl[1], Sharareh Salar-Behzadi[1], Detlev Haack[2], Kai Koch[3], Stephan Sacher[1], Jose C. Menezes[4] and Johannes G. Khinast[5,6], (1)Research Center Pharmaceutical Engineering, Graz, Austria, (2)Hermes Arzneimittel GmbH, Munich, Germany, (3)Innojet Herbert Hüttlin, Steinen, Germany, (4)Research Center for Pharmaceutical Engineering, Graz, Austria, (5)Institute for Process and Particle Engineering, Graz University of Technology, Graz, Austria, (6)Research Center Pharmaceutical Engineering GmbH, Graz, Austria

**Fluid Bed in a Flash**

O. Scheibelhofer[1], Roland Hohl[1], Sharareh Salar Behzadi[1], Detlev Haack[2], Kai Koch[3], Stefan Sacher[1], Jose C. Menezes[1], J. G. Khinast[1,4]

[1] Research Center Pharmaceutical Engineering GmbH, Graz, Austria

[2] Hermes Arzneimittel GmbH, Munich, Germany

[3] Innojet Herbert Hüttlin, Steinen, Germany

[4] Institute for Process and Particle Engineering, Graz University of Technology, Graz, Austria

Email for correspondence: otto.scheibelhofer@rcpe.at (O. Scheibelhofer), khinast@tugraz.at (J.G. Khinast)

Fluid Beds are a versatile and widely used unit operation in the pharmaceutical industry. Putting to use their property of very large particle to air contact surface, they are an advanced step for drying, coating and more. A great deal of interest has already been invested, concerning the properties of fluid beds, especially their geometry and flow pattern. Mostly, this is a field for simulations.

Nonetheless, the efficiency of fluid beds is often driven by the experience of their operators. This is a result of fluid beds being very sensitive to subtle changes, which can affect the flow pattern significantly, even dis-enabling the fluid bed process as a whole. Even more so, this increases the difficulty of placing sensors in such environment, making systematic experimental investigations an exhaustive work.

To overcome some of these difficulties, near-infrared spectroscopy seems a natural choice, for two reasons. First, the method is non-invasive, or even contactless, so to avoid the disturbance of the established flow pattern. Secondly, the method can be very fast and can therefore cope with the time-scale of flow inside the bed.

A single NIR probe is used for monitoring the fluid bed and following the trajectory of our process, for this a process NIR from Sentronic is used.

However, as there are different zones of flow and progress inside the bed, a single probe is not enough, to get an idea of the overall maturity inside the fluid bed.

Here we used a new spectrometer prototype provided by EVK, Raaba, Austria in order the monitor a coating process. The prototype is based on a push-broom principle, but instead of a line, a standardized port plate enables the mapping of different near-infrared probes, via a prism-grating-prism system onto the detection chip. As a consequence, several positions inside the fluid bed can be monitored simultaneously, with a frequency of several Hz. even more, the records of the probes can be directly compared, as they are guaranteed to be recorded at the same time, and there is no spectrometer drift in between probes, etc.

The probes are distributed inside the fluid bed vessel at carefully chosen positions, at the wall, in order to not disturb the flow pattern, but give at the same time the most relevant information.

The information gained is of value twofold. First, it tells about the process itself, and can be used to estimate necessary process time and quality. Second, it tells about the mechanisms inside the fluid bed, spatial differences, etc.; which is very valuable for final probe positioning, the task of up-scaling, and more.

Estimation of the sample positions and the process is done with risk analysis, and design of experiments is also used for generating an experimental setup pattern. Within the multiprobe measurement system, we will also evaluate the sample position of the single probe.

*Figure 1: The Ventilus 2.5 fluid bed reactor is shown., with the Sentronic NIR probe attached.*

# (146c) From Experimental Data to Building Integrated Dynamic Flowsheet Models for Pharmaceutical Processes

**Fani Boukouvala**[1], Fernando J. Muzzio[2] and Marianthi Ierapetritou[2], (1)Chemical and Biochemical Engineering, Rutgers University, Piscataway, NJ, (2)Department of Chemical and Biochemical Engineering, Rutgers, The State University of New Jersey, Piscataway, NJ

FDA harmonization guidelines [1] aim to promote building quality into design of pharmaceutical processes through mechanistic understanding of powder process behavior. Based on this paradigm, the industry is moving with a fast pace towards a model-based process and product design, which will undoubtedly lead to competitive manufacturing and better product quality [2]. However, process modeling is highly dependent on experimental or computational data, either in the form of empirical correlations or during the stage of parameter estimation. At the same time, there has been a high interest in the population of a flowsheet model library for solids-based pharmaceutical processes. Flowsheet modeling has been an imperative process engineering tool for the design and optimization of liquid-based processes [3-4]. This work aims to summarize all of the identified challenges that originate from the incorporation of experimental and computational data into a dynamic flowsheet model library for pharmaceutical processes, in order to develop a generic model library which may be used for a wide range of raw materials and processes, integrated to form complete production lines.

Many challenges arise due to the nature of pharmaceutical materials and processes compared to other industries, such as increased variability of powder material properties and operating conditions, uncharacterized materials, lack of well-established correlations of material properties and process operation and oftentimes lack of predictive models for certain processing steps. However, during the recent years where there has been a strong focus in enhancing process understanding, large amounts of data for the characterization of processes or materials have been collected. In certain cases, these data have provided insight about the governing underlying mechanisms and have lead to the development of first-principle models, while in other cases data have been used to build empirical response surface models, or multivariate latent variable models in cases of high dimensionality. In fact, even in the well-established liquid-based flowsheet models, certain physical phenomena which are insufficiently understood, such as chemical reaction kinetics or heat and mass transfer effects are represented through empirical correlations which are incorporated into first-principle models to create hybrid or semi-empirical modules [5].

Moreover, the complex nature of powder based-processes has also lead to the development and use of high-fidelity spatially distributed simulation models that aim to capture the complex behavior of interacting particles in order to provide insight into detailed critical material attributes such as particle size distributions. This introduces another challenge towards the final goal of flowsheet model building, which is the incorporation of multiscale computational data, derived from very expensive computer models. Fortunately, however, through this work it will be shown that the handling of such computational data is very similar to that of experimental data.

Finally it should be mentioned that the link between data and models is always a closed-loop process, since even after the development of the necessary models, the need of experimental data is once again imperative for the estimation of the model parameters. In this case the experimental design will be guided by the models, in order to ensure that they contain the necessary information needed for the parameter estimation.

A great opportunity arises in order to efficiently incorporate the information obtained from large data-bases and build and/or refine the flowsheet model library following well-established techniques of the process systems engineering community. This work aims to discuss the state of the art in first-principle models used to simulate pharmaceutical manufacturing and their proposed combinations and modifications in order to maximize the captured information through incorporation of useful data-based correlations. Most importantly, the need to consider each process as part of a large integrated puzzle should be considered even from the early stages of experimental design, since due to the need for simulation of continuous integrated lines, the propagation of a common set of material properties of powder mixtures is essential. The challenges and proposed remedies will be shown through specific case studies for a variety of manufacturing scenarios of pharmaceutical tablets such as direct compaction, dry and wet granulation.

References:


1. *Q8 Pharmaceutical Development*. 2006; Available from: http://www.fda.gov/downloads/RegulatoryInformation/Guidances/ucm128029.pdf.

2. Oksanen, C.A. and S. García Muñoz, *Process modeling and control in drug development and manufacturing.* Computers and Chemical Engineering, 2010. 34(7): p. 1007-1008.

3. Boukouvala, F., et al., *An integrated approach for dynamic flowsheet modeling and sensitivity analysis of a continuous tablet manufacturing process.* Computers and Chemical Engineering, In Press, 2012.

4. Werther, J., C. Reimers, and G. Gruhn, *Design specifications in the flowsheet simulation of complex solids processes.* Powder Technology, 2009. 191(3): p. 260-271.

5. Kahrs, O. and W. Marquardt, *The validity domain of hybrid models and its application in process optimization.* Chemical Engineering and Processing: Process Intensification, 2007. 46(11): p. 1054-1066.

# (146d) Modeling and Simulation of a Wurster Column for Particulate Coating Using Population Balance Moldels and Thermodynamics

Jianfeng Li, Chemical Engineering, Purdue University, West Lafayette, IN, Salvador García-Muñoz, Process Modeling & Engineering Technology, Pfizer Worldwide Research & Development, Groton, CT, William R. Ketterhagen, Pfizer Global Research and Development, Pfizer Inc., Groton, CT and James D. Litster, School of Chemical Engineering, Department of Industrial and Physical Pharmacy, Purdue University, West Lafayette, IN

The Wurster column coating process of pharmaceutical solids has been studied for many years and a considerable body of literature has been produced covering various aspects of the process. To achieve uniform active coating content is one of the key interests for the pharmaceutical manufacturing industry. Thus, a process model for predicting the evolution of a granule size distribution from first principles is desired.

In this work, a compartment based population balance model, combined with thermodynamic coating model has been developed to model a pharmaceutical coating process in the Wurster column coater. The model can predict certain key process parameters and the dynamic particle size distribution, and it can also be used to simulate responses of the system to a change in the working conditions.

In many industrial practices, the coating equipment has been identified into coating/non-spray regions. It is believed that both the variation of residence time in the spray region and the variation of cycle time lead to the non-uniform coating mass distribution among individual particles. Crites and Turton (2005) have developed a compartment based model to incorporate the cycle time distribution feature for the coating process in the Wurster column coater. The model was composed of fundamental flow types such as plug flow and well-mixed flow regions to describe phenomena including dead zone and side pass. Other instances include a series of well-mixed tanks model structure for a top-spray fluidized bed coater (Ronsse 2005), a twin region compartmental model for a rotating drum coater (Dennis et al. 2001).

In this work we use a population balance (PB) approach to model the particle size distribution of the coated particulates as a function of time. While the particle growth due to coating strongly relates to the number and duration particles passing through the spray region, a practical way to constructing PB models is based on the decomposition of the coater into spray/non-spray compartments. Thermodynamics of the spray are evaluated with a simple energy balance (see am Ende et al. 2005)

In terms of the flow of particles across the unite, the Wurster column coater is divided into three compartments including an ideal plug flow region, an ideal fountain CSTR, and a non-ideal annulus CSTR (in a similar way to that proposed by Crites and Turton (2005)).

1. The plug flow compartment refers to the region enclosed by the draft-tube, where particles are highly fluidized and heated by the upflow hot air. The coating solution is assumed to be delivered on to these particles from the spray nozzle centered in the middle of the distribute plate.

2. The ideal CSTR compartment represents the fountain region, in which particles enter into the chamber and continuously receive spray.

3. Particles are then settled down towards the bottom of the coater in the annulus region, described by the non-ideal CSTR. Particles spend most of the time in this region. The non-ideality refers to the factor that smaller particles have longer recycle time compared to larger particles (Liu and Litster 1993), so the characteristic time of the CSTR for the annulus region is a function of particle size.

Experimental data is used to determine the thermal properties of the column as well as growth kinetics. Predicted values of particle size as a function of time were found in good agreement with particle size distribution measurements (by laser diffraction) of samples taken throughout the process.

**References**

am Ende, M.T. and Berchielli, A., 2005. A thermodynamic model for organic and aqueous tablet film coating. Pharmaceutical Development and Technology 1, 47–58.

Crites, T and Turton, R., 2005. Mathematical model for the prediction of cycle-time distributions for the Wurster column-coating process. Ind. Eng. Chem. Res. 44, 5397-5402.

Denis, C., Hemati, A., Chulia, D., Lanne, J.Y., Buisson, B., Daste, G., Elbaz, F., 2003. A model of surface renewal with application to the coating of pharmaceutical tablets in rotary drums. Powder Technology 130,174-180.

Liu, L. X., Litster, J. D., 1993. Coating mass-distribution from a spouted bed seed coater - experimental and modeling studies. Powder Technology 74, 259-270.

Ronsse, F., Pieters, J.G., Dewettinck, K., 2007. Combined population balance and thermodynamic modelling of the batch top-spray fluidised bed coating process. Part I—Model development and validation. Journal of Food Engineering 78, 296–307.

# (146e) Exceptional Event Managemnt Applied to Continuous Pharmaceutical Manufacturing

**Anshu Gupta**, Gintaras V. Reklaitis and Venkat Venkatasubramanian, Chemical Engineering, Purdue University, West Lafayette, IN

One of the important challenges in effective real time process management is the implementation of intelligent systems that can assist human operators in making supervisory control decisions, instead of simply sounding an alarm when process variables go out of range. Operator failures to exercise the appropriate mitigation actions often have an adverse effect on product quality, process safety, occupational health and environmental impact. The economic effect of such exceptional events is immense; an estimated $20billion/year losses in petrochemical industry have been reported. The challenges and opportunities for improvements are even larger in the pharmaceutical manufacturing domain because so much of the processing involves particulate and granular systems whose characteristics tends to be more problematic than that of fluids. The development and manufacturing of pharmaceutical products are governed by strict safety regulations but with the advent of Process Analytical Technology (PAT) initiative advanced by the FDA; gives the pharmaceutical industry an opportunity to apply various systems engineering tools. Early detection and diagnosis of process faults while the plant is still operating in a controllable region can help avoid abnormal event progression, production disruptions and productivity losses. An EEM framework has been developed to deal with fault detection, diagnosis and mitigation of conditions that result from process anomalies. The framework developed uses a combination of Principal Component Analysis (PCA), Signed Digraphs (SDG) and Qualitative Trend Analysis (QTA) in applications involving continuous pharmaceutical product manufacturing line. An ontological database has been created to maintain records of the signatures of a number of exceptional events typical of a dry granulation line and their associated mitigation strategies. In trials on pilot scale equipment we have found that the EEM framework was able to detect and diagnose several types of faults within a few seconds of their inception and to provide mitigation advisories to the operator.

# (146f) Modelling of a Filter Drying Process

Giovanni Giorgio[1], Jason Mustakis[2], Mark A. Pinto[3] and Sean K. Bermingham[3], (1)Particle Generation and Control Engineering, GlaxoSmithKline Ltd., Stevenage, United Kingdom, (2)Pfizer Global Research & Development, Pfizer Inc., Groton, CT, (3)Process Systems Enterprise, London, United Kingdom

Filter drying, also referred to as vacuum contact drying, is a process widely used in the pharmaceutical industry to remove water or organic solvents. The advantage of this process over other drying operations is that it can be used to dry solids that are sensitive to oxygen and temperature.

The entire drying process is typically carried out in two stages in a vessel. Typically a cylindrical vessel is used although conical vessels are also encountered. In the first blowdown stage, pressure in the form of nitrogen is applied to remove excess solvent by convective drying. At the end of this stage, the wet filter cake typically has around 45% solvent on a wet basis.

The second, longer, period involves heating the walls of the filter cake while vacuum is applied to dry the cake to about 1% to 0.1%. During this stage, heat conduction from the wall to the interior of the cake is often the rate limiting step. While the process can be carried out without agitating the bed, this can lead to excessively long drying times and heterogeneity of the cake. On the other hand, continuous long agitation often results in agglomeration and/or breakage of the solid. As a compromise, intermittent periods of agitation are carried out to redistribute solvent and heat, resulting in faster drying times but without significant agglomeration and/or breakage of solid taking place.

A detailed model of the filter dryer unit operation has previously been published in the literature (Murru et al, 2011). The key features of the model are:

1. The second drying phase is modelled and drying is assumed to take place primarily due to heat conduction from the equipment walls. The blowdown phase where convective drying takes place is not accounted for.

2. A single solvent is considered. This allowed for process optimisation by allowing different solvents to be examined for the effects on total drying times.

The model was successfully used to predict drying times for various equipment sizes and solvents at both pilot plant and manufacturing scales.

In this work, the model is further extended to consider the following phemonena.

1. Convective drying is considered. This allows the modelling of the initial blowdown drying phase. The effects of the applied suction during the second drying phase are also explicitly accounted for. The application of hot nitrogen flow over the bed can significantly reduce the time of drying.

2. The model is extended to non-ideal mixtures of solvents. The presence of solvent mixtures in the drying operation has a significant effect on the drying time but most importantly lead to change of composition of the residual solvent during drying. This can lead to stability issues (hydrates / solvates) or can lead to particle bridging and agglomeration. Understanding the dynamics of the solvent composition is crucial to avoid these situations. The ability to handle multiple solvents simultaneously will allow for greater flexibility in selecting solvents for process optimisation.

3. Modelling of the vapour processing step is undertaken. This allows for data from downstream condensation and scrubbing unit operations to be used in model validation.

References

M Murru, G Giorgio, S Montomoli, F Ricard, F Stepanek (2011) Model-based scale-up of vacuum contact drying of pharmaceutical compounds. *Chemical Engineering Science* **66**, 5045-5054

# Techno Economic Assessment of a Direct Route to Propylene Oxide

*Salvador Aldrett, Shell Global Solutions International B.V*
*Arian van Mourik, Shell Global Solutions International B.V*
*Richard Sebesta, Shell Checmicals*

# Abstract

Production of propylene oxide (PO) is currently carried out commercially through indirect processes that involve extra purification steps. There is however, a potential process to produce PO similarly to ethylene oxide. The main barrier is in finding a catalyst with a selectivity high enough to make the economics favorable. This paper evaluates a process for the direct production of PO.

A process comprising reaction, absorption, CO2 removal, and purification is being proposed as a first pass for this process. Based on the concept, the equipment sizing and costing was performed using a capital estimation proprietary tool. Working together with the capital estimation team an estimate was put together for the capital required under such process. The estimate is in the range of $900-1000 mln for a nominal capacity of 300 kta placed on a Greenfield site in 2015.
Several patents have reported advances in catalyst development for this system. Our evaluation shows that a higher selectivity that currently demonstrated is needed to make the process economically attractive. In addition, developments in other process areas may also change the view from what is represented in this paper.

## Background

SMPO is currently one of the leading technologies for PO manufacture. However, alternative technologies, producing PO without co-product have been commercialized over the past year, despite this, they still require a 2-step process with H2O2 as an intermediate to be produced in a separate plant.

The "holy grail" is direct propylene oxidation technology (Ethylene Oxide look-alike). Analysis on the basis of previous studies indicates that such technology might be an attractive candidate with game-changing prospects, though there are significant technical hurdles to overcome.

The present report contains the results of our work:

- Process description and $CO_2$ footprint (t $CO_2$ / t PO) based on energy consumption for an 80% selectivity case.
- Material balance, equipment list and capital estimate for the PO-direct 80% selectivity case.
- Review of solvent selection for PO absorption by reaction-mapping based on thermodynamic behavior.
- Explore process design alternatives to the base case (membranes, other separation techniques, distillation sequence).

## Process Description

A relatively new PO only process is the Hydrogen Peroxide to Propylene Oxide (HPPO) technology. The HPPO process comprises in fact two separate plants: a hydrogen peroxide plant and a propylene oxidation plant. This technology has meanwhile been demonstrated at large scale in a 300kt/a HPPO plant in Antwerp (Belgium) and a 390 kt/a plant in Thailand. Further, South Korea's SKC has constructed a 100 kt/a HPPO plant at Ulsan. Apart from this HPPO other new PO only routes are under development, including:
• Epoxidation of propylene with in-situ generated hydrogen peroxide. This technology is similar to the HPPO process, but now the hydrogen peroxide is generated in situ from co-fed hydrogen and oxygen.
• Direct oxidation of propylene with oxygen (PO-direct)

Due to the large similarity of the direct oxidation route to the well-known and widely used direct ethylene oxide (EO) production from ethylene, the PO-direct route seems to be an attractive alternative, but the selectivity and activity of currently available direct PO catalysts are still much lower than for existing EO catalysts.

Patent US20050027134 discloses an example of typical state of the art catalyst used in the research for direct oxidation to produce PO. The selectivity of this PO catalyst is only about 60% and this would certainly not have been sufficiently high for a commercial process. In this study therefore a higher selectivity of 80% has been assumed although so far higher selectivity catalysts with acceptable workrates have not been disclosed in public literature.

The following section briefly describes a proposed process that helps us evaluate the process economics for this technology and determine the point at which catalysts selectivities become attractive from a commercial perspective.

## Reaction and main gas recycle loop

The fresh make-up feeds ($O_2$, $CH_4$ and chemical grade propylene, at 40°C and 15 bar) are mixed with the combined gas recycle (50 °C, 15 bar) and heated to approximately 200 °C by heat integration with the reactor effluent, before being introduced in the reactor. The propylene will react to PO but the propane could build up in the recycle and needs to be purged to avoid excessive build up. The reactor of choice is a multitubular reactor, very similar to the reactor of the EO process. The catalyst is packed in the tubes where PO is formed by partial oxidation of propylene ($C_3$=) at 15 bar according to the following gas phase reaction:

$CH_3\text{-}CH{=}CH_2 + 1/2\ O_2 \rightarrow CH_3\text{-}CH\text{-}CH_2$ (with O bridging the two CH carbons: \ / O) $[\triangle H_r = -156\ \text{kJ/mol}]$ (1)

Simultaneously, propylene may also be fully oxidized, to result in $CO_2$ and $H_2O$ according:

$$CH_3\text{-}CH{=}CH_2 + 9/2\ O_2 \rightarrow 3\ CO_2 + 3\ H_2O \qquad [\triangle H_r = -1968\ \text{kJ/mol}] \quad (2)$$

In the Aspen model used for the simulation the RSTOIC reactor, only reaction (1) and (2) are included.

Note that the heat of reaction from reaction (1) and (2) is -156 kJ/mol and -1968 kJ/mol respectively. This strongly implies that in case of low catalyst selectivity, the net heat generation in the reactor will increase significantly. The maximum gas temperature in the reactor is set at 250 °C. The excess of heat of reaction is removed by steam generation (evaporation of water) at the shell side of the reactor.

The reactor effluent is then fed into the PO absorber column. Where 99% of the PO is absorbed from the gas stream at 40 °C and 15 bar. The amount of required absorbent is largely depending on the PO concentration and the size of the gas stream. Apart from PO, also some propane, propylene and $CO_2$ is absorbed by the liquid.

The vapour from the absorber is led through a $CO_2$ absorption section where 78 % of the $CO_2$ in the gas stream is removed per pass. The $CO_2$ absorber has been represented in the flowsheet by a separator. After leaving the $CO_2$ absorber the gas stream is combined with the residual gases from the work-up section and recompressed into the main gas loop compressor at 15 bar.

## PO-Absorber/stripping section

In the PO-absorber, approximately 99% of the PO will be absorbed from the reactor effluent. To facilitate efficient PO absorption, the absorber column is operated at 40 °C, 15 bar. A mixture of mono propylene glycol (MPG) and water could be used as absorption fluid. As said in the previous section, a disadvantage of using this high pressure is that apart from PO also other species are absorbed in the MPG/water, which have to be separated downstream. Also the absolute size of the required MPG/water flow is a function of the pressure in the absorber (solubility of PO in MPG/water is a function of pressure).

To facilitate the stripping of the PO, the PO rich absorbent (MPG/water) needs to be heated again from 40°C to approximately 150 °C (at 3 bar). This may be achieved by heat integration, using the hot PO lean absorbent that leaves the stripper.

Prior studies indicate that MPG might be a good absorbent for the PO. At the low PO concentrations expected, it turns out that the advantages of MPG like good solubility of PO, have to be weighed against disadvantages like a higher solubility of propylene and a lower heat capacity compared with water. Calculations indicate that MPG will only have a strong advantage if the recovery of PO in the absorber has to be high (i.e. when PO recycled to the reactor is mostly destroyed). MPG as solvent has an advantage over water (lower capex & opex) if:

- The recovery of PO in the absorber has to be high.
- The PO concentration in the reactor effluent is high
- The temperature of the reactor effluent is low.
- The production of water as by-product in the reactor is low.
- Hydrolysis of PO in the absorber will result in net MPG make, which can be recovered as a by-product

Disadvantages of MPG as solvent are:

- The need for steam at a higher pressure level than with water or a water/MPG mixture.
- The extra investment needed (refrigeration or compressor) to recycle dissolved propylene

Even though the MPG or MPG/water mixture seems to be the most favourable option from a technical/thermodynamical perspective, it may turn out to be economically more favourable to use pure water, similar to the EO process. The use of pure water may for instance reduce the cost for make-up absorbent, absorbent purification and the energy consumption in the reboiler of the stripper but increases MPG make.

The stripper column is operated at 3 bar. The distillate rate is controlled/ adjusted such that 99% of the PO is carried over into the first overhead condenser. Then after leaving the stripper, the stripper bottoms are cooled in the absorbent heat integration unit from approximately 150°C to 50 °C, while simultaneously heating the PO rich absorbent from 40 to 140 °C. The PO lean absorbent needs to be pressurized again to 15 bar and is further cooled to 40 °C before being introduced into the absorber again. Again, also this size and duty of the heat integration equipment for the absorbent loop depends to a large extend on the size of the main gas loop and thus on the performance of the catalyst.

## PO work-up and residual gas recycle.

After leaving the top of the stripper, the vapour passes the first overhead condenser (40 °C, 3 bar) where approximately 70%wt of the PO is recovered via condensation from the gas stream. The resulting condensate contains mainly PO, but also contains 10-30 %wt of $H_2O$, which is condensed along with the PO. This condensate is combined with the condensate from the reactor effluent cooler and after preheating this combined stream is fed into the dehydrator column. In the second -chilled water- overhead condenser (5°C, 3 bar) again PO is condensed from the gas stream. The condensate mainly contains PO (approx 80%wt), but because of the lower temperature also some propane and propylene may be condensed along with the PO.

The PO rich condensate streams from the stripper vent gas condenser and the dehydrator overhead condensers are then pressurized to 15 bar again, combined and introduced into the light ends (LE) removal column. This column operates at 15 bar because of the high $C_3/C_3=$ concentration in the top of the column and to allow condensation thereof at approximately 20 °C. All vapours in the condenser are sub-cooled and the separated water phase is decanted. As a consequence of the applied higher pressure the bottom temperature of this column is approximately 130 °C.

The vapours from the top of the light end removal column are introduced into the PP splitter that is also operated at 15 bar to allow condensation of $C_3/C_3=$ in the top using cooling water. Apart from $C_3$ also $C_3=$ is removed by the PP splitter from the residual gas recycle, which is undesired because the $C_3=$ can be re-used as feedstock again.

## Catalyst

Multiple catalysts have been tested by various research groups with the main parameter being the selectivity to convert propylene to propylene oxide. Most catalysts contain silver, with other optional promoters. The catalyst used in evaluation was taken as in patent US2005/0027132. The catalyst once produced is impregnated in a solid matrix. The space velocity and recycles compositions can be tuned in order to improve the selectivity of such catalyst. The catalyst may contain optional promoters such as potassium, chlorine, molybdenum, rhenium, tungsten, gold, thallium, niobium, cobalt, or cerium.

## Modeling the PO-Direct Process in Aspen Plus

For this work we used an Aspen Plus model in order to derive the yields and energy requirements that feed the OPEX calculations. The model contains reaction and main purification sections, with recycles. Some sections such as the $CO_2$ absorption section were not modeled rigorously. At this point, it is assumed that the PO direct process will use an absorption system similar to that of the current EO technology as practiced in industry and will produce a similar separation. The purification section was designed to produce raw PO with approximately 2%mol water to be further purified in an azeotropic distillation column, propane with approximately 1% propylene to be burned as fuel, and waste water (with traces of organic impurities).

Given the nature of the distillation section and some of its complexities, we opted not to model rigorously this section but rather use the yields that have been reported elsewhere (refxx).

The model was tuned in order to improve convergence, to make the model more robust and to reduce the losses of PO and propylene in waste streams.

The results from the Aspen model were used to obtain consumption and generation of products, by-products, utilities, etc. An Excel spreadsheet is used to present the results in a tabular form, and to calculate the $CO_2$ footprint as well. An explanation of each category considered in the excel sheet follows:

Propylene Usage: This is a straight reading from the aspen model on the propylene required to produce the 300 kta of PO; the amount of propylene was adjusted according to a design specification to keep a propylene concentration of 10%mol in the stream fed to the reactor.

- Oxygen Usage: Oxygen flow required from Aspen to react the propylene to PO, $CO_2$, and water. Another design spec kept the $O_2$ concentration at 8%mol in the entrance to the reactor, as would roughly be expected to be the maximum in view of flammability.
- $CH_4$ Usage: Methane is a diluting gas and its contribution to the process ($CO_2$ footprint) is negligible.
- Electricity: This value includes the energy needed to operate pumps, compressors, and the water chiller that cools the overhead effluent from the stripper. An improvement in $CO_2$ footprint might be obtained if a steam-driven LiBr machine would be used, rather than an electrically-driven system, depending on the use options for excess steam produced in the plant.
- Steam: This accounts for the steam consumption in the purification section from the different columns.
- Condensate: This is the amount of condensate needed in the process, which is negligible.
- Chilled water: This is the chilled water consumption, but given that the electricity for the chiller is already accounted for, this can be counted as regular cooling water.
- Cooling Water: This is the water needed for column condensers, reactor, and reactor outlet stream.
- Process $CO_2$: Carbon dioxide directly generated by the process according to the conversion.
- Waste Water: Water from the dehydrating column that has to be treated accordingly.
- Energy Output: This line includes the heat generated from the reaction.

## Capital Cost Estimate

Based on the flowsheet developed by the team, the capital estimate was developed accordingly based on equipment sizing and other specifications. A summary of the capital estimate is shown in Table 1.

When financial modeling of the project cash flows is required, a project schedule would be proposed with a "ready for start-up" (RFSU) date premised. The capital costs would then be inflated and escalated to that RFSU date and the cash flows calculated based on assumptions of the cash spend of the project. The capital cost presented for the RFSU date would be Money of the Day (MOD); i.e. the costs at that future date and be significantly higher than the installed ISBL costs

**Table 1:** Capital Cost Estimate for PO direct Process

| Description | Cost (MMUSD) |
|---|---|
| ISBL direct costs: | 300 |
| OSBL direct costs | 150 |
| Project Costs | 400 |
| Contingencies | 75 |
| **Total Capital Cost** | **925 +40/-25%** |

## Main Equipment Design and Sizing

Design and sizing of the equipment for this process was based on a converged run from our Aspen Plus model. This work is based on a plant capacity of 300 KTa, reactor selectivity of 80% towards PO, a propylene concentration in the feed to the reactor of 10%vol, and a PO diluted concentration of 2%vol in the reactor outlet. The metallurgy assumed is typical of this industry.

## Operating Costs

Variable operating costs include raw materials, disposal/waste handling charges, co-product credits, periodic major replacements, and electricity, and are incurred only when the plant is in production. Raw material requirements were based on quantities generated by the Aspen Plus mass and energy balance.

The costs of raw materials and utilities were summed up to derive the total variable cost. The fixed cost is taken as a percentage of the total fixed cost. The major cost contributors are listed below:

- Propylene
- Oxygen
- Methane
- Process Water
- Fuel Gas
- Electricity
- Steam (HP, MP, LP)
- Cooling Water
- Catalyst
- Absorbant

## Process Economics

Given the total project investment, variable operating costs, co-product revenues and fixed operating costs, the economics of the process can now be evaluated. Two methods are frequently used for displaying economic results and valuing a project; the Profited Manufactured Cost (PMC) of a given product and/or a Discounted Cash Flow (DCF) analysis of the project. The PMC is a snapshot in time capturing both cash costs and capital charge. DCF analysis is a method using the concepts of the time value of money. All future cash flows are estimated and discounted to give their present values. The profited manufactured cost is used more often in the scouting phase of a project given its relative simplicity and the fact that it can be used as a comparison tool for different processes or process improvements.

In the current scouting phase of the project, the accuracy of the profited cost estimate is limited, for example it may vary somewhat from case to case due to the ongoing improvements and adjustments of the preliminary Aspen model during the progress of the work. Therefore, the absolute value of the profited costs should be regarded as indicative at this stage. Not modeled, but required units were included on the basis of estimates from existing processes.

The process economics are presented in this report as cash cost. This avoids having to assume a propylene oxide price since such price could vary depending on commercial terms and conditions.

**Table x:** *Cost Summary for PO Direct Process*

| | | |
|---|---|---|
| **Performance Data** | | |
| Propylene consumption | 271 | kta |
| Profited manufacturing cost (PMC) | 2225 | $/tonne |
| | | |
| **Capital Cost** | | |
| Capital cost (2015) | $925 | MMUSD |
| | | |
| **Operating Costs** | | |
| Variable cost | $400 | $MM/yr |
| Fixed cost | $35 | $MM/yr |
| **Total Operating Cost** | $435 | $MM/yr |
| **Cash Cost** | **$1455** | $/tonne |

## Conclusions

The present work show the evaluation of a propylene oxide direct process using a published process patent as the basis for the catalyst yield. The catalyst selectivity was set at 80% since this is believed to be a number that would make the economics attractive. A process flow was devised with help from various sources. Based on such process an Aspen model was put together to estimate the variable cost as well as the sizing of the equipment. The equipment size was used to estimate the capital required for a 300 kTa plant. Having the variable cost and the CAPEX one can calculate the cash cost to produce propylene oxide via direct oxidation. The cash cost calculated in this work is in the order of $1460/tonne.

## References

## Appendix A. Process Simulation Diagram

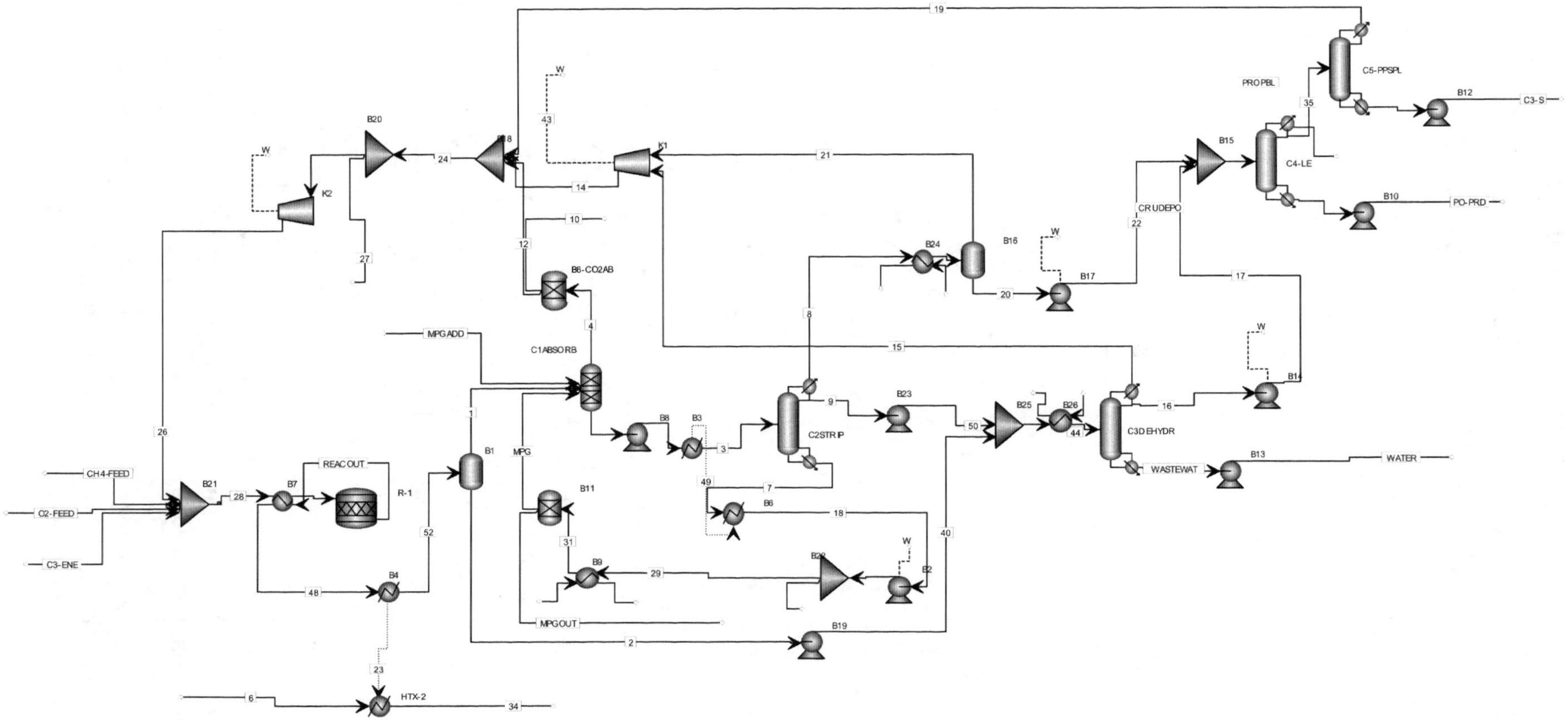

# PARTIAL OXIDATION OF O-XYLENE TO PHTHALIC ANHYDRIDE INSIDE OF THE EXPLOSION REGIME USING A MICRO STRUCTURED REACTOR

T. Lange[1], S. Heinrich[2], C. Liebner[2], H. Hieronymus[2], E. Klemm[1]

[1] University of Stuttgart, Institute of Chemical Technology, Faculty of Chemistry, 70550 Stuttgart, Germany

[2] BAM Federal Institute for Materials Research and Testing, Chemical Safety Engineering, 12200 Berlin, Germany

## Summary

The heterogeneously catalyzed gas phase oxidation of o-xylene (oX) to phthalic anhydride (PA) was investigated above the lower explosion limit of appr. 2 vol.% o-xylene in air using a micro structured reactor. For these investigations a $V_2O_5/TiO_2$ grafting catalyst was applied as micro fixed bed and as wall catalyst. Up to 7 vol.% o-xylene, selectivities as well as space-time-yields to PA could be increased siginificantly. At o-xylene concentrations higher than 7 vol. %, the selectivity and space-time-yield to PSA decreased, probably due to a hotspot, which was predicted at high o-xylene concentrations with an approximation and could be indirectly approved by thermal deactivation of the used catalyst.

## *Keywords*

o-xylene oxidation, $V_2O_5/TiO_2$ catalyst, micro structured reactor, explosion regime

## Introduction

Heterogeneously catalyzed selective oxidations in the gas phase are an important class of reactions which are usually carried out outside the explosion, i.e. at rich or lean feed compositions being above the upper explosion limit or below the lower explosion limit. Due to safety issues this is necessary, since a heterogeneous catalyst can always act as an ignition source because a hot spot may occur. Due to their small channel dimensions, micro structured reactors were successfully used in the past to investigate safety and reaction engineering aspects with explosive gas mixtures. It is known that in micro structured reactors both types of explosions, deflagrations as well as detonations, of flammable gases can be suppressed due to the flame-arrestor effect. This was shown for example for the hydrogen and ethylene oxidation [1-3]. The main goal of this work is to transfer the highly exothermic, heterogeneous catalyzed gas phase oxidation of o-xylene to phthalic anhydride in a micro structured reactor, to take advantage of the high surface/volume ratio to quench heat and chain explosions.

The micro structured reactor used for these investigations was derived from the so-called DEMIS-reactor concept [4] and is shown in figure 1. The micro channel was of a slit-like shape and the cross-section of the slit was 20 mm in width and 0.25 to 1.0

mm in height. A sapphire vision panel allowed visual, thermographic and spectroscopic analysis of the working catalyst. The $V_2O_5/TiO_2$ catalyst was prepared with the grafting technique, which is known to give highly active and selective catalysts [5]. The prepared catalyst was extensively characterized ($N_2$-adsorption, XRD, ICP-OES, IR, Raman) and subsequently coated to the wall of the slit as a catalyst coating or filled into the slit as a fixed bed. In this work the performance of a micro fixed bed was compared with that of a catalyst coating within the explosion regime. Also investigations on mass and heat transport limitations were carried out with high o-xylene concentrations.

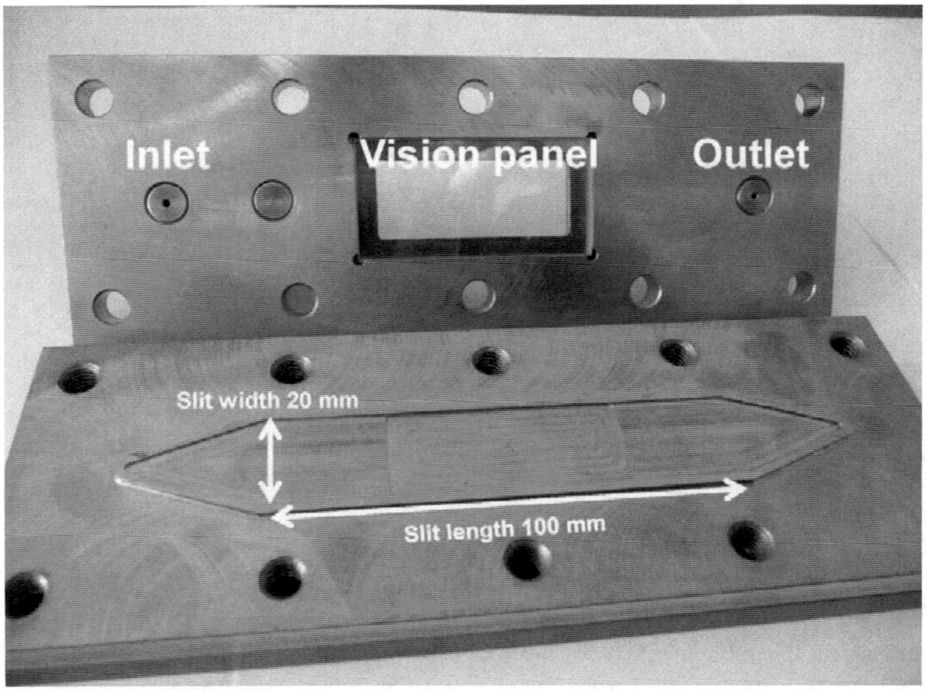


**Fig. 1:** Foto of the reactor showing the slit-like micro channel and its dimensions in width and length. The channel height is 0.25 to 1.0 mm.

## Results

The present work has shown that a highly exothermic, heterogeneously catalyzed reaction like the o-xylene oxidation to phthalic anhydride can be handled safely in the explosion regime using a micro structured reactor. We were able to operate the reaction with all the chosen feed compositions shown in the ternary diagram in figure 2 without any hazard, even if we used 25 vol. % o-xylene in pure oxygen. For feed compositions with up to 7 vol.% o-xylene, no significant losses in selectivity to phthalic anhydride (PA) occurred both in the micro fixed bed and the coated wall catalyst. The space-time yield for both catalyst systems was up to 2.3 times higher compared to conventional reaction conditions.

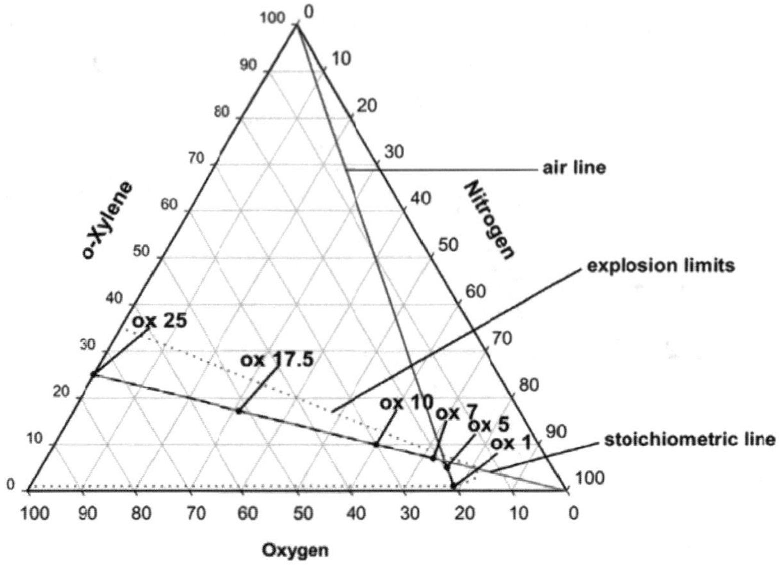


**Fig. 2:** The ternary diagram with the explosion limits of o-xylene in oxygen at standard temperature and pressure. The dashed line illustrates the adjusted feed compositions for the experiments in the explosion regime.

At feed concentrations exceeding 7 vol.% o-xylene, the selectivity to PA decreased due to increasing total oxidation which is shown in figure 3. Additionally, the maximal achievable conversion of o-xylene decreased due to the high oxygen comsumption through the total oxidation. However, the loss in selectivity to PA was for the micro fixed bed about 3 times higher than for the coated wall catalyst. A hot spot estimation according to Frank-Kamenetskii revealed the possibility of a hot spot at o-xylene concentrations above 7 vol.%, especially for the micro fixed bed due to the lower effective thermal conductivity [6]. The expected thermally induced deactivation of the catalyst was verified by a characterization of the used catalyst with XRD, FTIR and $N_2$-adsorption. In contrast to the fresh catalyst, the used catalyst contained crystalline vanadium oxide and higher amounts of rutil, which may also cause a favored total oxidation and may contribute also to the decreasing partial oxidation selectivity [7]. The thermographic investigations on the coated wall catalyst with a NIR camera proved the existence of a hotspot and revealed, that the hot spot temperature depends not only on the o-xylene concentration, reactor temperature and coating thickness, but also on the flow rate. Hot spots of more than 100 K were observed.

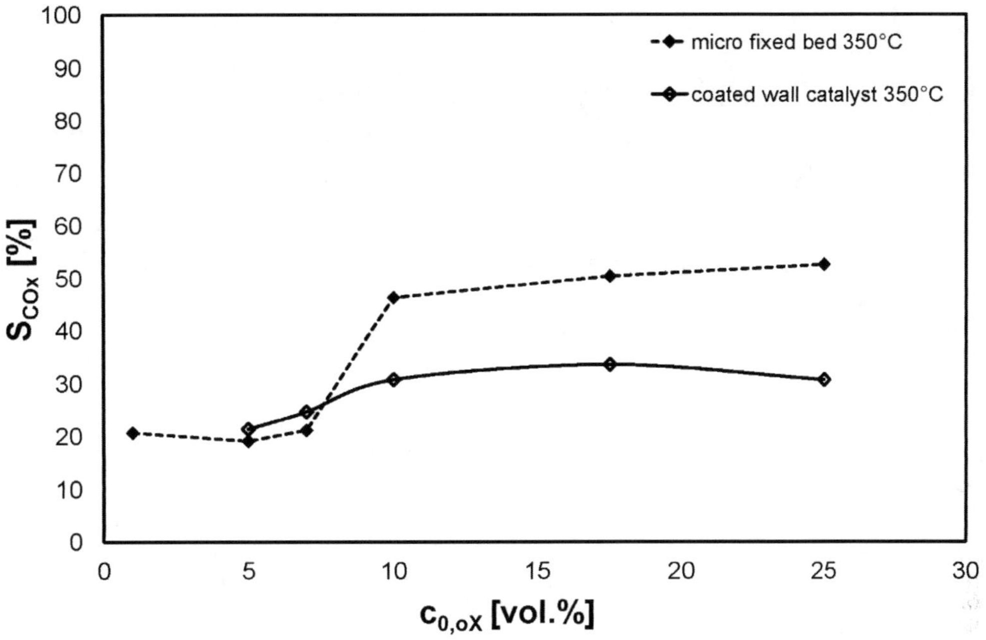


**Fig. 3:** Selectivity to total oxidation products ($CO_x$) versus the initial concentration of o-xylene at 30% conversion and 350°C.

## References


[1] G. Veser, Experimental and theoretical investigation of $H_2$ oxidation in a high-temperature catalytic microreactor, Chem. Eng. Sci. 56 (2001) 1265–1273.

[2] J. Fischer, C. Liebner, H. Hieronymus, E. Klemm, Maximum safe diameters of micro capillaries for a stoichiometric ethene/oxygen mixture, Chem. Eng. Sci. 64 (2009) 2951–2956.

[3] S. Heinrich, F. Edeling, C. Liebner, H. Hieronymus, T. Lange, E. Klemm, Catalyst as ignition source of an explosion inside a micro reactor, Chem. Eng. Sci., accepted.

[4] E. Klemm et al., Direct Gas-Phase Epoxidation of Propene with Hydrogen Peroxide on TS-1 Zeolite in a Microstructured Reactor, Ind. Eng. Chem. Res. 47 (2008) 2086-2090.

[5] G.C.Bond, Preparation and properties of vanadia/titania monolayer catalysts, App. Cat. A: General 157 (1997) 91-103.

[6] Frank-Kamenetskii, D., Stoff- und Wärmebertragung in der chemischen Kinetik. Springer Verlag GmbH (1959).

[7] T. Lange, S. Heinrich, C. Liebner, H. Hieronymus, E. Klemm, Reaction Engineering Investigations of the heterogeneously catalyzed Partial Oxidation of o-Xylene in the explosion regime using a micro structured reactor, Chem. Eng. Sci. 69 (2012) 440–448.

# (164c) Scale-up Approaches to Preserve Particle Properties of Api's in Agitated Filter Driers

**David am Ende** and Mark T. Maloney, Chemical R&D, Pfizer, Inc, Groton, CT

Agitated Filter Driers (AFD's) are commonly used in the pharmaceutical manufacturing for performing both filtration and drying operations. Successful scale-up from lab to manufacturing AFD equipment requires that the physical properties be met. These properties include product quality such as purity and particle size. Depending on the API-solvent system and other operational parameters, undesired attrition or agglomeration may occur impacting the desired particle properties. In such cases mitigation strategies are needed for scale-up that minimize agitation during drying while maintaining acceptable drying rates and ultimately preserving particle properties. The focus of this presentation will be to describe our modeling and scale-up approaches into agitated filter driers with an emphasis on the challenging API's with propensity for agglomeration or attrition during drying.

# (164d) Kilogram-Scale Production of Energetic Materials in a Remote-Controlled Microreactor Plant

Stefan Loebbecke, Energetic materials, Fraunhofer Institute for Chemical Technology (ICT), Pfinztal, Germany

Today, microstructured reactors having internal fluidic structures in the sub-millimeter domain are well known to provide far better heat exchange characteristics and mass transport performance than attainable in macroscopic batch or continuous reactors. In the last decade, a large number of studies have impressively demonstrated that the accumulation of strong reaction heats and hot spots, which result in unwanted side, subsequent and decomposition reactions, can be successfully suppressed in microstructured reactors. Hence, the hazardous potential associated with reactions that are strongly exothermic or potentially explosive can be significantly reduced. Greater safety is also attained with handling thermolabile substances due to the small hold-up of microfluidic reactors.

Here we report on our recent developments of using microstructured reactors for the safe synthesis and subsequent purification of different liquid nitrate esters classified as energetic materials. The hazardous potential of these reactions is due to the huge reaction exothermicity (reaction of alcohols with nitrating agents) and a considerable thermolability of the reaction products and intermediates. A continuous production plant has been developed which is operated in a completely remote-controlled way. At the heart of this technical plant are microreactors made of glass, which have been specially developed for high-throughput applications for both synthesis and downstream processing (washing and extraction). Due to its modular design, the plant allows the production of different nitrate esters at different process conditions. For example, microreactors and pumps can be quickly exchanged for a new production campaign depending on the viscosities of the starting materials and product. The appropriate microreactors are integrated into the production plant to match the throughput, mixing intensity, hold-up, and pressure drop that are required. The entire plant has a wide range of safety features, is under permanent video surveillance and every technical aspect is controlled.

The microreactor plant has been particularly developed for the production of the following nitrate ester oils: ethylene glycol dinitrate (EGDN), butanetriol trinitrate (BTTN), methyl nitrate (MN), and trinitroglycerin (NGL). The oils are synthesized in the microreactor by continuously mixing the corresponding alcohol with a nitrating agent such as HNO3/H2SO4 under precisely controlled stoichiometric and thermal conditions. Afterwards, the energetic raw products (raw oil) are automatically transferred to the downstream process where they are purified and stabilized in further tailor-made microstructured reactors by continuous washing and extraction steps.

The produced nitrate esters fulfill all necessary specifications with respect to (pharma-grade) purity as well as thermal, impact and friction stability.

# (164e) Scale-up of Primary Explosives Using Reaction Engineering Tools

**Jerry S. Salan**, Nalas Engineering Services Inc., Salem, CT

Scale-up is a relative term to the material being produced. In some cases scale-up challenges may include heat and mass transfer. In other cases the scale-up challenges may include the safety of the equipment, facility and personnel. Primary explosives are the most sensitive explosives produced and require great care with respect to processing and handling. These materials are not forgiving and typically detonate if insulted through stimuli such as friction, impact, electrostatic discharge or heat. Scaling up these types of explosives requires a 'hands-off' approach while still being able to acquire all the necessary chemical engineering data for safe processing that produces a high quality product. Primary explosives are first synthesized by the process chemist and eventually transitioned to the chemical engineer. This presentation focuses on how we utilized a combination of in-situ and off-line tools to develop and transfer a 'chemical process' to a plant environment for a primary explosive. The primary explosive was produced in multi-gram quantities.

# (164f) Scale-up of LLM-105 Involving Online Analytics

Andrew G. Pearsall, Chemical Development, Naval Surface Warfare Center, Indian Head, MD

The development of continous processes for the production of energetic materials has been proven several times within the Chemical Development Branch at the Naval Surface Warfare Center, Indian Head Division (NSWC IHD). The advantages of producing these materials continuously lies in the fact that processing and product costs can be lowered, potential hazards can be reduced, and batch to batch variability can be minimized. Safety is at the forefront of the production of energetic compounds, and the ability to minimize the amount of hazardous reaction solution that is being processed at any given time in a reaction is one method of increasing safety. The avoidance of multiple equipment setups between batches and the increased efficiecy that can be achieved with a properly designed continuous process will reduce product costs. Performing a process continuously at a smaller scale as compared to a batch process also increases the ease of controlling reaction temperature due to the smaller volumes in the continous process and also can help in avoiding some of the mixing challenges that can be experienced at larger scales. NSWC IHD has developed continuous processes for several energetic compounds, and details on the modeling, simulations, and reaction information gathered in the laboratory to develop these processes will be presented.

# (176a) Molecular Design with Simultaneous Topological, Topographical, and Electrostatic Characterization

Robert Herring III[1], Subin Hada[1], J. Colin Haser[1], Nishanth G. Chemmangattuvalappil[2] and Mario Richard Eden[1], (1)Department of Chemical Engineering, Auburn University, Auburn, AL, (2)Department of Chemical and Environmental Engineering, University of Nottingham - Malaysia, Semenyih, Malaysia

Computer-aided molecular design (CAMD) is emerging as a very powerful tool in chemical and pharmaceutical industries today. The ability to design products, which meet a set of specified target properties, exemplifies the applicability of CAMD. With the advent of new computing technologies, a more detailed and comprehensive description of molecular structures is possible. This growth in the ability to model and predict a wide range of molecular properties should parallel the increased stringency of regulations on environmental, health and safety standards. Thus, more tools must be developed which take advantage of relationships between molecular structure and the macro scale properties realized for these systems. It is well understood that most properties of interest are not solely a function of the topological arrangement and connectivity of the molecules under investigation. The spatial arrangement, e.g. steric hindrance or ligand binding, and electrostatic nature of these structures are also necessary to reveal more optimal candidate molecules. It has been suggested that searching the chemical space with a combination of 2D and 3D descriptors could prove more effective than employing single methods [1].

This contribution details a systematic methodology for doing just that, and it is based on the signature molecular descriptor [2]. One difficulty in the application of reverse molecular design using geometric based descriptors is that the minimum energy molecular conformation is a function of the complete structure. The most stable conformation is ultimately determined by interactions among subatomic particles through the force fields they generate, which can be accurately modeled by quantum mechanics. The effect of these force fields diminishes as a function of increasing inter-particle distance. Thus, a reasonable estimation of local geometry can be made by considering atoms extending radially, up to fixed distance, from a central atom. The signature descriptor is ideal for this approach as it details the extended valence sequence of an atom up to a pre-defined height or distance away from the central atom. Previous studies have proven the applicability and effectiveness of atomic signature descriptors utilized in molecular design with topological and topographical considerations [3-5]. These signatures can be geometry optimized through various quantum mechanical approaches, and this information is conveniently stored in the form of a geometry matrix. Also, the electron densities of these signature "fragments" can be generated through *ab initio* calculations, given the previously obtained geometry information, and recombined to produce complete molecular electron distributions. This approach, known as MEDLA (Molecular Electron Density Lego Approach) [6], utilizes a "fuzzy" electron density description so that boundary adjustments upon fragment recombination are unnecessary. The nature of these signature descriptors and boundary-less electron cloud fragments allow for a linear optimization approach and are consequently much less computationally demanding than other *ab initio* quality calculations. The framework and mathematical techniques necessary for formulation and solution of property or activity based molecular design problems, with 2D, 3D and electrostatic characterization, are presented here along with a case study which illustrates the benefits of such an approach.

[1] Nettles J.H., Jenkins J.L., Bender A., Deng Z., Davies J.W., and Glick M. (2006) Bridging Chemical and Biological Space: "Target Fishing" Using 2D and 3D Molecular Descriptors. *J. Med. Chem.*, **49**, pp. 6802-6810

[2] Faulon J.-L., Visco D.P. Jr., and Pophale, R.S. (2003) The Signature Molecular Descriptor. 1. Using Extended Valence Sequences in QSAR and QSPR Studies. *Journal of Chemical Information and Computer Sciences,* **43**, pp. 707-720.

[3] Herring R.H., Rudolfs N., Chemmangattuvalappil N.G., Roberts C., and Eden M. (2012). Molecular Design Using

Three-Dimensional Signature Descriptors. *Computer Aided Chemical Engineering*, **31**, (In Press).

[4] Chemmangattuvalappil N.G., Solvason C.C., Bommareddy S. and Eden M.R. (2009) Incorporating Molecular Signature Descriptors in Reverse Problem Formulations. *Computer aided Chemical Engineering*, **27**(1), pp. 73-78.

[5] Chemmangattuvalappil N.G., Solvason C.C., Bommareddy S. and Eden M.R. (2009) Novel Molecular Design Technique using Property Operators based on Signature Descriptors. *Computer aided Chemical Engineering*, **27**(2), pp.897-902

[6] Walker P. D., Mezey P. G. (1993) Molecular Electron Density Lego Approach to Molecule Building. *J. Am. Chem. Soc.*, **115**, pp. 12423-12430.

# (176b) Amodeo: Automated Molecular Design Using Optimization

**Apurva Samudra** and Nick Sahinidis, Chemical Engineering, Carnegie Mellon University, Pittsburgh, PA

We describe and demonstrate a new software that automates, augments, and accelerates steps in computer-aided molecular design (CAMD). Our overall approach extends the scope of CAMD methods beyond group contribution models by employing accurate and extensive property models with a wide application domain. The molecular design problem is tackled in three stages: 1) composition design, 2) structure determination, and 3) extended design. Composition identification and structure determination are decoupled to achieve computational efficiency. Using approximate group contribution methods in the first stage, molecular compositions that fit the relaxed design targets are identified. In the second stage, isomer structures of solution compositions are determined systematically and structure-based property corrections are utilized to refine the solution pool. In the final stage, the design is extended beyond the scope of group contribution methods by utilizing problem-specific property models. At each stage, optimization algorithms generate a large and diverse pool of candidate designs using an assortment of property models. The molecule generation machinery and external property models are bundled together to create an easy-to-use software.

The use of integer optimization techniques instead of enumeration at each stage, leads to an efficient algorithm, unrestricted in terms of the number of solutions sought and the maximum size of the molecule sought. Unique solutions are generated effectively with cuts to handle redundancy in chemical space. These optimization-based models lead to orders of magnitude decrease in solution time over enumeration or MINLP approaches. The software includes interfaces for external property models. The software also provides automated database searches and computational chemistry calculation options. We demonstrate the flexibility of the software using algebraic correlations, optimization-based models, and simulations as property estimation techniques.

# (176c) Design of an Emulsified Hand Wash Through a Systematic Model-Based Methodology

Michele Mattei, Department of Chemical and Biochemical Engineering, Technical University of Denmark, Kgs. Lyngby, Denmark, Michael Hill, Chemical Engineering, Columbia University, New York, NY, Georgios Kontogeorgis, Chemical and Biochemical Engineering (CERE), Technical University of Denmark, Kgs. Lyngby, Denmark and Rafiqul Gani, CAPEC, Department of Chemical and Biochemical Engineering, Technical University of Denmark, Kgs. Lyngby, Denmark

**Design of an Emulsified Hand Wash through a Systematic Model-based Methodology**

Michele Mattei[1] (e-mail: *micu@kt.dtu.dk*), Michael Hill[2] (*mih16@columbia.edu*), Georgios M. Kontogeorgis[1] (*gk@kt.dtu.dk*), Rafiqul Gani[1] (*rag@kt.dtu.dk*)

1 *Department of Chemical and Biochemical Engineering, Technical University of Denmark, Søltofts Plads, Building 229, DK-2820, Kgs. Lyngby, Denmark*

2 *Department of Chemical Engineering, Columbia University, Rm. 801 Mudd, 500 West 120th Street, New York, NY 10027, USA*

A general concern shared by everyone, wherever they live, whatever their social rank, whatever their age, is to be neat and to live in a clean environment. Maintaining or restoring cleanliness is usually among the top priorities of everyone and a significant amount of time and money is involved in the cleaning processes. The real breakthrough for soaps and detergents appeared during the early 1940s with the introduction of "built" detergents in the United States; since then, a great variety of products appeared on the market to help people address the cleaning tasks. Today, cleaning products are becoming increasingly more sophisticated, for various reasons; mainly the technological progress (dishwashers, washing machines, *etc.* are nowadays in almost every house in the developed world) and the evolution in lifestyle (consumers need more refined products to deliver superior performances on a wider spectrum of soils with a minimum of efforts and time commitment)[1]. Among the many raw materials used in a formulated cleaning product like liquid soaps and detergents, surfactants are by far the most important and needed ingredients. They are present in every cleaning or conditioning product and several factors are important in the choice of the surfactant system[2]. Some of the applications of surfactants in cleaning products are improving the cleaning activity, improving mildness, boosting foaming properties, making the product antimicrobial active and in forming the emulsion[3]. The design of these emulsified liquid formulated products has quite a few degrees of freedom: in addition to the nature of the three main components (water, oil and surfactant), decisions also need to be made on the influence of electrolytes, alcohols and other additives on the final product. As a consequence of this and of consumer assessments, the number of components in a commercial emulsified product is often quite high. Consequently, the design of these formulated products has been considered more of an art, in which a successful "magic recipe" is worth a lot of development time and money[4].

The above problem definition matches perfectly with the general definition of a product (formulation) design problem: given a set of desired (target) product attributes, determine a set of blends (a mixture of chemicals) that satisfy the desired attributes, and, from them, select the most appropriate candidates for final validation by experiments[5]. The development of a systematic model-based methodology to quickly and reliably reduce the search space and within it, to identify a set of candidates (to be afterwards experimentally validated and, if necessary, refined for further improvement) can then make the design and development of these products less costly, so that the product itself can be introduced to the market earlier, while using the experimental resources mainly for verification of the product.

A systematic model-based step-by-step methodology for design of emulsion based chemical products has been developed, which employs a hierarchical approach consisting of seven sequential steps, starting with the identification of the needs to be satisfied by the product; then building up the formulation by adding one-by-one the different classes of chemicals needed; and finally determining the composition of the formulated product. Based on the ingredients chosen and the target properties, their compositions on the formulation are adjusted. The choice of ingredients is ordered, so that the most advantageous active ingredients are identified first, followed by the dispersed and continuous phase solvents, then the emulsifiers and finally the additives. The selection is carried out considering effectiveness, safety, toxicity and cost of each candidate. A specially developed database of different types of chemicals (ingredients) and a library of property models help in evaluating, screening and defining the candidate formulations. This model-based methodology is intended to be generic, in the sense that many different emulsified products can be designed through this procedure once the needs-property relations are established, appropriate structured databases for all the categories of ingredients are accessible and the property prediction models are applicable (to estimate the needed properties when their data are not available in the database). The concepts of the methodology have been tested on the design of an emulsified sunscreen lotion[6].

In this work, an extended and more generic version of the methodology is presented and the main steps are illustrated through a case study involved with the design of an emulsified hand wash product. The focus of the application of the methodology presented here is on the definition of consumer assessment and their translation into measurable target properties, as well as on modelling of some of the critical properties, such as, properties of surfactants (acting in a hand wash product both as active ingredients and emulsifiers at the same time) in order to estimate their effectiveness so that the most effective can be selected. These properties have been modelled through the development of a classical group-contribution (GC) method, based on the $GC^+$ approach. The work-flow methodology and all the related tools will be implemented into "The Virtual Process-Product Design Laboratory[7] software.

---


[1] Broze G., *Handbook of Detergents, Part A: Properties*; 1999: 1-5.

[2] Schramm L.L. et al, *Annu. Rep. Prog. Chem., Sect C: Phys. Chem.*; 2003, 99: 3-48.

[3] Polefka T.G., *Handbook of Detergent, Part A: Properties*; 1999, 433-468.

[4] Salager J.L., *Progr. Colloid Polym. Sci.*; 1996, 100: 137-142.

[5] Gani R., *Chem. Eng. Res. Des.*; 2004, 82 (A11): 1494-1504.

[6] Mattei M. et al, *11$^{th}$ International Symposium on Process System Engineering*; 2012, in press.

[7] Conte E. et al, *Fluid Phase Eq.*, 2011; 302: 294-304.

# (176d) Strategic Search for Experimental Conditions for Efficient Product Design

**Takuya Kishio**, Hiromasa Kaneko and Kimito Funatsu, Department of Chemical System Engineering, The University of Tokyo, Tokyo, Japan

In experimental design of materials and products, experimental results are affected by various parameters, for example, chemical constitution and composition of raw materials, and synthesis condition. There are many parameter combinations and the relationships between parameters and physical properties are very complicated. Hence, parameter search for the candidate having the desired properties takes cost and time.
Until now, experimenters have chosen the parameter combination from a large choice with their experience and/or intuitions. However, this inefficient search method increases experimental number of times and development cost.
For understanding relationships between parameters and physical properties, prediction models with regression methods are useful. Traditionally, predicted values of physical properties with a regression method were used for the parameter search. However, in regions where data density is low, the predictive ability of a regression model is also low and the reliability of predicted values is not ensured.
In this study, we use regression methods, and evaluate candidates with predicted physical properties and data distribution as evaluating criteria. We can get the material with desired physical properties with less experimental number of times by using the proposed method.

First, a physical property prediction model is built between experimental parameters (X) and a physical property (y) from database by using regression methods such as gaussian process (GP), partial least squares (PLS), and support vector regression (SVR). Second, values of new X of candidates are input into the model and values of y are predicted. Third, these candidates are evaluated with the following criteria. After evaluating the candidates, the physical property of the best ones is measured. If the measured property is insufficient, the prediction model is updated with these data. This flow is repeated until objected values of y are obtained.

Two evaluating criteria are used in this method. One is probability (P) to get the material which has the objective y ($y_1 \leq y \leq y_2$). P is obtained by integration of the normal distribution, whose average is a predicted value of y ($y_{pred}$) and variance is estimated variance of prediction errors. This variance can be calculated with GP.
Another criterion is data density (DD) calculated with one-class support vector machine (OCSVM). In regions where data density is high, the predictive ability will be high and these regions are called as applicability domains of a regression model.
After range-scaling of P and DD, a weighted average of these two criteria is regarded as an overall criterion. We changed the weight of the criteria variously and investigated the experimental number of times to get the objective solution with each regression method.

The proposed method was applied to various types of data. Simulation data were generated with Goldstein-Price Log Function[1]. The number of input variables (X) is two and the number of output variables (y) is one. The minimum value was 1.10 when $x_1$ equals to -1 and $x_2$ equals to 0. We got 1,681 data by changing each x ($-2 \leq x \leq 2$) in steps of 0.1. The objective region of y was set as $-5 \leq y \leq 3$, and the first 10 data for constructing prediction model were selected randomly from the candidates whose values of y were larger than 7.
The weight of P for a weighted average was set as 0.5, which meant that P and DD were dealt as equivalent. When GP or SVR which is a nonlinear regression method was used, the experimental number was smaller than that of PLS which is a linear regression method. When search for the candidates in the objective region of y was performed 50 times, the average experimental number with the proposed method was 13.6 when GP was used. Meanwhile, when only predicted value was used as a traditional evaluating criterion, the average number was 27.9. Herewith, the effectiveness of our method was confirmed.
Next, we applied this method to actual data which were the logarithmic water solubility values of 1,290 chemical compounds obtained from the literature[2]. X-variables were 187 structural descriptors and y-variable was water

solubility (logS). The objective region of y was $1.15 \leq y \leq 2.00$, and the first 10 data were selected randomly from the chemical compounds whose values of y were smaller than -5. In this case, when we used PLS the experimental number was smaller than that of GP and SVR because this data had strong linearity between X and y. The average experimental number was 4.98 when the proposed method was used and 5.60 when only predicted value was used as traditional evaluating criterion.
By using the proposed method, the decrease of the experimental number of times was achieved, and therefore more efficient experimental design of materials and products can be expected.

[1]G. Li, V. Aute, and S. Azarm, *Structural and Multidisciplinary Optimization,* **40**, 137-155, 2010.
[2] I. V. Tetko, V. Y. Tanchuk, T. N. Kasheva, A. E. P. Villa, J. *Chem. Inf. Comput. Sci.*, **41**, 1488-1493, 2001.

# (176e) Data-Driven Optimization of Product Blends Using Multivariate Property Clustering Techniques

**Subin Hada**[1], Robert Herring III[1], J. Colin Haser[1], Nishanth G. Chemmangattuvalappil[2] and Mario Richard Eden[1], (1)Department of Chemical Engineering, Auburn University, Auburn, AL, (2)Department of Chemical and Environmental Engineering, University of Nottingham - Malaysia, Semenyih, Malaysia

Mixture processes play an important role in today's manufacturing of value-added chemical products such as commodity chemicals, food, cosmetics, oil and pharmaceuticals. In classical non-mixture designs such as factorial and response surface designs, all the factors are orthogonal or independent. This means that it is possible to freely choose the level of a factor regardless of the other factors' levels. However, this freedom does not exist for mixture designs, because each component in a mixture is dependent upon the settings of the other component settings. In mixture design, the factors are interdependent and the effects of the factors on the responses are not separable. In addition, traditional mixture models are usually employed to investigate the relationships between the blend ratio matrix (**R**) and the final blend product property matrix (**Y**) only, given that the properties of the pure raw materials (**X**) and the process conditions used to manufacture them (**Z**) are already chosen. However, in the development of formulated products that meet target properties with minimum experiments and minimum total material cost, it is important to simultaneously account for all three degrees of freedom (**X**, **R**, and **Z**) available in the blending operations.

In this paper, projection methods like PCA and PLS are employed to extract and utilize the necessary information from all three degrees of freedom that are available to control the properties of any product manufactured in blending operation. This information is used to build models to simultaneously explore variations in the raw material properties data (**X**), their blend ratios (**R**), and process conditions (**Z**) and the resulting effect on the product quality (**Y**). These methods capture information about the variance and correlation structure within predictor and quality variables that were present during the production of previous product grades and that are important for consistency of the desired process conditions with past operating procedures. An integrated solution approach is employed that formulates two reverse problems, one for determination of design targets and another for matching of the design targets. The empirical models are used in a reverse problem formulation to ensure that a complete set of candidate materials and their mixtures are found which match the design target subject to the predictive power of the model. Mapping the design problem onto the lower dimensional cluster space utilizing property clustering algorithm facilitates visualization and simultaneous solution of the problem and also avoids combinatorial explosion when dealing with multi-component mixtures.

The method and concepts will be illustrated using a case study describing the design of polymer blending problems in a latent variable space.

# (176f) Computer Aided Ionic Liquid Design (CAILD) Method for Development of Green Ionic Liquids

**Amirhossein Mehrkesh** and Arunprakash T. Karunanithi, Department of Civil Engineering, University of Colorado Denver, Denver, CO

Solvents comprise two thirds of all industrial emissions. Traditional organic solvents easily evaporate into the atmosphere as they have high vapor pressure and are linked to a host of negative environmental effects including climate change, urban air-quality and human illness. Room temperature ionic liquids (RTIL), on the other hand, have low vapor pressure and are rarely flammable or explosive, thereby presenting fewer environmental risks and health hazards. For this reason they are considered as 'green' solvents. RTILs are molten salts that exist as liquids at relatively low temperatures and have unique properties. Ionic liquids consist of a large organic cation and charge-delocalized inorganic or organic anion of smaller size and asymmetric shape. The organic cation can undergo unlimited structural variations through modification of the alkyl groups attached to the side chain of the base cation skeleton (Fig. 1) and most of these structural variations are feasible, from chemical synthesis point of view, due to the easy nature of preparation of their components. Functionally, ionic liquids can be tuned to impart specific desired properties by switching anions /cations or by incorporating functionalities into the cations/anions. It is estimated that theoretically more than a trillion ionic liquid structures can be formed. Due to their tunable nature, these molten salts have the potential to be used as green solvents for a variety of applications.

This paper presents a computer aided molecular (IL) design (CAILD) methodology with an aim to design optimal task specific ionic liquid structures for different applications. We utilize group-contribution based ionic liquid property prediction models (forward problem) within a mathematical programming framework to reverse engineer functional ionic liquid structures. The CAILD model is formulated as a mixed integer non-linear programming (MINLP) optimization problem and the solution results in optimal ionic liquid cation-anion combination for a given application. The formulation of the optimization model including representation of IL structural constraints will be discussed in detail. Specific case study results on applications will be presented.

# (187a) Layer Melt Crystallization of Long-Chain Aldehydes From Hydroformylation Processes

**Thorsten Beierling**, Department of Biochemical and Chemical Engineering, Laboratory of Thermodynamics, Technische Universität Dortmund, Dortmund, Germany and Gabriele Sadowski, Department of Biochemical and Chemical Engineering, Laboratory of Thermodynamics, Technische Universität Dortmund, Dortmund, Germany

The hydroformylation of long-chain olefins, an industrially highly relevant reaction, produces a mixture of isomeric long-chain aldehydes. The purified substances are used as intermediates for further bulk and fine chemicals, such as diluent or for the flavour and fragrance industry. Conventional methods like stand-alone distillation tend to be costly or even infeasible, due to a decreasing difference in boiling points with increasing chain lengths. Layer melt crystallization in contrast is a technically proven and selective method for separating isomeric compounds and has generally the advantages of an easy scale up, facile solid-liquid separation and process handling. Despite these advantages, the promising method of separating isomeric long chain aldehydes with melt crystallization has not been investigated yet.

In this work the applicability of layer melt crystallization for various isomeric long chain aldehyde systems with different chain lengths was evaluated. Necessary parameters like related phase-equilibria were measured and modeled. Layer melt crystallization experiments were conducted, in which achievable growth rates and the corresponding purities were studied. The growth rates were modeled using the physical properties of the aldehyde systems as a function of process conditions like concentration, temperature and cooling rate. Additionally, the efficiency of sweating as a post-purification step was investigated. The sweating behavior can mathematically be described in high accordance to the experimental values as a function of temperature, crystal purity and layer thickness.

Summing up we show that melt crystallization should, due to its high efficiency, be considered as the method of choice for the separation of isomeric long-chain aldehydes. Furthermore our results reveal useful methods for the extrapolation of melt crystallization including the sweating step to systems with different chain-lengths, notably without conduction of further experiments.

# (187b) Hybrid Filtration-Drying-Dissolution Device for Compact Pharmaceutical Manufacturing System

**Shin Yee Wong**, Jie Chen and Allan S. Myerson, Department of Chemical Engineering, Massachusetts Institute of Technology, Cambridge, MA

Pharmaceutical manufacturing is typically done in batch vessels. The recent interest in continuous pharmaceutical manufacturing had triggered a lot of interest in both academic and industrial communities, driven by the many benefits of the transformation. To bring the idea of continuous manufacturing further, we are developing a bench-top sized Pharmacy-on-Demand (PoD) system, capable of producing formulated drugs, corresponding to a daily production rate of few kilograms of Active Pharmaceutical Ingredients (API). The realization of the compact PoD system would require innovation in all unit operations (chemical synthesis, crystallization, and downstream processes), and high level of controls and automations. In this study, a hybrid device that integrated all post-processing operations after the crystallization process was developed.

This device consisted of five major components: 1) top cover consisted of inlets for slurry (from crystallizer) and liquid excipients; 2) steel mesh (0.5μm) holder; 3) lower barb end connected to vacuum; 4) wall heating component; 5) side drain valve. The filtration, drying and dissolution operations were conducted in semi-batch manner. Once the crystallization process was completed, the slurry was pumped into the hybrid device. Then, the refined API was separated from the mother liquor through the steel mesh under vacuum condition. After the filtration, the API was dried under vacuum condition at the temperature controlled by power supplied to the wall heating element. Once dried, the API was re-dispersed in the liquid excipients and drained from the side valve. Initial testing had shown high separation and drying efficiency. More than 98% of the crystallized API was successfully recovered after re-dissolution with the liquid excipients. The hybrid device had successfully transformed the slurry obtained from the crystallization process into formulated drug products in a single integrated device.

In short, the invention of this hybrid device had essentially eliminated the need for multiple stand-alone units for filtration, drying, blending and dissolution. On top of its integrated function, the compact (max dimension = 90mm) device had also significantly simplified the implementation of the downstream purification process of the PoD system.

# Design and Characterization of a Continuous Stirred Tank Crystallizer

G. Morris, G. Hou, M. Barrett, B. Glennon

Solid State Pharmaceutical Cluster, The School of Chemical and Bioprocess Engineering, University College Dublin, Belfield, Dublin 4, Ireland

**Keywords:** Mixed Suspension Mixed Product Removal Crystallizer, Benzoic Acid, Cooling Crystallization, Process Analytical Technology

**Paper ID#:** 270744

## Introduction

This study presents the development of a continuous crystallization platform consisting of a single stirred tank Mixed Suspension Mixed Product Removal (MSMPR) crystallizer. A photo and schematic drawing of the MSMPR setup appear in Figures 1 and 2 respectively. A recirculation loop is incorporated such that the withdrawn product is recycled to a dissolution vessel which serves as the feed tank to the crystallizer. Two OptiMax$^{TM}$ workstations (Mettler-Toledo) are used as both the feed/dissolution tank and the MSMPR crystallizer. Difficulties can often arise in the continuous removal of representative, non-classified product from lab-scale MSMPRs as a result of the relatively low volumes involved and consequently slow flowrates and removal velocities observed **[1]**. Intermittent withdrawal, whereby a certain percentage volume of the MSMPR is periodically removed at a high velocity so as to aid the potential for iso-kinetic removal and mitigate classification has been applied previously and is adopted in this work **[2, 3]**. This quasi-continuous transfer of the MSMPR product to the feed/dissolution vessel is achieved using vacuum with 7.1% of the suspension volume removed per transfer.

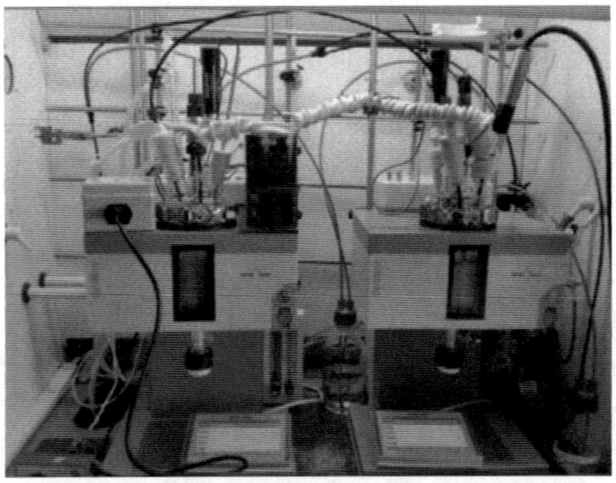

*Figure 1: Image of the MSMPR setup showing feed/dissolution vessel on the left and MSMPR on the right with FBRM in-situ*

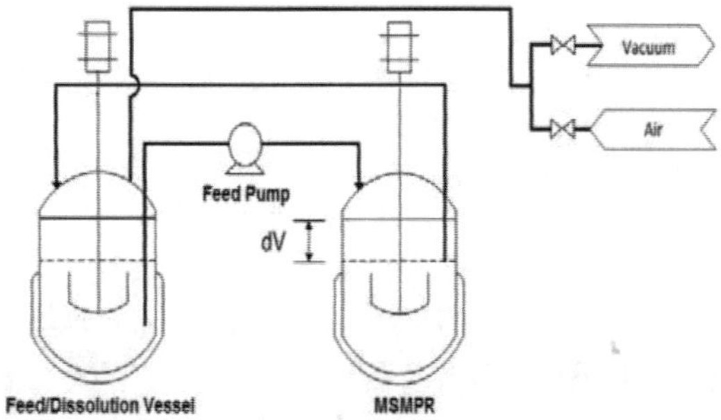


*Figure 2: Schematic drawing of the MSMPR setup showing application of vacuum in recycling material from MSMPR to feed/dissolution vessel*

In-situ process analytical technologies (FBRM, PVM), together with periodic sampling of the withdrawn product at steady-state and during the dynamic period of the process, have been applied in order to characterize the continuous cooling crystallization of benzoic acid from water and ethanol. Comparison is drawn between *high magma density* crystallizations (feed saturated at 40°C with the crystallizer at 18°C) and *low magma density* crystallizations (feed saturated at 15°C with the crystallizer at 0°C). For the low magma density runs the influence of residence time on affecting the steady-state particle size was investigated. For each of these studies, continuous operation was started with a suspension in the MSMPR at time zero by first batch cooling the saturated feed to the desired MSMPR temperature. In addition, a further set of experiments was carried out to investigate whether the mode of start-up bears any influence on the final steady-state product obtained and the route to steady-state.

## Characterization with FBRM and Periodic Sampling

The crystallization is monitored in-situ using FBRM and PVM. This allows real time tracking of the changes in the particle size distribution and attainment of steady-state. Figure 3 shows the dynamic response in the FBRM total counts and mean chord length trends for a high-magma density crystallization proceeding to steady-state.

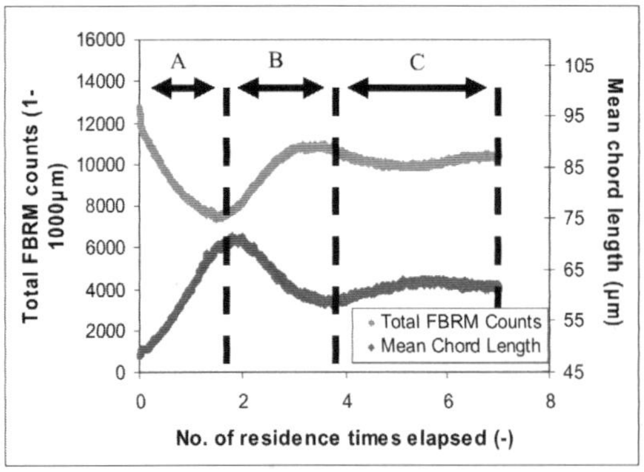


*Figure 3: Tracking MSMPR to steady-state with FBRM for a high magma density run of 60 minute residence time. Showing regions: (A) wash-out period, (B) nucleation response and (C) attainment of steady-state*

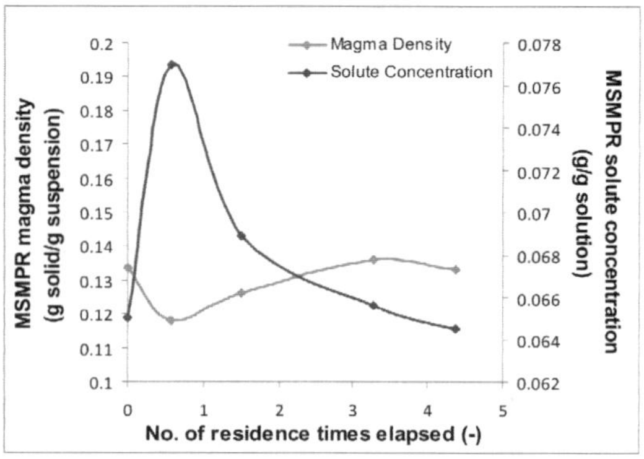


*Figure 4: MSMPR solute concentration and magma density trends from start-up to steady-state for a 15 minute residence time high magma density run*

As the crystallization proceeds initially there is a wash-out period (A) where particles are withdrawn in the absence of nucleation as the supersaturation builds. Over this period crystal growth is observed which is confirmed by an increase in the FBRM size trends and in-situ PVM images (Figure 5). Supersaturation increases to a maximum just before region (B), where the kinetics of the system force a period of constant nucleation which steadily consumes some of the supersaturation and slightly shifts the CSD back towards the fine end. Eventually the rate at which new crystals are formed by nucleation balances the rate at which they are withdrawn from the MSMPR and the system tends towards a steady-state (C).

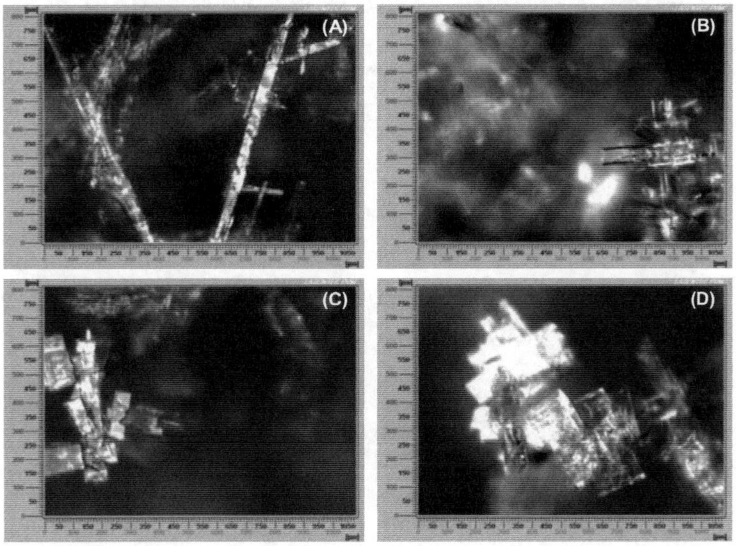


*Figure 5: In-situ PVM images from a low magma density run of 15 minute residence time showing the change in particle size during MSMPR start-up. Taken at: (A) time zero after batch cooling at 1.5°C/min, (B) at 1 residence time, (C) at 2 residence times and (D) at 3 residence times*

## High & Low Magma Density Crystallizations

Comparing high and low magma density crystallizations it was observed that the level of the nucleation response in the dynamic route to steady-state is reduced when the slurry density in the crystallizer is relatively low (Figure 6), demonstrating the dependency of the nucleation rate on magma density. Consequently for the high magma density process the steady-state chord length distribution is shifted more towards the fine end (Figure 7).

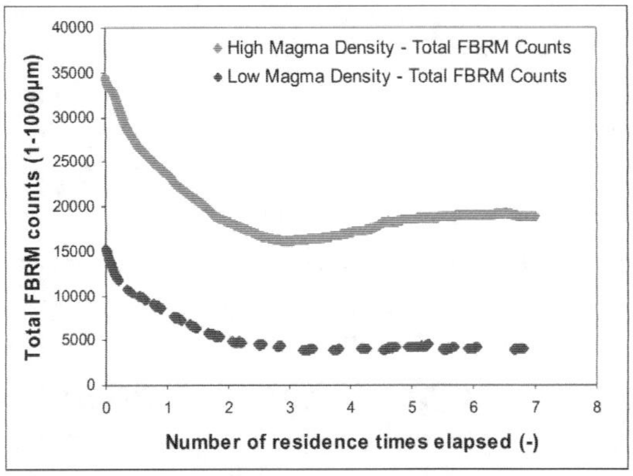


*Figure 6: FBRM total counts trend for high and low magma density MSMPR runs of 15 minute residence time. For the low magma density run FBRM fouling was an issue so periodic cleaning of the probe was adopted*

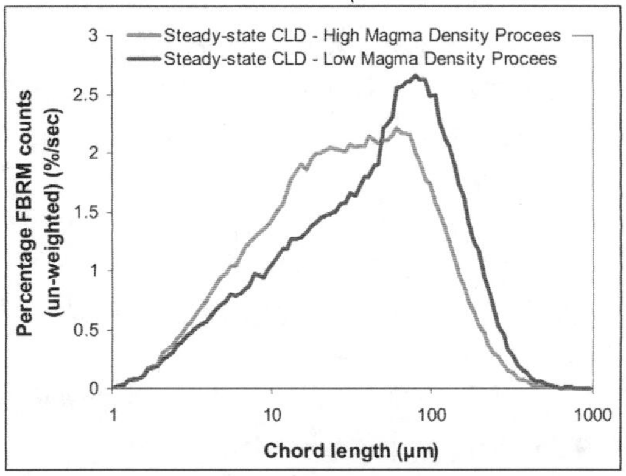


*Figure 7: Final steady-state (un-weighted) chord length distributions (CLDs) for both high and low magma density processes*

### Evolution of Particle Size with Increasing Residence Time

In an MSMPR, for a given temperature drop between feed and crystallizer, it is possible to access different positions on the phase diagram by varying the residence time and as a result manipulate the steady-state supersaturation. The speed of the crystallization kinetics and the meta-stable zone width will ultimately dictate the supersaturation range accessible, and whether changing residence time will be favourable in terms of manipulating particle size advantageously. The benzoic acid system used in this study has a small meta-stable zone width and is more growth dominated. Consequently, the accessible range of supersaturation is limited but particle size has been shown to grow with residence time (Figure 8). Figure 9 demonstrates that steady-state particle size increases whilst MSMPR solute concentration and hence supersaturation deceases with increasing residence time.

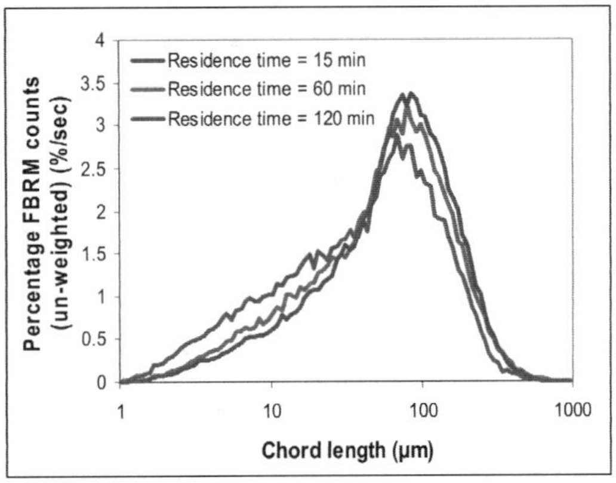


*Figure 8: Steady-state CLDs (un-weighted) for low magma density runs of varying residence time*

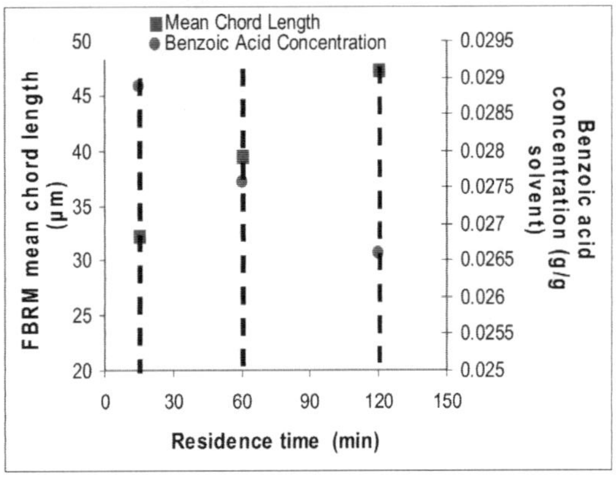


*Figure 9: Demonstrating increase in particle size with residence time whilst solute concentration decreases*

### Influence of Start-Up Mode on Attainment of Steady-State

Several start-up regimes for running an MSMPR could potentially be employed. To examine whether particular choices in this area can have a significant bearing on the steady-state particle size obtained by the system, three methods of starting the MSMPR were investigated whilst keeping all other conditions the same:

(i) *Batch cooling the saturated feed to the desired MSMPR operating temperature, therefore allowing continuous operation to be started with an equilibrium suspension in the MSMPR at time zero*

(ii) *Starting with a clear solution which is saturated at the MSMPR operating temperature*

(iii) *Seeding a solution that is saturated at the MSMPR operating temperature with isolated steady-state product from a previous run, therefore creating a suspension of the final product at time zero*

All start-up methodologies were assessed through high magma density benzoic acid crystallizations, where the feed to the MSMPR is saturated at 40°C and the crystallizer is operated at 18°C. The resulting steady-state chord length distributions from each start-up regime are seen to be similar in all cases (Figure 10). This suggests that the final steady-state condition attained by the system is largely unaffected by the start-up methodology adopted.

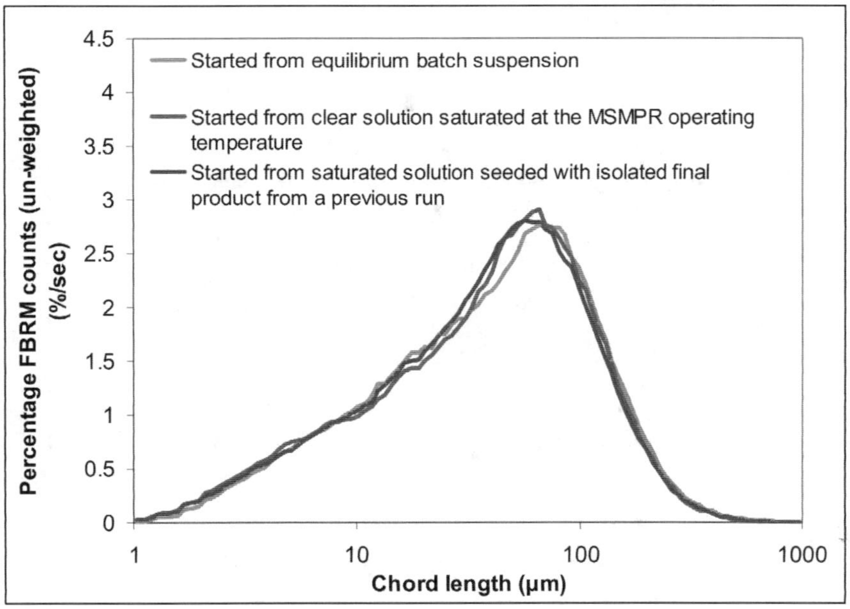


*Figure 10: Steady-state CLDs (un-weighted) for high magma density crystallizations with different start-up regimes*

Comparison of the FBRM total counts trends during the dynamic route to steady-state suggests that starting with saturated solution that is seeded with final product material may offer the quickest route to steady-state (Figure 11). Unlike the equilibrium batch start-up, in this case the initial particle size distribution is the same as the final and as a result the system does not require periods of significant oscillation in the size distribution to develop the steady-state CSD for the continuous process. This therefore may represent an optimum start-up strategy for MSMPRs, as the time required to reach steady-state is reduced and hence wastage of material whilst the final product is potentially out of spec is minimised.

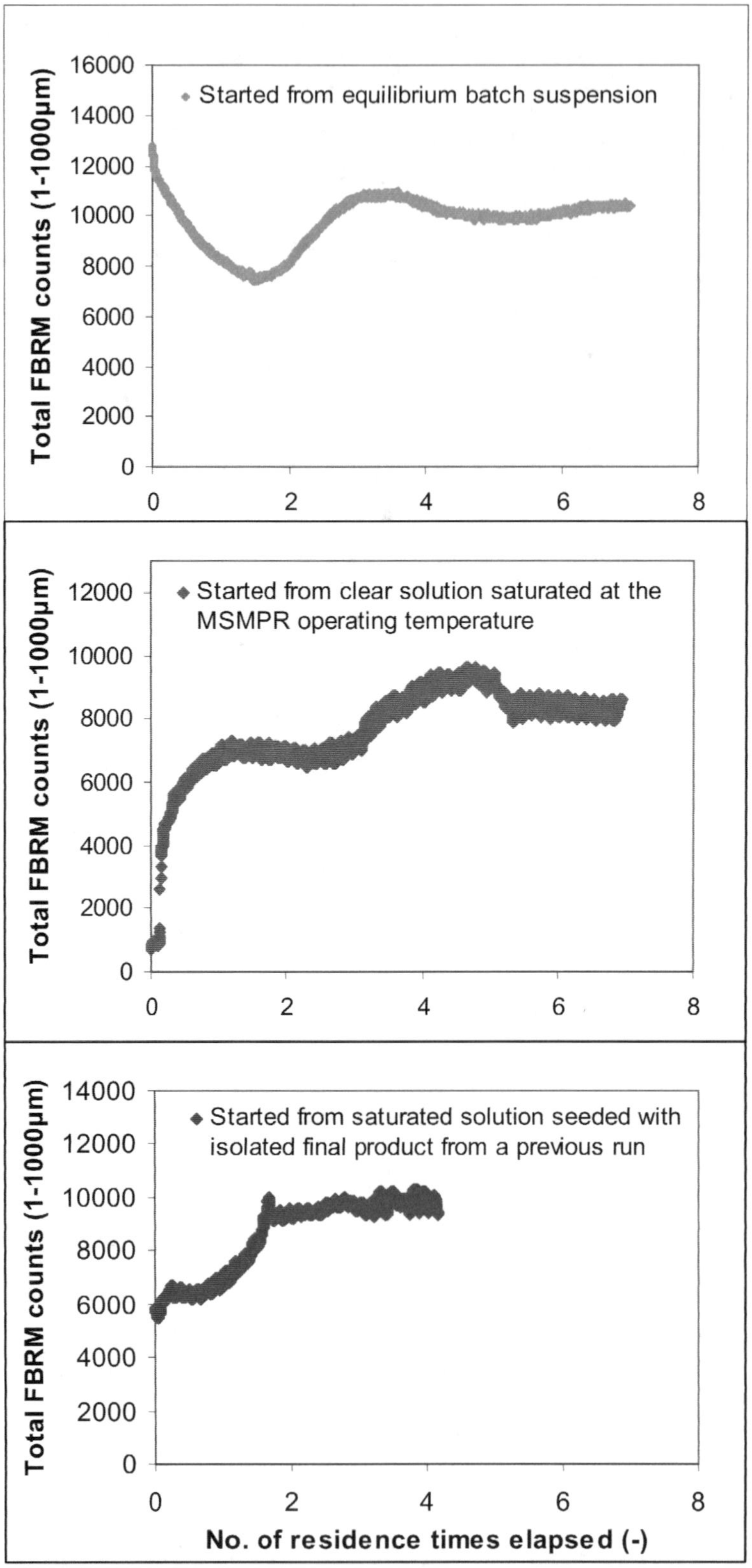


*Figure 11: Comparison of the dynamic response in the FBRM total counts trends as the system proceeds to steady-state for all three start-up methodologies investigated*

## References


**[1]** A.D. Randolph, M.A. Larson, Theory of Particulate Processes, 2nd edition, Academic Press, New York, 1988.
**[2]** V.A. Chan, H.M. Ang, Journal of Crystal Growth 166 (1996) 1009-1014
**[3]** E. Kougoulos, A.G. Jones, K.H. Jennings, M.W. Wood-Kaczmar, Journal of Crystal Growth 273 (2005) 529–534

# (187d) X-Ray Compatible Microfluidic Platforms for On-Chip Structure Determination of Membrane Proteins

**Sudipto Guha**, Ashtamurthy Pawate, Sarah L. Perry and Paul J.A. Kenis, Department of Chemical & Biomolecular Engineering, University of Illinois at Urbana-Champaign, Urbana, IL

Microfluidics has found extensive use in fields as diverse as energy conversion, medical and clinical diagnostics, point-of-care medical devices, proteomics and protein crystallization. Microfluidics allows for ease of sample preparation, small sample sizes, ease of scale-up, as well as integration of various sample analysis techniques on a single platform[1,2].

Here, we focus on the application of microfluidic platforms for membrane protein crystallization. Membrane proteins (MPs) play a crucial role in many important biological processes including energy and material transduction across cellular membranes, molecular recognition and immune response. They are amphiphilic, structurally diverse, and extremely responsive to the surrounding environment. Efforts to understand the structure-function aspects of these proteins through X-ray crystallography have been hampered by difficulty in obtaining high quality crystals[3]. Identifying the optimal crystallization condition(s) requires systematic screening of a huge chemical space (salts, buffers and precipitants), which requires a lot of protein sample that is generally not available. Additionally, current methods of structure determination are not optimized for handling small and fragile crystals that MPs tend to form.

We present an array-based, X-ray compatible microfluidic platform comprised of cyclic olefin copolymer (COC) and a thin polydimethylsiloxane (PDMS) membrane needed for valve actuation. The array-based design allows for screening for upto 96 conditions using < 6μL of protein solution. After screening, a similar chip is used for structure determination of the protein by merging data from multiple crystals grown on-chip. The use of X-ray compatible materials allows *on-chip* analysis, thus preventing damage to fragile protein crystals that might occur during manual harvesting and mounting of crystals. As proof of concept, we have screened crystallization conditions for a soluble protein, phosphonoacetate hydrolase (PhnA). After obtaining the optimal condition and using anomalous data collected solely on-chip, we determined its structure to a resolution of 1.99 Å.

We are now in the process of validating our platform by screening and solving the structure of two MPs - photosynthetic reaction center (RC) and cytochrome $ba_3$-type oxidase. We have collected data on RC to 3.0 Å and are currently optimizing the crystallization conditions and chip design to get higher quality data.

(1) Hansen, C.; Quake, S. R. *Current Opinion in Structural Biology* **2003**, *13*, 538-544.

(2) Li, L.; Ismagilov, R. F. *Annual Review of Biophysics* **2010**, *39*, 139-158.

(3) Stevens, R. C. *Current Opinion in Structural Biology* **2000**, *10*, 558-563.

# (187e) Investigation of a Counter-Current Crystallization Process for Separation of Systems Forming Solid Solutions

**Erik Temmel**[1], Sebastian Wloch[1], Uwe Müller[2], Detlef Grawe[3], Robert Eilers[4], Heike Lorenz[5] and Andreas Seidel-Morgenstern[6], (1)Max Planck Institute for Dynamics of Complex Technical Systems, Magdeburg, Germany, (2)Hapila Gera GmbH, Gera, Germany, (3)Jesalis Pharma GmbH, Jena, Germany, (4)Hapila Gera GmbH, Magdeburg, Germany, (5)Physical and Chemical Foundations of Process Engineering, Max Planck Institute for Dynamics of Complex Technical Systems, Magdeburg, Germany, (6) Physical and Chemical Foundations of Process Engineering, Max-Planck Institute for Dynamics of Complex Technical System, Magdeburg, Germany

***Investigation of a counter-current crystallization process for separation of systems forming solid solutions***

Erik Temmel[1], Sebastian Wloch[1], Uwe Müller[2], Detlef Grawe[3], Robert Eilers[2], Heike Lorenz[1], Andreas Seidel-Morgenstern[1,4]

[1]*Max Planck Institute for Dynamics of Complex Technical Systems, Magdeburg, D-39108, Germany*

[2]*HAPILA GmbH, Gera, D-07552, Germany*

[3]*Jesalis Pharma GmbH, Jena, D- 07745, Germany*

[4]*Otto von Guericke University Magdeburg, Chair of Chemical Process Engineering, Magdeburg, D-39108, Germany*

The occurrence of complete or partial solid solutions in solution crystallization is still the focus of several recent investigations. One reason is that the underlying mechanisms of the formation of mixed crystals are not clarified in every detail. On the other hand this special solid liquid equilibrium (SLE) limits the applicability of commonly used crystallization processes and complicates separation tasks. Especially for pharmaceuticals this can lead to a huge economical drawback.

Process design for crystallization from solution is mainly based on the solid-liquid equilibria (SLE) of the substances of interest [1]. For the, in case of ternary mixtures, commonly appearing simple eutectic and intermediate compound-forming systems several opportunities for crystallization based separation from a solvent are known and are well understood, e.g. processes based on applying preferential crystallization in case of two dissolved enantiomers. Besides, the tie lines of systems involving solid solutions do not lead to a corner of the pure substance in the ternary phase diagram (figure 1). Therefore it is not possible to separate these mixtures within one crystallization step and there is no option to apply the commonly used crystallization processes. For many examples such a miscibility behavior is published including inorganic systems like salts containing one common ion [1, 2] or simple organic substances like amino acids [3]. For more complex chiral molecules the occurrence of complete mixed crystal formation seems to be rare [4]. However, there are more cases reported where partial solid solutions are present at the pure enantiomer side of conglomerate and racemic compound-forming chiral systems.

Some theoretical studies suggest a separation opportunity based on fractional crystallization [5-7]. By means of repeated dissolution and crystallization it is possible to separate systems forming solid solutions. This principle can be realized by exploiting a cascade of batch crystallizers. Furthermore, it was shown that a counter-current between the solid and liquid phase in such a process would lead to optimal process performance in terms of yield and productivity. Nevertheless, realization and automation of a counter-current crystallization cascade is quite complicated. Between every crystallization stage a solid/liquid separation is required. Additionally, the automatic transport of the crystalline solid phase between the stages is difficult and would lead to huge instrumental effort.

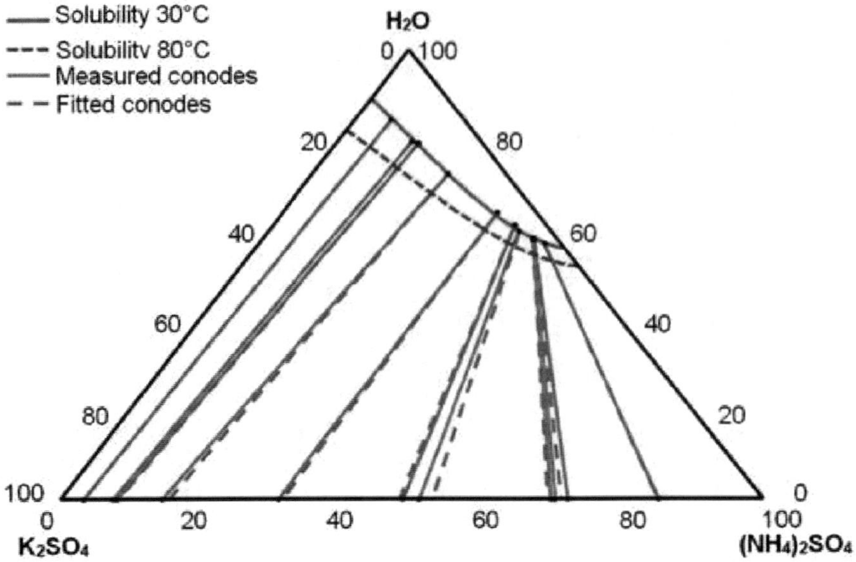


Figure 1: Ternary phase diagram of the system $K_2SO_4$ / $(NH_4)_2SO_4$ / $H_2O$ for 30°C and 80°C with conodes indicated.

Recently an automatic counter-current crystallization cascade was developed by the HAPILA Gera GmbH [8]. In this plant exclusively liquid phases are transported. The solid phase is dissolved after the crystallization to avoid the expensive transport of the crystalline phases in the solid state. Hence, it is possible to proof this up to now only theoretically considered separation opportunity for systems involving solid solutions.

In the contribution the SLE of two different systems forming complete solid solutions will be shown. Furthermore, the mathematical description of the process of interest will be explained and verified by a manually operated counter-current crystallization in a rotary evaporator. The validated model was then used to run a semi-automatic pilot plant followed by a comparison of theoretical and experimental results. Finally a partial solid-solution forming system was theoretically investigated in a simulation study.

References


[1] Mullin, J. W., *Crystallization*, Boston, Butterworth-Heinemann (1997)

[2] Dejewska, D.; Szymański, T., *Cryst. Res. Technol.*, 1998, 33, 757-765

[3] Kurosawa, I. W.; Teja, A. S.; Rousseau, R. W., *Fluid Phase Equilib.*, 2004, 224, 245-249.

[4] Jaques, J.; Collet, André; Wilen, *Enantiomers, racemates, and resolutions*, S. H., Florida, Krieger (1994)

[5] Lin, S. W; Ng, K. M.; Wibowo, C., *Comput. & Chem. Eng.*, 2008, 32, 956–970

[6] Balawejder, M.; Galan, K. ; Elsner, M. P.; Seidel-Morgenstern, A.; Piatkowski, W.; Antos, D., *Chem. Eng. Sci.*, 2011, 66, 5638–5647

[7] Matz, G., *Chem. Ing. Tech.*, 1980, 52, 562-570

[8] HAPILA GmbH, Offenlegungsschrift DE 10 2008 023 833 A1, 2009

# (187f) Designing a Fit-for-Purpose Agglomeration Method for Pharmaceutical Crystals

**Chenchi Wang**, Daniel Hsieh, Mario Hubert and Chiajen J. Lai, Research and Development, Bristol-Myers Squibb Company, New Brunswick, NJ

**Designing a Fit-for-Purpose Agglomeration Method for Pharmaceutical Crystals**

*Chenchi Wang, Daniel Hsieh, Mario Hubert, Chiajen Lai.*

Research and Development, Bristol-Myers Squibb Company, New Brunswick, New Jersey

In pharmaceutical development various particle engineering techniques are commonly employed to alter crystal morphology, improve crystal size, or increase purity of the active pharmaceutical ingredient (API). Crystal agglomeration is an alternative method to improve slurry and powder properties, especially for materials with a crystal habit or size that is difficult to change by classical crystallization techniques. In a crystal agglomeration process, the primary crystals can be directed to form agglomerates in a multi-solvent system. The formation of agglomerates can significantly improve slurry filterability or bulk powder properties compared to the primary crystals. A "fit-for-purpose" method will be presented that can be used to identify suitable conditions to encourage crystal agglomeration. Multiple case studies will be presented in detail to demonstrate how specific slurry or powder properties can be attained through the crystal agglomeration approach.

# (187g) Utilization of Population Balance Models to Develop a Continuous Crystallization Process; Determination of Crystallization Kinetics

Christopher L. Burcham[1], Michael A. Lovette[2], Christopher S. Polster[2], Sean K. Bermingham[3] and Hassan Mumtaz[3], (1)Chemical Product Research and Development; Particle Design Laboratory, Eli Lilly and Company, Indianapolis, IN, (2)Chemical Product Research and Development, Eli Lilly and Company, Indianapolis, IN, (3)Process Systems Enterprise, London, United Kingdom

Due to material and time constraints, experimentation alone often cannot provide a means for design optimization of continuous crystallization processes within the pharmaceuticals industry. While several conventional process design software packages offer the ability to handle process optimization, such software packages often lack a robust means for predicting the effect of changing process variables on the corresponding particle size distributions (PSD). In the crystallization of active pharmaceutical ingredients, specific PSD requirements are often a crucial to the manufacturability and bioavailability of the final drug (solid dosage form). We use a combined population balance and process design software (e.g., gCRYSTAL, Process Systems Enterprises) as a means for process optimization for continuous crystallization. In performing this process optimization, we use rate constants for nucleation and growth that were obtained from regressed experimental data using the same software package. Furthermore, we are able to perform this process optimization with the assertion of constraints on the PSD for the optimal solution.

In this talk, experiments are conducted to provide the necessary data needed to regress kinetic parameters associated with crystallization; nucleation, growth and agglomeration. This talk will discuss both the experimental and computation aspects of the work and subsequent refinements as continuous crystallization results are collected later in development. This is the first of two discussions at this meeting, illustrating our proposed "universal" work flow for the design of continuous crystallizations.

# (187h) Utilization of Population Balance Models to Develop a Continuous Crystallization Process; Process Design Using Optimization

Christopher L. Burcham[1], Michael A. Lovette[2], Christopher S. Polster[2], Sean K. Bermingham[3] and Hassan Mumtaz[3], (1)Chemical Product Research and Development; Particle Design Laboratory, Eli Lilly and Company, Indianapolis, IN, (2)Chemical Product Research and Development, Eli Lilly and Company, Indianapolis, IN, (3)Process Systems Enterprise, London, United Kingdom

Due to material and time constraints, experimentation alone often cannot provide a means for design optimization of continuous crystallization processes within the pharmaceuticals industry. While several conventional process design software packages offer the ability to handle process optimization, such software packages often lack a robust means for predicting the effect of changing process variables on the corresponding particle size distributions (PSD). In the crystallization of active pharmaceutical ingredients, specific PSD requirements are often a crucial to the manufacturability and bioavailability of the final drug (solid dosage form). In this talk, we use a combined population balance and process design software (e.g., gCRYSTAL, Process Systems Enterprises) as a means for process optimization for continuous crystallization. In performing this process optimization, we use rate constants for nucleation and growth that were obtained from regressed experimental data using the same software package (the details of this regression are contained in the talk "Utilization of Population Balance Models to Develop a Continuous Crystallization Process; Determination of Crystallization Kinetics," which is also being given at this conference). Furthermore, we are able to perform this process optimization with the assertion of constraints on the PSD for the optimal solution. This talk concludes with our insights regarding the implementation of process optimization as a part of a proposed "universal" work flow for the design of continuous crystallizations.

# (206a) Controlling the Composition of a Mixture of Polymorphs by Examining Seeded Smpt Kinetics

Ramya Kumar, Venu Nalivela and **Srividya Ramakrishnan**, Dr. Reddy's Laboratories Ltd., Hyderabad, India

Solvent mediated phase transformation (SMPT) occurs in a variety of organic and inorganic crystals. In this study SMPT kinetics were investigated in the context of generating polymorphic mixtures. It was required to crystallize an anti-epileptic drug as a binary mixture of forms with consistent composition for various reasons.

This API exhibited polymorphism with three forms, A, B, and C, which were found to exist as three pairs of enantiotropes. A mixture of Forms A and C was selected for development based on the thermodynamic order of stability which indicated that Form A was the most stable below 55°C, Form C at intermediate temperatures (between 56°C to 65°C) and Form B at high temperatures. Their enantiotropic relationship was exploited by heating a slurry of Form-A to a temperature slightly above the transition temperature, and seeding with Form C to trigger partial transformation to the marginally more stable Form-C.

Controlled implementation of SMPT to produce a mixture of the desired polymorphic composition required an investigation into the impact of various crystallization parameters such as transformation temperature, stirring rate (RPM), crystal size distribution of Form-A on the rate and extent of conversion to Form-C. This study was performed systematically through Design-of Experiments. Increasing the temperature was expected to provide a greater thermodynamic driving force and increase the rate of conversion of Form A to C. Though the DoE model confirmed that temperature was indeed the most significant variable, it was observed that kinetic parameters such as RPM and CSD of Form-A gained greater prominence close to the transition temperature. These were significant enough to even cause dissolution of Form-C seeds, thus preventing conversion of Form A to Form C. This interesting reversal of the thermodynamic script demonstrates that dissolution and growth kinetics in SMPT are determined by the dynamic equilibrium that exists between the solute in the liquid phase and the two solid phases, not just temperature and solubility. The ratio of the surface area between the metastable and stable crystals and the hydrodynamic environment in which this equilibrium operates were found to be as important as the thermodynamic driving force provided.

The second part of our work focuses on elucidation of the mechanism of nucleation and growth of the stable Form-C. Differences in conversion behavior were observed when the particle size of Form C seed used to trigger the conversion was changed. Using conversion-time profiles in combination with microscopic observations, two mechanisms of conversion were identified. With coarse Form-C seeds, cross-nucleation of Form-C out of the metastable Form-A crystal faces was observed which could be attributed to insufficient surface area for growth provided by the coarse seed. On the contrary, when micronized seed was used, the extent of surface cross-nucleation was considerably lessened as growth of the micronized seed was the primary route by which transformation to Form-C occurred. Thus the conversion mechanism followed was found to be dependent on the surface area of the seed.

**Communication Number: IPDO IPM-00330**

# (206b) Amine-Functionalized Metal-Organic Framework Crystallization Process Design

**Paul M. Schoenecker**, Grace A. Belancik, Bogna Grabicka, William J. Koros and Krista S. Walton, Chemical and Biomolecular Engineering, Georgia Institute of Technology, Atlanta, GA

Metal-organic frameworks (MOFs) are porous crystalline structures consisting of metal centers connected via organic ligands. MOF crystals are typically formed via solvothermal methods, which consist of heating a mixture of metal salt, organic ligand, and solvent(s) in a sealed vessel. Most current MOF research is devoted to crystal structure identification and application specific experimentation. However, we examine the scalability of amine-functionalized UiO-66, a water-stable MOF with direct application potential, via batch crystallization vessels as well as a novel continuous-flow crystallization process. The flow-through crystallizer utilizes a single draft-tube in an attempt to preferentially produce larger MOF crystals and agglomerates. A recycle stream is also implemented to minimize solvent usage and provide seed crystals to the crystallizer feed. Characterization of the products is accomplished by determining the product yield as well as quality. In particular, crystal phase, size, and porosity are examined via pXRD, SEM, and BET modeling of $N_2$ adsorption at 77 K, respectively.

# (206c) The Role of Meso-Mixing in Anti-Solvent Crystallization Processes

Des O'Grady, METTLER TOLEDO, Columbia, MD, Mark Barrett, School of Chemical and Bioprocess Engineering, University College Dublin, Dublin 4, Ireland, Brian Glennon, School of Chemical & Bioprocess Engineering, University College Dublin, Dublin 4, Ireland and Eoin Casey, School of Chemical and BioProcess Engineering, University College Dublin, Dublin, Ireland

This paper establishes links between the available turbulent mixing capability in a fed-batch reactor and important crystallization characteristics such as the Metastable Zone Width (MSZW) and product nucleation during the anti-solvent crystallization of benzoic acid. As demonstrated an increase in the locally dominant meso-mixing timescale, $\tau_D$, facilitates the premature and subsequent sporadic nucleation of the solute. An alternative well mixed addition location, defined via Computational Fluid Dynamics (CFD) simulations and turbulent mixing theory, allows this dispersion limiting timescale to be reduced by over 50% across all anti-solvent addition rates assessed. Additionally, a more direct pathway from the feed point to the well mixed region of the impeller is also presented. These more favorable hydrodynamic conditions improve supersaturation dispersion and blending which in turn eliminate the premature and additional nucleation of the solute throughout the batch. Correlations between the population of fine particles and $\tau_D$ are also presented. These relationships demonstrate the sensitivity to the systems crystal size when hydrodynamic properties are improved.

# (206d) Crystallization of Active Pharmaceutical Ingredients in Scrapped Surface Crystallizers

**Jie Chen**[1], Shin Yee Wong[1] and Allan S. Myerson[2], (1)Chemical Engineering, Massachusetts Institute of Technology, Cambridge, MA, (2)Department of Chemical Engineering, Massachusetts Institute of Technology, Cambridge, MA

Scraped-surface heat exchanger (SSHE) (Härröd, M. 1986) is used widely in the chemical and food industries, for example the separation of para-xylene from xylene mixtures, the separation of chlorinated benzenes, the production of ice cream and margarine and etc. The scrapping motion minimizes fouling and improves the heat-transfer rates by removing the solid layer on the wall and transferring solid to the bulk. Despite SSHE's proven efficiency, its application in pharmaceutical crystallization is very limited.

To explore the concept of SSHE, a laboratory scale scraped surface crystallizer (SSC) was designed and tested using several active pharmaceutical ingredients (API) of different crystal characteristics (needles, rods, fines, and aggregates). The crystallizer was designed using a MultiMax$^{TM}$ 50 mL reactor with a custom made PTFE anchor impeller. The close clearance (~ 1 mm) between the impeller and the reactor's interior wall prevented crystal encrustation on both the side and bottom wall. The temperatures of all crystallization experiments were controlled using the MultiMax$^{TM}$ station.

The performance of SSC was investigated for cooling and anti-solvent crystallizations. Crystallization kinetics, mixing efficiency, slurry loading tolerance of SSC were evaluated and compared to a conventional stirred tank crystallizer. The results show that SSC can operate at a significantly lower stirring speed for suspension of thick slurry. In addition the crystals produced from SSC are significantly larger, and higher yield was observed due to minimal encrustation on the crystallizer wall. This study showed that SSC is suitable for pharmaceutical crystallization applications, especially those processes with high solid loading of fines and needle-like crystals.

Härröd, M., "Scraped Surface Heat Exchangers", *J. Food Proc. Eng.,* **9**, 1, 1986.

# (206e) Continuous Enantioseparation in a MSMPR Crystallizer – Experimental and Theoretical Investigation of Different Process Conditions

**Kamila Galan**, Physical and Chemical Foundations of Process Engineering, Max Planck Institute for Dynamics of Complex Technical Systems, Magdeburg, Germany, Martin Peter Elsner, Department of Applied Chemistry, Georg Simon Ohm University of Applied Sciences, Nuremberg, Germany and Andreas Seidel-Morgenstern, Physical and Chemical Foundations of Process Engineering, Max-Planck Institute for Dynamics of Complex Technical System, Magdeburg, Germany; Chemical Engineering, Otto von Guericke University, Magdeburg, Germany

Growing demands for products with constant high quality make continuous crystallization very attractive. Carrying out a process at a steady state can guarantee the continuity of product properties. This might be especially important, when one considers drugs produced in an enantiopure form.

Within this contribution the idea of a Mixed Suspension Mixed Product Removal (MSMPR) crystallizer for preferential crystallization is analyzed both experimentally and theoretically [1]. Preferential crystallization is applied to resolve one enantiomer from a racemic mixture of both. The separation process takes place in the metastable zone, where spontaneous nucleation is suppressed kinetically. Preferential crystallization is initiated by introducing seeds of only one enantiomer to the solution. The surface of these seeds drives resolution due to their growth. Executing preferential crystallization in a continuous mode requires a continuous supply of seeds together with racemic feed solution and a continuous withdrawal of the product suspension containing the seeded target enantiomer.

To gain insight into the process a mathematical model was developed assuming an ideally mixed vessel, uniform conditions inside the tank, identical crystals size distribution of crystals in the vessel and in the product stream. Population balance and mass balance equations as well as kinetics for crystal growth, primary and secondary nucleation were used to simulate the behavior of the solid and liquid phases. A simulation study using parameters of the enantiomer of threonine [2] was done to investigate the effects of different conditions on the process performance, including the saturation and crystallization temperature as well as mass flow rates of seeds and the feed solution. An optimum set of process conditions was chosen considering also the aspect of reducing the duration of the dynamic start-up phase.

The experimental part of the contribution will focus on the following aspects: first, observed resolution results of L- and D-threonine will be compared with model predictions. Effects of selected conditions will be discussed to identify the influence of key process parameters on the length of the dynamic start-up phase and the product quality (purity and mean characteristic size of the product). A comparison of the continuous process with the performance of a corresponding batch process will conclude this contribution.

[1] Jones A.G., *Crystallization Process Systems*, Butterworth-Heinemann, Oxford, 2002

[2] Qamar S., Elsner M.P., Hussain I., Seidel-Morgenstern A. *Seeding strategies and residence time characteristics of continuous preferential crystallization* Chemical Engineering Science, 71, 5-17, 2012

# (206f) Continuous Crystallization Process Development

**Haitao Zhang**[1], Justin Quon[2], James M.B. Evans[3], Allan S. Myerson[4] and Bernhardt L. Trout[2], (1)Chemcial Engineering department, Massachusetts Institute of Technology, Cambridge, MA, (2)Massachusetts Institute of Technology, Cambridge, MA, (3)Department of Chemical Engineering, Massachusetts Institute of Technology, Cambridge, MA, (4)Chemical Engineering, Massachusetts Institute of Technology, Cambridge, MA

Batch-based manufacturing system currently employed by the pharmaceutical industry is costly and inefficient. A drug's active ingredients are synthesized in a chemical manufacturing plant and then shipped to a separate facility where they are converted into large batches of pills, liquids, or creams. With multiple interruptions, including transport to separate locations, the production of just one batch of drugs can take weeks. What's more, the manufacturing design and scale-up required to produce a new drug can be financially unsustainable and exceedingly time-consuming. Continuous Processes/Manufacturing offers a paradigm shift that will enable the industry to produce quality products more efficiently. Continuous crystallization has long been a standard method for the crystallization of bulk commodity chemicals and large-volume specialty chemicals, but it has only recently become a focus in the pharmaceutical and fine chemical industries. This presentation will:

• Highlight the current approaches to continuous crystallization

• Assess the advantages of a continuous crystallization process

• Analyze the key challenges and draw on real life case studies to present the key points

# (243a) To Teach or Not to Teach, What Is the Answer?

Santiago Faucher, Technologies, Hatch, Mississauga, ON, Canada

The typical university graduate will have spent 17 of their 21 years being taught a prescribed education regimen. The retention of this education is tested through written exams which, in the best of cases, will permit the student to elaborate on the answer and outline the process they took to achieve it. In the worst of cases, the evaluation is reduced to a multiple choice format due to the limited capacity of the education system to do otherwise. In all cases, there is typically only one correct answer or approach thereto.

Fortunately the next 60 years of one's life and career are not so simple and much richer: there is no one right answer, but rather a multitude of possible acceptable solutions to choose from. For Chemical Engineers who have been subjected to problem-based and/or self directed learning, this realization may come early during their college education; for others, the capstone project may present the first opportunity toward this realization. Ultimately it is the professor's approach to learning that dictates when students begin to think more richly towards varied and novel solutions. Should learning be pushed onto the student through professing or be pulled by the student through the establishment of the appropriate problem and reward? Which of these approaches will lead to a deeper retention of the curriculum? As professor of a capstone course, I chose a problem-based learning approach thus taking on the role of mentor and client rather than professor. I responded to student questions with questions. Each question designed to guide but not direct; no answers were provided. My students' response to my unconventional approach fitted well with the model of Kübbler-Ross describing grief management, while their group dynamics nicely fit Tuckman's Group Development Model.

This talk will report on the pros and cons of a problem-based capstone project, the educator and student roles, student and group dynamics, mentoring mechanisms and effectiveness, and finally on an unexpected team performance. As is often the case, both in this course and in life, the question asked in this educational experiment has lead to learning and to a more profound question: should one teach or mentor the chemical engineers of the future?

# (243b) Helping Students to Learn by Doing - Capstone Design Experience At Columbia University

Michael Hill and Stanley A. Leshaw, Chemical Engineering, Columbia University, New York, NY

Columbia has transformed its two-semester Capstone Design sequence from a traditional lecture-based format to one which imitates the real-world dynamics of industry. Students meet in small groups or teams for up to one hour each week with an advisor/technical manager/coach who critiques progress, offers suggestions for improvement, and provides direction for the following week. Student teams complete their design projects on a semi-independent basis, receiving individual guidance, but not complete answers, from their advisor. The format also plays to the strengths of adjunct faculty, who have found it easy to adopt. As a result, students "learn by doing" as they address three significant process or product design problems of increasing complexity.

Projects are selected by the instructors to provide the students with sufficient hands-on exposure to several important pedagogical issues in design, such as the interaction of components within a system and the impact of constraints on optimization in design. These projects are also chosen to reflect an awareness of broader societal issues, such as environmental relevance, green energy, and sufficient water.

Student feedback has expressed appreciation for the format, with students frequently citing Design as their favorite course.

# (243c) Chemical Process Design and Projects Two Semester Sequence

**Alan W. Weimer**, Chemical and Biological Engineering, University of Colorado at Boulder, Boulder, CO

The Fall semester 3 Credit course, “Chemical Process Synthesis”, along with the Spring semester 2-Credit "Design Projects" course, is the *capstone* course sequence in chemical and biological engineering at the University of Colorado. The 1st semester utilizes a design text for approximately 8 weeks and ends with a "mini-design" team project which serves as the final exam. Topics covered from the text include: (1) an introduction to the design method; (2) health, safety, environmental, and ethics; (3) process creation/simulation, (4) process synthesis heuristics; (5) design methods reviewed for major unit operations; (6) heat and power integration; (7) ideal and non-ideal separation trains; and (8) process economics and profitability analysis (3 weeks of economics). Every class starts with 10 minutes dedicated to current events. The mini-design project is a well-defined problem that can be simulated and students are organized into typically 3-person teams. All teams work on the same design problem. Grading of the mini-design projects is taken very seriously (not hand-waved) in order to "set the bar" for the 2nd semester. For unsatisfactory mini-design projects, the teams must submit a satisfactory "retread" prior to the drop/add deadline of the 2nd semester. First semester grading is 30% homework, 30% two midterm exams, and 40% mini-design project, however, a passing grade is required on the mini-design project independent of all other grades in order to pass the 1st semester class. A peer-review counts as 20% of the mini-design project grade, but does not impact the requirement that the team project itself requires a passing grade - independent of the peer-review. The 2nd semester "Design Projects" course is supported by external typically industrial liaisons. The 3-person team projects are organized in the fall by the instructor and the projects are selected during the first week of class. Students work very closely with their external liaisons. Problem statements are typically very open-ended and may be only one or two paragraphs in length. Teams are required to research the project and to write a project proposal due within 3 weeks. The proposal needs to incluide all required "battery limits". The proposal is graded by the instructor with input by the external liaisons. Teams give two oral presentations during the remaining 8 weeks of class and are required to make an oral presentation at their liaison's facility, if local, during the last week of class, or during final exam week. GoToMeeting is used for non-local liaison presentations. Industry liaisons typically attend the oral presentations and comprise an external panel that asks quesions along with students in the class. Oral grading is done based on certain metrics with input from both the students and the external liaisons. Liaisons that cannot attend are engaged using GoToMeeting. A final written report is submitted and graded with input from the external liaisons. Organizing the external projects is challenging, but the instructor has been able to organize 36 different projects over each of the last two years. Student evaluations are typically excellent regarding the capstone sequence. Many of the industry liaisons are former students who want to contribute to the educational process. Many also recruit at the university and even hire students that they are engaged during the course. The instructor has spent half of his career in industry and half in academia and also directs a very active research program.

# (243d) Senior Design At UC Irvine: Process Development and Economics Analysis with PRO/II

**Gang (Gary) Xu**, R&D, Invensys/SimSci-Esscor, Lake Forest, CA

Industrial standardized simulator PRO/II is applied in UC Irvine senior design class. Students are explore to real time simulation and project design assignments.

With fundamental knowledge of process design and economic analysis learnt from classroom, students get to develop projects by literature searching, PRO/II flow-sheet exercising, data reconciliation, and high-level process optimization from economic and environmental consideration.

Students were also stimulated by choosing their own projects reviewed by instructor. Interesting ones including generation of long-chain hydrocarbon (as fuel easy to store and transport) from natural gas, bio-energy generation from grass, supercritical CO2 decaffeination, sulfuric acid production from SO2, alcohol fermentation from potato (sugar) ... From which practice of fundamental engineering and economic understanding were encouraged.

With all the activities, popular topics such as

1. Azeotropic distillation

2. process optimization by controlling of design variables

3. economic model analysis such as Pareto optimal with engineering invest vs. environmental impact

were also emphasized as key portion of the design projects. Some of these topics, such as Pareto optimal, exceeded undergraduate level. However, basic knowledge of such topics were well understood with reasonable scale of exercise directly connected with projects chosen by the students.

# (243e) Conduct Your Capstone Design Class As a Consulting Company

**Richard L. Zollars**, Voiland School of Chemical Engineering and Bioengineering, Washington State University, Pullman, WA

Capstone design classes are often taught in a similar manner – design a process (ammonia plant, methanol plant, etc.) to meet some production goal. By selecting an appropriate plant the instructor can be insured of covering almost any unit operation to which the students have been exposed at some time during their education. There are a number of shortcomings to such an approach, however. One of these is that the details of such plants, which unit operations are used, in what order, etc., are readily available in the literature. Students will repeat the standard design, go through a laborious series of calculations to size and cost equipment, perform an economic analysis and report their findings. In terms of promoting any in-depth thinking, there is little to offer. In addition, only a small fraction of students will ever be employed in such a situation.

A more common situation for students upon graduation is to be employed as a process or unit engineer. As such, in addition to keeping their unit operating, they have the responsibility of upgrading or debottlenecking their operation. This smaller scale, more focused, design work typically involves only a few unit operations. While few in number the unit operations that are involved are highly variable depending upon the unit in which the engineer is employed.

In an academic setting trying to operate a design class using this narrower focus is difficult without having the problems seem contrived. Thus, eleven years ago, we began operating our capstone design class in a different manner. The instructor for the capstone design class solicited regional industries for projects. With this set of projects available the design class is then conducted like an engineering consulting firm with teams of students vying for various projects. During the design process the students must keep in contact with their "clients" to insure that they are addressing the client's concerns. At the end the student groups must satisfy not only the class instructor, but also the industrial contact person who originally proposed the problem.

The capstone design class at Washington State University is a two semester sequence of classes. The fall semester course covers economic analysis, equipment sizing, safety and team work in addition to starting the design project. The spring semester class is devoted exclusively to the design project. In preparation for the academic year regional industries are contacted for proposed projects. For a variety of reasons (cost, laboratory facilities, safety) the projects are to be paper studies, not requiring work in a laboratory or shop. The projects should require approximately 500 person-hours to complete. In addition to developing such a project the industry is expected to provide part of a person's time to serve as the point of contact for the student design team as well as some financial remuneration to cover expenses for the project.

During the fall semester the students select the group in which they want to work. Ideally, each group is three people but no group will be larger than four people. Once the groups are formed one-page descriptions of the projects that are available for the year are given to each group. Each group then goes through a bidding process for the selection of the project on which they will work. At the completion of this bidding/selection process each group will visit the industrial site to meet with their contact as well as get a fuller understanding of the project. Prior to the end of the fall semester but after the initial site visit each group will conduct a brainstorming session to develop a slate of potential ideas for the solution of the problem they have been assigned.

During the spring semester the students work exclusively on the design project. They develop Gantt charts to help them monitor their progress. These are updated formally three times during the semester. They are expected to maintain regular communications with their industrial contact. A second site visit about half way through the semester is conducted to insure that both the students and industry contact are in agreement with the direction of the project. Final written reports are due after 12 weeks. This is followed by a day of formal oral presentations the following week. Because of the number of projects in any given semester the time for these presentations is short. Thus we have gone to a poster/oral presentation format. The student design groups participate in a poster session first followed by 20 to 25 minute oral presentations. Finally, before the end of the semester, each group is to hand-in a final revised written report which will be sent to the industrial contact.

To date a total of 58 different design projects have been conducted for 15 different industrial sponsors. The industries have ranged from major corporations, to government labs, to state government agencies, to small companies (5 – 10 employees). The range of topics is also quite broad with few projects covering similar processes. There have been some very traditional chemical engineering projects (heat exchanger optimization, hydrogen recovery in a refinery, debutanizer optimization), some others that are still definitely chemical engineering in nature (active metal reduction of nitrates, oxidation of uranium using Fenton's Reagent, purification of soy and palm oils, water treatment/recovery, and copper recovery) while a few others stretch the boundaries of the discipline (recycling of carbon fiber reinforced composites, recovery of neutraceuticals from fruit waste, and converting municipal and/or food waste into biofuels).

As measured by student involvement conducting the design course in this manner has been a success. Student evaluations of the course are uniformly high. More than one alumnus has reported that their experience in the design course was the closest thing they had to actual practice than any other course in the curriculum. The industries that have participated have also expressed their support. Not only have they continued to provide projects, manpower and financial support but also have hired students as a result of the student's work on a project supported by the industry.

Another way to assess the success of conducting the design course in this manner is the quality of the work performed by the students. The quality of the written reports has gone up, in part due to the fact that the reports are going to people who might be hiring the student. The oral/poster presentations have been excellent. More than one person from industry has commented that they meet or exceed the quality that they would expect from their engineers.

The quality of the designs is also outstanding. Approximately 20% of the student design suggestions have been implemented in some from by the company that supplied the project. The quality of the student's work was verified by a situation that occurred recently. One of the companies offered a project for which they had already signed a contract with an engineering consulting firm. They received both the professional firm's report at about the same time as they received the student's report. While the student's report lacked the detail of the professional report (types of welds, fasteners, support structures, etc.) the student design and the professional design contained almost identical unit operations, arranged in the same order, with equivalent material and energy balance calculations.

# Integrating Active Research on CO2 Capture in Traditional Process Design

Omkar Namjoshi[1], Paul Nielsen[1], Matthew Walters[1], Bo Lu[1], Siyun Wang[1], Gary Rochelle[1]

[1]The University of Texas at Austin, Austin, Texas, USA

The typical capstone design course in most chemical engineering curricula integrates material from heat and material balances, unit operations, engineering economics, and reaction engineering. At the University of Texas at Austin, we also incorporate aspects from sustainable engineering and from our research program in the design class. Students who take this class are tasked with completing a preliminary design of a post-combustion carbon capture process (capturing $CO_2$ from a point source) using amine-based solvents or physical solvents.

Themes from sustainable engineering are incorporated at the beginning of the course and reinforced throughout the semester. Students are first given an introduction to climate change, how capturing $CO_2$ might help mitigate climate change, a brief overview on $CO_2$ storage sites. These themes are reinforced throughout the class as detailed design progresses and economics are finalized. Students must answer what price $CO_2$ must be "sold" for the plant to make a profit and also must discuss the environmental impacts of $CO_2$ capture facility. The link to sustainability gives students valuable insight regarding energy and environmental policy, which is not commonly encountered in a traditional "core" chemical engineering class.

Within the first three weeks of class, students are expected to form teams of three students each and submit a proposal outlining their choice of $CO_2$ source, capture solvent, and storage site, justifying their choices with quantitative arguments. The teams are expected to use as many resources as possible and are encouraged to get information and guidance from graduate students, research sponsors (which include large oil and gas companies, amine technology suppliers, and power plant operators), and old student reports (which are posted on the course website) to help strengthen their arguments for their proposal (as well as for general help, if it is needed). Students – past and present – have proposed a variety of $CO_2$ sources, including (but not limited to) coal-fired power plants, integrated combined-cycle power plants, fluidized catalytic crackers, and LNG treating plants. Example of capture solvents include aqueous amine solvents, such as monoethanolamine, concentrated piperazine, and activated MDEA. Students are encouraged to design plants using unconventional sources and solvents; those who do so are evaluated more on effort and a good initial, "first-draft" design. Students who have chosen solvents and sources that have been selected in prior years are evaluated on producing an optimized design that represents an improvement over the prior year's design.

The integration with our research program is achieved by incorporating thermodynamic models built using the AspenPlus simulation software / framework and developed by graduate students into the student projects. Our research group has developed rigorous models for activated MDEA and concentrated PZ and has developed preliminary frameworks for other amine systems; the use of these models by the undergraduate students is a helpful validation tool for the graduate students and helps in improving future revisions of the model.

Assignments are generally assigned each week, with a few exceptions for some of the lengthier and more involved assignments. Each assignment can be best described as "goal directed, mission oriented" – students will have a set number of deliverables for each assignment cycle and are expected to complete them (or explain why they were unable to) and are held accountable to these deliverables. Table 1 gives a list of assignments along with deliverables (note that the deliverables can change from year to year and are group specific) that are assigned to students. A few assignments are individual assignments, and the assessments from each individual assignment are used to differentiate student performance within the group.

The weekly assignment cycle ensures that students do not fall behind on their reports. The final report essentially consists of students putting all of their reports together and addressing any areas for improvement that TA feedback has identified. The grade for the final report is 50% of the overall course grade. This performance bias ensures that students who fall behind are able to catch up (should they choose to do so) and that students who earned good evaluations on prior reports will continue to maintain their performance through the end of the semester.

One unique aspect about course management is that highly motivated graduate students serve as teaching assistants (TA's) for the class. Each teaching assistant supervises three to four student design groups and is responsible for assigning deliverables to the students, mentoring student design groups and ensuring that they have the resources to be successful in completing the assigned deliverables, and holding the students accountable to those deliverables (addressing feedback). A head teaching assistant (generally the most experienced amongst the TA's) is also appointed and, in addition to having his/her own student groups to supervise, mentors and advises the other teaching assistant and also manages course coordination such as AspenPlus tutorial development and ensuring that student evaluation is completed in a way that is fair, consistent, and free from individual TA bias. The supervising professor (currently Dr Gary Rochelle) delivers the lectures during the classroom sessions, is responsible for supervising each TA, and holds a standing weekly "alignment" meeting to collectively set goals and deliverables for upcoming assignments with the TA's as well as address any TA concerns. The teaching assistant experience for this course is an excellent way for motivated graduate students to obtain "real-world" management skills that are useful in either a professional or academic setting. In recognition of this fact, UT's chemical engineering department grants three hours of Supervised Teaching course credit to each student who serves as a process design TA.

The emphasis on sustainability, the integration of research into the student design projects, coursework organization, and emphasis on graduate TA development make UT's design course unique.

**Table 1. Plant Design Assignments at UT-Austin**

| Assignment / completion time | Individual or Group? | Deliverables |
|---|---|---|
| Proposal / 2 weeks | Individual | Pick a $CO_2$ source, capture solvent, and storage site |
| Proposal / 1 week | Group | Finalize source / solvent / storage site decision, submit one proposal, make improvements to proposal |
| PFD / 1 week | Group | Develop preliminary flow diagram, outline equipment required, model specifications (e.g. $CO_2$ compressor pressure) |
| Simple Heat and Material Balance / 1 week | Group | Develop basic heat and material balance, model $CO_2$-amine interactions based on an equilibrium model |
| Rigorous Heat and Material Balance / 2 weeks | Individual | Develop rigorous heat and material balance on one unit operation, model $CO_2$-amine interactions using rate-based models |
| Rigorous Heat and Material Balance / 1 week | Group | Combine reports, address feedback |
| Equipment Specifications / 1 week | Group | Size all equipment (absorption / regeneration columns, heat exchangers) and specify materials (e.g. stainless steel) |
| Equipment Costs / 1 week | Group | Cost all sized equipment using Aspen Economic Toolkit |
| Project Economics / 1 week | Group | Find price of CO2 required for plant to be profitable, develop project economics – cash flows, IRR, DCF |
| Process Safety and Environmental Assessment / 1 week | Group | Identify environmental impacts that the process might have |
| Draft Final Report / 1 week | Group | Combine all previous assignments into one report, address all feedback. No new information is needed beyond the environmental assessment section. |
| Final Final Report / 2 weeks | Group | Address feedback on draft final report, also present oral presentation on design to class |

# (243g) Incorporating Process Simulation Across the Chemical Engineering Curriculum to Improve Student Performance On the Capstone Design Project

David A. Rockstraw and Martha C. Mitchell, Chemical Engineering, New Mexico State University, Las Cruces, NM

The New Mexico State University Chemical Engineering Department recently made changes to the core curriculum to accommodate increased training in the use of chemical process simulation (Aspen Plus®) in parallel with the fundamental topics of the discipline. Over the course of five consecutive semesters, students learn aspects of the simulator, solving the same problems in simulation lab that were previously solved by hand calculation in their courses. The 2012 graduating seniors received twice the exposure hours to the simulator as previous graduating classes, and the 2013 graduating class will have received twice the hours of exposure of the 2012 graduating class. The increased level of understanding of how to properly use the simulator for process design has led to significant improvement in student performance on their capstone design projects in 2012. This presentation will describe the curriculum changes made that have permitted this improvement in student work, and the content and methodologies of the design courses in the sequence.

# (300b) Crystallization Kinetics Identification within a Generic Modeling Framework

**Kresten Troelstrup Meisler**[1], Noor Asma Fazli Abdul Samad[1], Nicolas von Solms[2], Krist V. Gernaey[3] and Rafiqul Gani[4], (1)CAPEC, Department of Chemical and Biochemical Engineering, Technical University of Denmark (DTU), Kgs. Lyngby, Denmark, (2)CERE, Deparment of Chemical and Biochemical Engineering, Technical University of Denmark (DTU), Kgs. Lyngby, Denmark, (3)Department of Chemical and Biochemical Engineering, Technical University of Denmark (DTU), Kgs. Lyngby, Denmark, (4)CAPEC, Department of Chemical and Biochemical Engineering, Technical University of Denmark, Kgs. Lyngby, Denmark

Crystallization kinetics identification within a generic modeling framework

*Kresten Troelstrup Meisler*[1] (kretm@kt.dtu.dk), *Noor Asma Fazli bin Abdul Samad, Krist V. Gernaey*[1] (kvg@kt.dtu.dk), *Nicolas von Solms*[1] (nvs@kt.dtu.dk) and *Rafiqul Gani*[1] (rag@kt.dtu.dk)

*(1) Department of Chemical and Biochemical Engineering, Søltofts Plads, Building 229, Technical University of Denmark, DK-2800 Lyngby, Denmark*

Crystallization serves as an efficient separation process for compounds, which are solid in their pure form at the given separation conditions. Crystallization is an essential step in the production of many pharmaceutical products as the active pharmaceutical ingredients (APIs) are often separated efficiently through this operation. The monitoring and analysis of crystallization operations has recently received increased attention due to the growing need to control the final crystal size distribution (CSD) in a relatively narrow range as well as to measure and monitor the final product characteristics CITATION EAa10 \l 1030 (1). This monitoring and analysis requires models that describe the multiple phenomena and their interplay encountered in crystallization operations in order to have the ability to design effective control strategies and the possibility of planning crystallization operations. To this end, models of the phenomena (kinetics, solubility, etc.) are needed within a generic framework that allows the generation of the needed models to achieve a more complete description of specific crystallization operations. For example, certain size distributions can be desired in order to obtain specific properties for the product such as rate of dissolution. Growth of the crystals occurs in multiple dimensions and the growth rates of the facets determine the shape of the crystals. The size distribution is obtained because of different competing phenomena such as nucleation and growth, each with different kinetics. A full representation of a crystallizer requires models for all these kinetic phenomena and saturation descriptions (constitutive equations) coupled with descriptions of the equipment and an operational policy. Such a description is possible within a generic framework where the models can be combined, reused and identified. This framework has been established CITATION Noo11 \l 1030 (2) and tested for simulation of selected crystallization processes.

The objective of this work is to provide the ability to establish the kinetics of a crystallization operation systematically and efficiently. Initially an operational scenario is defined for which the specific balance equations are set up. The constitutive models are chosen based on the system description and the objectives of the model. Once the model is ready, that is, all model parameters are available, it can be used for simulation of the corresponding crystallization operations. If an established model for the desired kinetics is found within the model library, then this model is used. If not available, or in case a new model is desired, then identification of model parameters is performed. This identification step requires measured data, which may be available in different forms such as single crystal growth data or chord length measurements (for example, data from Focused Beam Reflectance Measurements, FBRM). Use of such data requires appropriate translational policies to convert measurements into one or more variables that are described by the model. A preliminary version has been developed and tested to obtain information about the development of crystal size distribution (CSD) in a given operation. The measurement types that can be handled through the modeling framework are continuously being expanded to accept more data types with information relevant for the crystallization operations. The measurements can be used for offline analysis and parameter regression. With an established kinetic model it is thus possible to translate the predicted model results into a form, which is directly comparable (and visualized) to the data available for evaluation of the model. Furthermore, if there is model – data mismatch, the measured data can be used for online parameter estimation.

The expanded model framework combined with the systematic approach to establish the kinetic models for use in general crystallization operations in combination with monitoring tools will be presented using case studies involving different scenarios for crystallization operations.

<>Works Cited

BIBLIOGRAPHY 1. *Optimal seed recipe design for crystal size distributionn control for batch cooling crystallisation processes.* **E. Aamir, Z.K. Nagy, C.D.Rielly.** 2010, Chemical Engineering Science, Vol. 65, pp. 3602-3614.

2. *A generic multi-dimensional model-based system for batch cooling.* **Noor Asma Fazli Abdul Samad, Ravendra Singh, Gürkan Sin, Krist V. Gernaey, Rafiqul Gani.** 2011, Computers and Chemical Engineering, Vol. 35, pp. 828–843.

# (300c) Development of an Empirical Framework for Monitoring CSD Using FBRM in Batch Crystallization

**Huayu Li**, Martha Grover, Yoshiaki Kawajiri and Ronald W. Rousseau, School of Chemical & Biomolecular Engineering, Georgia Institute of Technology, Atlanta, GA

Focus Beam Reflectance Measurement (FBRM) allows in-situ monitoring of crystallization processes. In order to utilize FBRM for control of crystallizers, accurate and robust measurement of crystal size distribution must be realized. To fulfill this goal, models for crystallization kinetics and the sensor are needed; the former is a population balance equation, while the latter converts the measured chord-length distribution (CLD) to the crystal size distribution (CSD) [1-3]. In our previous study, we proposed an empirical model for suspensions of paracetamol crystals in a non-solvent toluene [4]. The CLD obtained for a population of crystals of a specific size is referred to as a fingerprint. By obtaining fingerprints for different crystal sizes in toluene, the model was validated by investigating the linearity and additivity of fingerprint vectors.

In the present work, we apply the empirical model to a crystallization process. As paracetamol crystals of a certain size range are added into the saturated solution, the increase in total counts and changes in the chord-length histogram are recorded and analyzed to build the empirical model. We use a microscope and ATR-FTIR to confirm there is no significant transformation of crystal shape and no dissolution. Our investigations include the following: (1) a regime where the linearity of number of crystals and chord counts holds is identified, (2) linear additivity among different sizes of crystals is verified, (3) detectability of different crystal sizes by the FBRM is investigated, and (4) the most appropriate inversion technique is identified.

With the crystal size distribution estimated by the empirical model, liquid concentration and temperature recording, the full scope of the crystallization process is obtained for simulation purposes, such as parameter estimation and control scheme development. Such a framework is directly generated from practical FBRM data and requires no complicated physical modeling. Therefore, the small-scaled model can be applied in real-time conveniently to monitor and control crystallization processes.

References

1. Yu, Z.Q., P.S. Chow, and R.B.H. Tan, "Interpretation of focused beam reflectance measurement (FBRM) data via simulated crystallization," *Organic Process Research & Development*, 2008. 12(4): p. 646-654.
2. Ruf, A., J. Worlitschek, and M. Mazzotti, "Modeling and experimental analysis of PSD measurements through FBRM," *Particle & Particle Systems Characterization*, 2000. 17(4): p. 167-179.
3. Kail, N., W. Marquardt, and H. Briesen, "Estimation of particle size distributions from focused beam reflectance measurements based on an optical model," *Chemical Engineering Science*, 2009. 64(5): p. 984-1000.
4. Li, H., Y. Kawajiri, M. Grover, and R. W. Rousseau, "Estimation of Crystal Size Distribution with an empirical model of FBRM," AIChE Annual Meeting, Minneapolis, MN, October 2011

# (300d) Apply PAT-Enabled Platform to Realize the QbD Control Strategy for Crystallization Processes

**George Zhou**[1], Aaron Moment[2], Shane T. Grosser[3], Paul Fernandez[2], Luke Schenck[4] and John Higgins[5], (1)Analytical Development and Commercialization, Merck & Co., Inc., Rahway, NJ, (2)Chemical Process Development and Commerciliazation, Merck & Co., Inc., Rahway, NJ, (3)Chemical Process Development and Commercialization, Merck & Co., Inc., Rahway, NJ, (4)Chemical Process Development and Commercialization, Merck & Co, Inc., Rahway, NJ, (5)Process Analytical Technologies, Merck and Company, Inc., West Point, PA

It is critical to achieve the desired crystal form for an active pharmaceutical ingredient (API) because polymorphic behavior may affect its chemical stability, solubility, morphology, density and bioavailability. The automated API platform enabled by process analytical technologies (PAT) provides the flexibility for crystallization process development. It can significantly enhance our capability to explore the design space of crystallization processes and improve the ability in maintaining the crystallization performance under different or unexpected scenarios of disturbances and scale-up variations. In this presentation, this enhancement of capability will be illustrated through several case studies. For example, risk assessment and corresponding DoE studies on crystallization and PSD indicate that the final PSD of a high volume API strongly depends on the seeding-point temperature which in turn relies on the solution composition. Therefore, by successfully utilizing this PAT-enabled automated API platform to map out the solubility for a through-process stream across the multi-dimensional solvent composition space as a function of temperature, control space of the crystallization process can be defined. It has enabled a QbD control strategy which incorporates the accurate measurement of batch composition in real-time for a robust process that reduces cycle time, and increases yield and productivity as well as realizes QbD benefits - greater process stability, process portability, and process flexibility.

# (300f) Theoretical and Numerical Analysis of Critical Seed Loading in Batch Crystallization

**Yu-Ti Tseng**, Chemical Engineering, National Taiwan University, Taipei, Taiwan and Jeffrey D. Ward, Department of Chemical Engineering, National Taiwan University, Taipei, Taiwan

Doki et al. (2002) introduced the idea of critical seed loading for batch crystallization. The idea is that for a given crystallization process there is a certain seed mass that will effectively suppress nucleation and lead to a product with a large averages crystal size and narrow size distribution. Furthermore, the critical seed loading ratio depends primarily on the average seed size and much less on other process variables. Doki et al. determined the critical seed loading chart experimentally and developed the following empirical expression for critical seed loading:

$$C_s^* = 2.17\times10^{-6} L_s^2$$

where is the critical seed loading (initial seed mass divided by theoretical crystal yield) and $L_s$ is the seed volume mean size in microns. Later authors (Hojjati et al. (2005) and Kubota et al. (2009)) determined critical seed loading ratios experimentally for other solute-solvent systems.

In this work, we show that critical seed loading charts can also be produced by numerical simulation based on kinetic models of crystallization processes. Results based on processes for the crystallization of potassium sulphate and potassium nitrate show good agreement with the correlation suggested by Kubota and coworkers. This further supports the utility of the critical seed loading concept for rapid process development and suggests that the critical seed loading can be determined for other solute-solvent systems using kinetic models already published in the literature.

Furthermore, we show that with certain reasonable assumptions it is possible to develop an implicit analytical expression for the critical seed loading as a function of seed mean size. These results are helpful for understanding the concept of critical seed loading intuitively and using it in process development.

**References**

Doki N, Kubota N, Yokota M, Chianese A. Determination of critical seed loading ratio for the production of crystals of uni-modal size distribution in batch cooling crystallization of potassium alum. *J Chem Eng Jpn.* Jul 2002;35(7):670-676.

Hojjati H, Rohani S. Cooling and seeding effect on supersaturation and final crystal size distribution (CSD) of ammonium sulphate in a batch crystallizer. Chemical Engineering and Processing 2005;44(9): 949-57.

Kubota N, Onosawa M. Seeded batch crystallization of ammonium aluminum sulfate from aqueous solution. Journal of Crystal Growth 2009;311(20): 4525-29.

# (304a) Process Intensification of the Production of Di-Methyl Carbonate (DMC) Using a New Synthesis and Design Process Intensification Methodology

**Deenesh K. Babi**, Department of Chemical and Biochemical Engineering, Technical University of Denmark, Lyngby, Denmark, Johannes Holtbruegge, TU Dortmund University, Department of Biochemical and Chemical Engineering, Laboratory of Fluid Separations, D-44227 Dortmund, Germany, Philip Lutze, Department of Biochemical and Chemical Engineering, Laboratory of Fluid Separations, TU Dortmund University, D-44227 Dortmund, Germany, Andrzej Gorak, Dortmund University, Chair of Fluid Separation Processes, 44221 Dortmund, Germany, John Woodley, Center for Process Engineering and Technology, Department of Chemical and Biochemical Engineering, Technical University of Denmark, Lyngby, Denmark and Rafiqul Gani, CAPEC, Department of Chemical and Biochemical Engineering, Technical University of Denmark (DTU), Kgs. Lyngby, Denmark

**Process Intensification of the production of Di-Methyl Carbonate (DMC) using a New Synthesis and Design Process Intensification Methodology**

Deenesh K. Babi[1], Johannes Holtbruegge[2], Philip Lutze[2], Andrzej Górak[2], John M. Woodley[1], Rafiqul Gani[1]

*dkbabi@kt.dtu.dk, johannes.holtbruegge@bci.tu-dortmund.de, philip.lutze@bci.tu-dortmund.de, andrzej.gorak@bci.tu-dortmund.de jw@kt.dtu.dk, rag@kt.dtu.dk*

[1] *Department of Chemical and Biochemical Engineering, Technical University of Denmark (DTU), DK-2800 Kongens Lyngby, Denmark*

[2] *Department of Chemical and Biochemical Engineering, Technical University of Dortmund (TU Dortmund), D-44227 Emil-Figge-Str. 70 Dortmund*

**Abstract**

Process Intensification (PI) is a means by which processes, whether new or existing, can achieve a more efficient and sustainable chemical process through the improvement of, for example, energy efficiency and waste generation. PI can be defined as the improvement of a process at the phenomena level which ultimately has an impact at the higher levels of a process, that is the functional and unit operations (Unit-Ops) levels. More specifically PI is the enhancement of the involved phenomena through the integration of Unit-Ops, functions and phenomena. Except for reactive distillation and dividing wall columns, the implementation of PI still faces hurdles [1]. For achieving PI via a systematic approach different methodologies exist, of which two are mentioned here, the task-based means-ends analysis developed by Sirrola [2] and the Unit-Ops based synthesis and design method for achieving PI developed by Lutze et al. [1]. These methods, which work at the Unit-Ops level, are limited to PI equipment already in existence and are therefore able to generate new integrations/combinations of intensified existing equipment but may not generate new/innovative PI solutions.

The objective of this work is threefold: (1) the development of a new/innovative systematic PI synthesis and design methodology, which includes model based synthesis/design and experimental based validation; (2) the development of a phenomena based synthesis and design method (PBS) which in addition to the knowledge-based (KBS) and Unit-Ops based (UBS) methods (thereby extending further, the application range) is one of the methodologies within the new PI methodology; and (3) the application of the new PI methodology to a case study - the production of di-methyl carbonate (DMC).

The new PI methodology has the following characteristics: (a) flexible for a wide range of applications (b) finds innovative (new) and predictive (existing) solutions and (c) model and experimental based validation of intensified options. The new PI methodology follows a systematic decomposition approach starting with a base case as a reference point. In Step 1 a general problem definition is determined. In Step 2 system information is collected based on thermodynamic data and the type of chemical/process systems that are identified. Using this data, the process is analysed and in Step 3, it is possible to use one or two or all of the PI methods. The three synthesis and design methods considered are the KBS, the UBS and the PBS. The KBS and UBS methods use knowledge stored in a knowledge-based tool on implemented PI equipment used for overcoming certain limitations/bottlenecks (LB's), for example, the use of reactive distillation for overcoming unfavourable equilibrium in a reaction. The PBS methodology not only uses the knowledge of the existing methodologies at the Unit-Ops level but also operates at a lower level of aggregation (i.e phenomena level), where the apriori knowledge of existing Unit-Ops is needed to some degree but also where new Unit-Ops, in principle, can be designed. In Step 4 the feasible flowsheets obtained from Step 3 are validated/ verified either through a model or experimental based analysis. At the end of Step 4 the best PI process candidate and the other feasible options are used to update the systems information. It should be noted that the degree of complexity increase from the KBS to PBS while the possibility of obtaining/generating novel PI equipment decreases from PBS to KBS.

A preliminary version for a phenomena-based synthesis/design (PBS) methodology has been developed [3] and consists of a six step workflow: Problem and objective function definition→ Process analysis using thermodynamic insights→ Identification of LB's of the process together with the desirable and accompanying phenomena to overcome these LB's→ Phenomena building blocks are connected to form simultaneous phenomena building blocks (SPBs) which are screened for the most feasible connections using for example connectivity rules. The SPB's are then connected to form operations which are then connected to form flowsheets just as atoms are combined to form molecules→ Flowsheets are first screened using for example logical constraints and performance metrics to obtain feasible PI options→ The feasible flowsheets from are optimized with respect to the objective function with the end result being the identification of the best intensified process candidate. The emerging flowsheets consists of either novel or existing PI equipment. Defined phenomena are for example mixing, two-phase mixing, phase transition and phase separation. This preliminary version is now extended and implemented within the general PI synthesis design methodology.

In this presentation the PI synthesis and design methodology with focus on the PBS workflow will be highlighted together with the details of its application to the production of DMC. DMC is an important bulk chemical used as a solvent and fuel additive. It will be shown that the PBS methodology systematically generates, besides other options, the current intensified option proposed by the co-authors who have followed the UBS path of the combined methodology and is as follows: a reactive distillation column→ membrane separation →distillation.

References:

[1] P. Lutze, A. Román-Martinez, J. M. Woodley & R. Gani. A systematic synthesis and design methodology to achieve process intensification in (bio) chemical processes. Comput. Chem. Eng. 2012 (36) 189– 207

[2] Jeffrey J. Siirola. Strategic process synthesis: Advances in the hierarchical approach. Comput. Chem. Eng. 1996, Supplement 2 (20) S1637-S1643

[3] P. Lutze, R. Gani & J. M. Woodley. Phenomena-based Process Synthesis and Design to achieve Process Intensification. Comp. Aided Chem. Eng. 2011 (29) 221–225

# A NOVEL MILP APPROACH TO THE SYNTHESIS OF THERMALLY COUPLED DISTILLATION SEQUENCES.

*José A. Caballero*; Ignacio E. Grossmann***
**Department of Chemical Engineering. University of Alicante. Ap. Correos 99. 03080 Alicante. Spain*
*** Department of Chemical Engineering. Carnegie Mellon University. Pittsburgh, PA. USA.*

## Introduction

The distillation is still the most widely used separation technique in the chemical industry. It handles more than 90% of separations and purifications and this trend seems unlikely to change in the near future. The capital investment for these distillation systems was estimated in $8x10^9$ US$ [1]. Using data by Mix et al [2], Soave & Feliu [3] calculated that distillation accounts about the 3% of the total US energy consumption which is equivalent to $2.87x10^{18}$ J (2.87 million TJ) a year or to a power consumption of 91 GW or 54 million tons of crude oil.

A renew interest in Thermally Coupled Distillation (TCD) appeared in, say the last 15-20 years, due the important potential savings in energy: typical values around 10% to 40% (in some cases even larger) has been reported [4-9] when compared with conventional distillation sequences. An although the TCD can only be implemented in mixtures involving the separation of at least three components the reader can have an idea of the impact just looking at the numbers presented in the first paragraph.

The design of TCD sequences is much more complex than conventional distillation (each column with a condenser and a reboiler) by at least two reasons. First, the number of alternatives is much larger and second when we introduce a thermal couple we are introducing also two thermodynamically equivalent configurations (TEC) [10-12],. Therefore, the number of alternatives rapidly increases (i.e. in a five component mixture there are 203 basic configurations, more than 8000 if we consider internal heat exchangers and more than $2\cdot 10^5$ if we take into account TEC). In order to avoid the degeneracy created by TEC it is convenient to use a task based approach instead of a column based approach. –All TEC shares the same sequence of separation tasks-.

As noted by Andrecovich and Westerberg [13] in the 80's of last century, when we consider conventional separation sequences, we can, a priori, optimize each possible column of the sequence, and then use a MILP approach to extract the optimal sequence. But, when we deal with non-sharp separations we cannot know a priori the distribution of no-key components and therefore we must simultaneously optimize the columns and the sequence which produces a non-convex MINLP even with shortcut methods [4, 5, 14]. However, taking into account the following two facts, it is possible to develop a strategy that allows decoupling the column and sequence optimization in TCD systems.

1. Any distillation sequence of zeotropic mixtures have a structure without recycles (A thermal couple can be simulated with an equivalent system without recycles).

2. In TCD sequences the flows of some column sections are not in the optimal operation conditions because they must be modified in order to accomplish with the mass balance. But, the introduction of heat exchangers allows decoupling these sections and avoids the necessity of a simultaneous optimization.

   It is possible then generating a tree of possible separation tasks as follows:

1. First we consider, and optimize, all the initial separation tasks (those that have the initial mixture as feed).

2. For each of the products (distillate and bottoms of each column) we generate all the possible separations. At this point we consider that mixtures of the same components with different concentrations or different thermal states are different -explicitly takes into account the possibility of a heat exchanger or a thermal couple-.

3. Following recursively this approach we can optimize all possible separation tasks and, implicitly, generate a tree that includes all the separation tasks.

The number of possible separation tasks is large, but much smaller than the number of sequences. Using the Underwood-Fenske shortcut method the complete tree for a 5 component mixture can be generated in around 4-5 seconds of CPU time.

The final step consists of extracting the optimal separation sequence. However this is not straightforward. The simple connectivity, like in conventional columns, does not assure a feasible basic sequence:

a) It is necessary to include logical relationships accomplish with these task (a comprehensive description of these logical relationships can be found in the work by Caballero & Grossmann [5].

b) For some internal mixtures, there are more than one possible separation task that depends on their origin (which separation task generate it) and the thermal state (include or not a heat exchanger or a thermal couple). Therefore it is necessary to include a set of logical relationships to assure the feasibility and relates the tree of separations with the feasible alternatives.

c) Divided Wall Columns (DWC) are explicitly included. A DWC is generated by a subsystem of three fully thermally coupled tasks [15] and some logical relationships added to avoid only feasible configurations.

As all the data of all the separation tasks is calculated when each of the individual columns are calculated, the resulting problem is a MILP in where we select the best feasible sequence of tasks (i.e. the sequence with minimum Total Annualized Cost). Calculate the excess or defect of heat (or vapor flows) in the connectivity of each individual task and estimate the extra cost (energy).

**Example.**

As an example considers a mixture of 5 hydrocarbons (all relevant data and main results are showed in Table 1).

The optimal solution includes a DWC and a set of small heat exchangers in the final products that appears as a consequence of the differences in the flows in the different column sections –See Figure 1-. Alternatively it is possible to remove all these heat exchangers by adjusting the flows in the implied column sections. The net result will be an increase in the heats of reboiler (associated with product E) and condenser (product A) approximately in the same quantity of the heat removed in the intermediate products. This could have two opposite effects. In one side we remove three heat exchangers, but at the same time we lose the possibility of using utilities at intermediate temperatures.

It is possible to re-optimize the optimal sequence presented in Figure (1) taking now into account the possibility of removing the heat exchangers related to products B, C and D. Remove one of those heat exchangers save its investment cost but the energy consumed by the heat exchanger is transferred to the upper heat exchanger if the heat exchanger removed is a condenser, and to the lower heat exchanger if it is a reboiler. At the same time the internal flows in the corresponding column sections increases and then also the investment costs in the column. The optimal solution obtained varies in less than a 2% in TAC respect to that shown in Figure (1) but in this case, the reboiler C and the condenser B were removed. Removing reboiler D produce an important increase in the internal flows of the separation tasks (D/E) and then it is better to maintain the reboiler D.

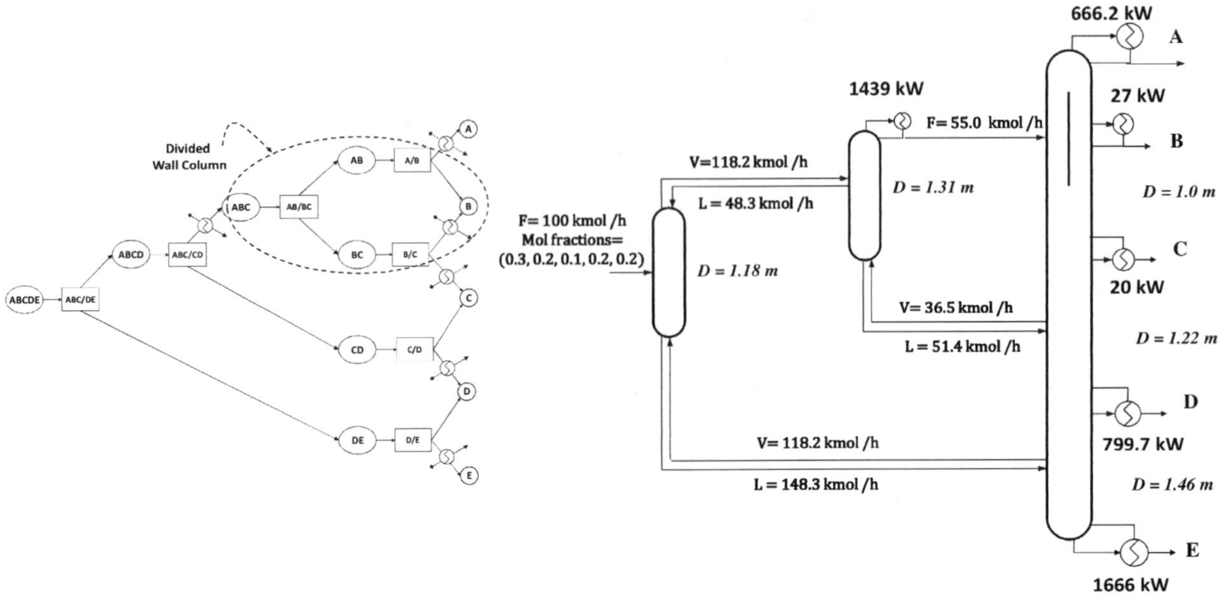


**Figure 1.** ***Sequence of separation tasks and a possible arrangement in actual columns.***

*Table 1. Data and results for example 1.*

| Component | Feed composition (mol fraction) | | |
|---|---|---|---|
| (A)- Benzene | 0.3 | Total Flow | 100 kmol/h |
| (B)- Toluene | 0.2 | Pressure | 1 atm |
| (C)- Ethylbenzene | 0.1 | Steam cost | 5.02 $/GJ |
| (D)- Styrene | 0.2 | Cold water cost | 0.19 $/GJ |
| (E)- α-methyl styrene | 0.2 | (8000 h/year) | |
| | | Capital charge factor 0.18 (9% interest ; 8 years) | |

| | |
|---|---|
| **Optimal sequence** | ABC/DE - ABC/CD - AB/BC – A/B – B/C – C/D – D/E |
| | Condensers: ABC; A; B |
| | Reboilers: C; D; E |
| *Vessels cost (k$)* | 574.27 |
| *Tray cost (k$)* | 256.51 |
| *Heat Exchangers cost (k$)* | 599.5 |
| *Total Investment (k$)* | 1430.03 |
| *Total utilities cost (k$/year)* | 568.42 |
| ***TAC ( (k$/year)*** | **825.8** |

*Table 1 (cont). Data and results for example 1.*

| | |
|---|---|
| ***Model Statistics*** | |
| *CPU time (s)* | 57 |
| *Nº equations* | 12064 |
| *Nº continuous variables* | 3142 |
| *Binary variables* | 2563 |

## Acknowledgements

The authors wish to acknowledge support from the Spanish Ministry of Science and Innovation under project (CTQ2009-14420-C02).

## References

1. Humphrey, J., *Separation processes: playing a critical role.* Chemical Engineering Progress, 1995. **91**(10): p. 43-54.
2. Mix, T., et al., *Energy conservation in distillation.* Chemical Engineering Progress, 1978. **74**(4).
3. Soave, G. and J.A. Feliu, *Saving energy in distillation towers by feed splitting.* Applied Thermal Engineering, 2002. **22**(8): p. 889.
4. Caballero, J.A. and I.E. Grossmann, *Design of distillation sequences: from conventional to fully thermally coupled distillation systems.* Computers & Chemical Engineering, 2004. **28**(11): p. 2307-2329.
5. Caballero, J.A. and I.E. Grossmann, *Structural considerations and modeling in the synthesis of heat-integrated-thermally coupled distillation sequences.* Industrial & Engineering Chemistry Research, 2006. **45**(25): p. 8454-8474.
6. Fidkowski, Z.T., *Distillation configurations and their energy requirements.* AIChE Journal, 2006. **52**(6): p. 2098-2106.
7. Fidkowski, Z.T. and R. Agrawal, *Multicomponent thermally coupled systems of distillation columns at minimum reflux.* AIChE Journal, 2001. **47**(12): p. 2713-2724.
8. Rudd, H., *Thermal Coupling For Energy Efficiency.* Chemical Engineer-London, 1992(525): p. S14-S15.
9. Triantafyllou, C. and R. Smith, *The Design And Optimization Of Fully Thermally Coupled Distillation-Columns.* Chemical Engineering Research & Design, 1992. **70**(2): p. 118-132.
10. Agrawal, R. and Z.T. Fidkowski, *More operable arrangements of fully thermally coupled distillation columns.* AIChE Journal, 1998. **44**(11): p. 2565-2568.
11. Caballero, J.A. and I.E. Grossmann, *Thermodynamically equivalent configurations for thermally coupled distillation.* AIChE Journal, 2003. **49**(11): p. 2864-2884.

12. Rong, B.G., A. Kraslawski, and I. Turunen, *Synthesis and optimal design of thermodynamically equivalent thermally coupled distillation systems.* Industrial & Engineering Chemistry Research, 2004. **43**(18): p. 5904-5915.
13. Andrecovich, M.J. and A.W. Westerberg, *An Milp Formulation For Heat-Integrated Distillation Sequence Synthesis.* AIChE Journal, 1985. **31**(9): p. 1461-1474.
14. Caballero, J.A. and I.E. Grossmann, *Generalized disjunctive programming model for the optimal synthesis of thermally linked distillation columns.* Industrial & Engineering Chemistry Research, 2001. **40**(10): p. 2260-2274.
15. Caballero, J.A. and I.E. Grossmann, *Logic-Sequential Approach to the Synthesis of Complex Thermally Coupled Distillation Systems*, in *Computer Aided Chemical Engineering*, M.C.G. E.N. Pistikopoulos and A.C. Kokossis, Editors. 2011, Elsevier. p. 211-215.

# (304c) Process Intensification in Ethylene Oxide Production: Optimal Reactor Design From a Process Point of View

**Hannsjörg Freund**[1,2], Andreas Peschel[2] and Kai Sundmacher[2,3], (1)Chemical Reaction Engineering, University of Erlangen-Nürnberg, Erlangen, Germany, (2)Process Systems Engineering, Max Planck Institute for Dynamics of Complex Technical Systems, Magdeburg, Germany, (3)Process Systems Engineering, University of Magdeburg, Magdeburg, Germany

In this work, a new method for the identification of the best reaction concept from an overall process point of view is proposed. Using this method the optimal reaction route is determined independently of existing apparatuses and the recycle affecting the reactor is fully considered. The reaction concept and the design parameters of the process are simultaneously optimized. For this purpose, a detailed model of the reaction system including reaction kinetics, thermodynamic relationships, and intrinsic bounds such as explosion limits as well as a complete model of the downstream process is set up. The potential of different reaction concepts including innovative process intensification alternatives can easily be determined by comparison with an optimized reference case.

The example considered in this work is the oxygen based ethylene oxide process, which is one of the most important chemical bulk processes. Sophisticated cooling in combination with distributed oxygen dosing is identified as best technical reaction concept with a significant potential of reducing the running expenses compared to an optimized reference case. In addition, the utility consumption (e.g. the consumption of electricity) can be drastically reduced.

In conclusion, the proposed method enables the predictive determination of the best reaction concept from an overall process point of view taking the full interconnections between the reaction concept and the process into account and considering highly innovative process intensification options. Thereby, this method provides a key stone for the development of more economical and more sustainable chemical reactors and processes of the future.

# (304d) Microwave Assisted Flow Processing: Coupling of Electromagnetic and Hydrodynamic Phenomena

Narendra Patil[1], Erik Esveld[2], Faysal Benaskar[1], Evgeny V. Rebrov[3], Lumbertus Hulshof[1], Jan Meuldijk[1], Volker Hessel[1] and Jaap C. Schouten[1], (1)Chemical Engineering and Chemistry, Eindhoven University of Technology, Eindhoven, Netherlands, (2)Food and Biobased Research, Wageningen UR , Wageningen, Netherlands, (3)School of Chemistry & Chemical Engineering, Queen's University Belfast, Belfast, United Kingdom

**Microwave assisted flow processing: coupling of electromagnetic and hydrodynamic phenomena**

Narendra G. Patil,[1] Erik Esveld,[2] Faysal Benaskar,[1] Evgeny V. Rebrov,[3] Lumbertus A. Hulshof,[1] Jan Meuldijk,[1] Volker Hessel[1] and Jaap C. Schouten[1]

[1]*Laboratory of Chemical Reactor Engineering, Eindhoven University of Technology, P.O. Box 513, 5600 MB Eindhoven, the Netherlands*

[2]*Wageningen UR Food and Biobased Research, P.O. Box 17, 6700 AA, Wageningen, The Netherlands*

[3]*School of Chemistry and Chemical Engineering, Queen's University Belfast, Stranmillis Road, Belfast, BT9 5AG, United Kingdom*

**Introduction**

Fast and volumetric heating behavior of microwaves is attracting attention for their application in continuous flow synthesis of specialty chemicals. However, the use of tuned microwave field even in small defined reactors does not always provide the expected temperature patterns.[1,2] This is majorly due to a lack of position specific information with respect to velocity profile of liquid. Additionally, the accuracy of these predictions in microwave heating is limited when no space distribution of volumetric heating source is considered.[1] Understanding the space distribution of microwave heating process with detailed modeling is a prerequisite for process development and control. In this work, we take a closer look at the influence of liquid velocity profiles on the axial and radial temperature profiles in a tubular microwave integrated millireactor combined with a heat exchanger (Figure 1a).

**Results and discussion**

Horizontal co-current flow of a reactant (ethanol) and a microwave transparent coolant (toluene) was established in a teflon supported quartz tube (i.d.: 3 mm, o.d.: 4 mm) and shell (i.d.: 7 mm, o.d.: 9 mm), respectively (Figure 1a). The co-current flow of coolant was assumed to avoid overheating of the reactant by inherently fast microwave heating. The reactor was inserted in a microwave cavity with a length of 47 mm. The physical model included electromagnetic interactions, fluid dynamics and heat transport (Figure 1b). The temperature and flow profiles were obtained in a 3-D domain using the COMSOL software. As the dielectric properties of the reactant were changing with temperature, it was expected that they should influence the heating rate. However, this effect was of minor importance (Figure 1c) and the remaining simulations were made in a 2-D domain. Effect of gravitation was observed due to horizontal arrangement of the reactor-heat exchanger. The effect of gravity was clearly pronounced in the coolant flow where a recirculation pattern was observed due to difference in density (Figure 2). The coolant flow rate had a minor effect on temperature profiles and energy dissipation at a high heating rate of 180 °C/s (Figure 3, a and b). Higher temperatures were observed at the wall (stagnant liquid films), leading to drastic drop in the energy dissipation (Figure 3, c and d). The laminar flow of ethanol was disturbed by a temperature probe (Figure 4). The probe was inserted either from the outlet (Figure 4a) or inlet (Figure 4b) side of the assembly. In the former case, measured temperature was lower as compared to the latter case (Figure 5). This effect could be explained by local hydrodynamics near the tip of the probe.

When the probe was inserted from the outlet, the highest velocities were found at the tip. Conversely, when the probe was inserted from the inlet, a stagnant film assured lowest velocity at the probe tip. These observations were then validated by experiments (data points, Figure 5). The temperature profiles obtained by the modeling study agreed with the experimental observations.

## Conclusions

The stagnancy of the flow of the microwave absorbing fluid influenced the temperature distribution in a microwave integrated flow reactor. The buoyancy influence (as a result of gravitational forces) was already visible for horizontal arrangement of reactor-heat exchanger assembly at millimeter sizes. The stagnant layer formation caused by any insertion or at reactor walls yielded higher temperatures and lower energy dissipation regions. The coolant flow, unless used as a heated jacket to minimize heat losses, was found to be ineffective. Lastly, complex 3-D geometry modeling efforts can be eliminated at very early stages unless there are specific disturbances in electric field pattern.

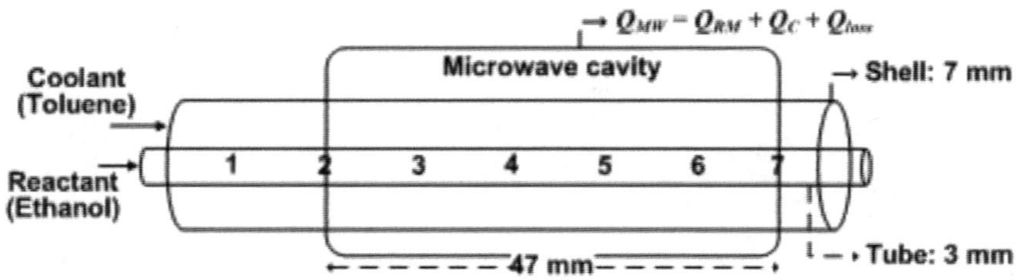


*(a)*

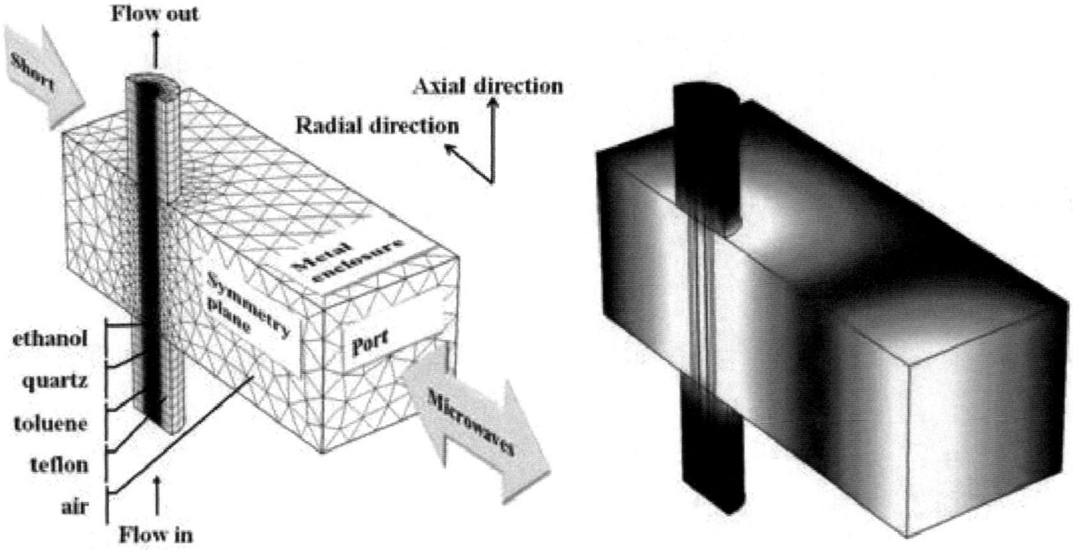


*(b) (c)*

**Figure 1:** *a)* Schematic view and process details of the microwave integrated reactor-heat exchanger assembly used for experimental validations, *b)* 3 D computational domain of the microwave cavity showing individual components of the assembly, *c)* Electric field intensity in and around the microwave integrated reactor-heat exchanger. Ethanol flow rate: 40 ml/min, toluene flow rate: 100 ml/min. Input microwave power: 55 W.

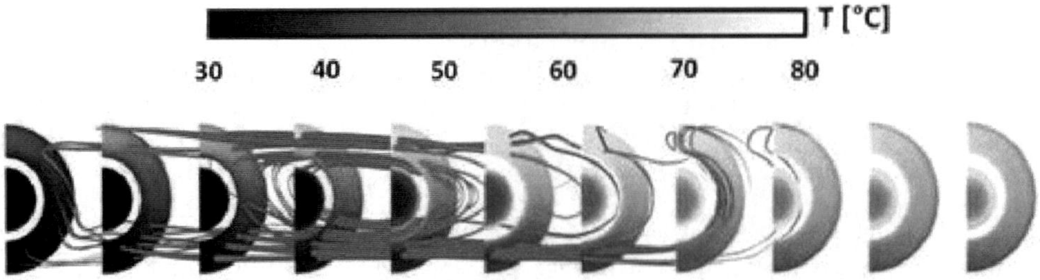


**Figure 2:** Effect of buoyancy in a horizontal arrangement of the microwave integrated reactor-heat exchanger assembly. Gray lines (with thickness indicating the strength) show velocity profiles of coolant. Axially sectioned colored pattern shows temperature in the reactant as well as the coolant section.

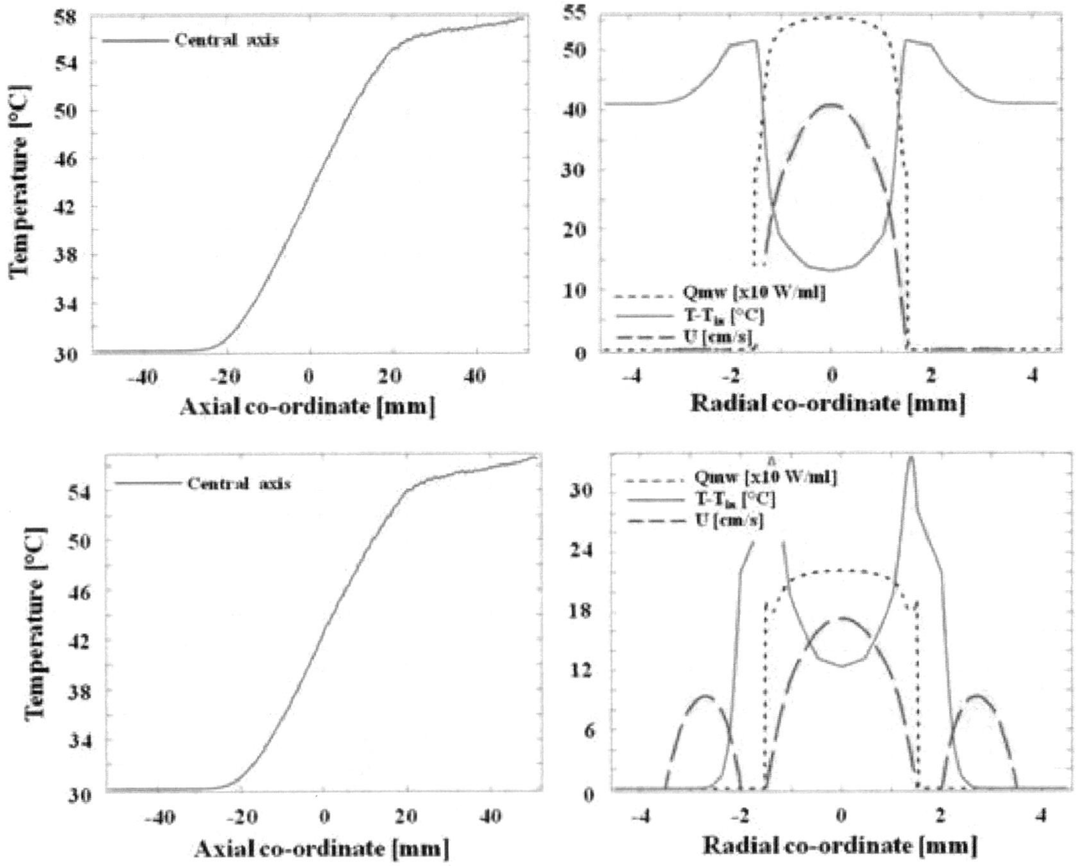


*(a) (b)*

*(c) (d)*

**Figure 3:** Temperature profiles *(a,c)* as a function of axial co-ordinate, and *(b,d)* energy dissipation of microwaves ($Q_{MW}$, W/ml), temperature rise ($T-T_{in}$, °C) and velocity profile (U, cm/s) as a function of radial co-ordinate. *(a)* near stagnant coolant flow, $F_{EtOH}$ = 100 ml/min, $F_{TOL}$ = 0.5 ml/min. *(b)* high coolant flow, conditions are the same as those in Fig 1c.

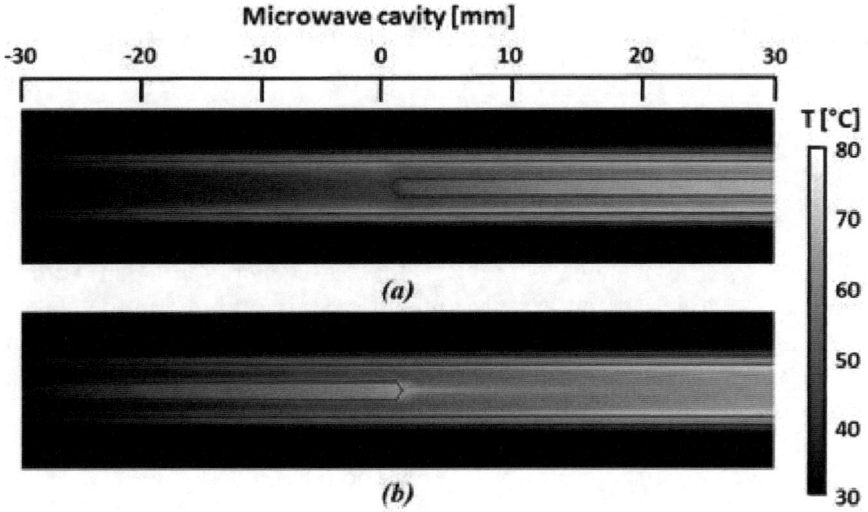


**Figure 4:** Influence of the direction of probe insertion on the temperature profiles obtained by modeling, a) probe inserted from the outlet, *b)* probe inserted from the inlet. Conditions are the same as those in Fig 1c.

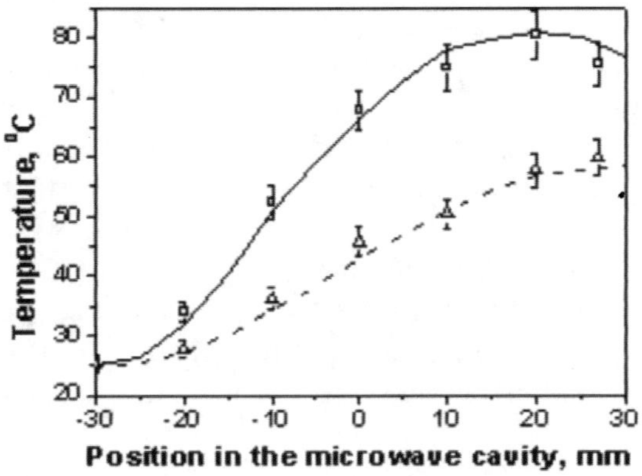


**Figure 5:** Influence of the direction of probe insertion on the temperature profiles obtained by modeling (lines) and experiments (data points). Solid line and squares: probe inserted from inlet, dotted line and triangles: probe inserted from outlet. Conditions are the same as those in Fig 1c.

**Reference:**

1. Patil, N. G.; Hermans, A. I. G.; Benaskar, F.; Rebrov, E. V.; Meuldijk, J.; Hulshof, L. A.; Hessel, V.; Schouten, J. C. Energy efficient and controlled flow processing under microwave heating by using a milli reactor-heat exchanger. *AIChE J.* **2011**, *Accepted*. DOI 10.1002/aic.13713.

2. Patil, N. G.; Benaskar, F.; Rebrov, E. V.; Meuldijk, J.; Hulshof, L. A.; Hessel, V.; Schouten, J. C. Continuous multi-tubular milli-reactor with a Cu thin film for microwave assisted fine-chemical synthesis. *Ind. Eng. Chem. Res.* **2012**, *submitted.*

# Process Intensification for Hydrogen Generation using Micro-Channel Steam Reforming

**Presentation ID# 281083**

**Peter Bossard, Jacques Mettes and Luis Breziner** — **Power & Energy, Inc.**
**Benjamin A. Wilhite** — **Texas A&M University**

**1. Abstract.** Today, the advantages of micro-channel technology for next generation steam reforming have been demonstrated on a pilot scale. Modeling and experimental efforts are reported in the context of Power & Energy, Inc's (P&E) development of a radial tube-in-tube microreactor (TTμR) for efficient, scalable production of hydrogen ($H_2$) and/or synthesis gas (CO, $H_2$) from natural gas at competitive costs. The novel radial micro-channel reactor allows manipulating heat- and mass-transport phenomena in catalytic-wall heat-exchanger microreactors to achieve increased catalyst utilization, process intensification and reactor stability. Using similar time proven production methods as used in hydrogen purifier manufacturing, a successful commercialization of the novel microreactor technology is expected with the potential to achieve transformative breakthroughs in system cost, durability and thermal efficiency as compared to existing planar μreactor systems.

**2. Introduction.** The efficient catalytic conversion of hydrocarbons, including natural gas or methane, to synthesis gas remains a critical challenge for both emerging clean energy industries and existing petrochemical and natural gas markets.[1,2,3,4,5,6,7,8] Hydrogen gas represents a universal energy currency, in that it may be harvested from myriad hydrocarbon resources. Hydrogen can be utilized in electrochemical power systems capable of exceeding Carnot-Cycle thermodynamic limits on efficiency. Process intensification strategies have been employed in the design of next-generation methane reformers where closely coupling of endothermic steam reforming with exothermic combustion,[9,10] yields breakthroughs in thermal efficiency. Potential integration with gas purification via membrane[11,12] or adsorption[13,14] systems further enhances the reaction conversion process while providing a pure hydrogen product. Thanks to pioneering efforts of companies like the Battelle, Velocys and the Institut fur Mikrotechnik Mainz (IMM), micro-reactor technologies have been identified as a potentially transformative technology for realizing next-generation methane reformers for several markets.

Microchemical systems have been demonstrated as a promising technology for chemical processing, with (i) improved heat and mass transport owing to reduced characteristic lengths in microchannels, (ii) resulting enhanced control of reaction temperature and hot-spot formation, and (iii) their potential for integrating multiple processes within one compact unit for unrivalled process intensification and portability.[15] Enhancement of heat and mass transport rates within micro-channels has fueled significant research towards developing thermally-integrated μreactor systems capable of combining multiple endothermic and exothermic catalytic and/or physical processes within a compact, efficient device. To-date, several μreactor designs have been reported for integrating catalytic combustion with endothermic steam reforming of hydrocarbons [16, 17] and subsequently with water-gas-shift [18] and/or membrane separations[19,20] for further cleanup of hydrogen product.

Fabrication of μreactors traditionally involved machining two-dimensional patterns into individual silicon,[18] metal[7,21] or ceramic[22,23] plates via chemical etching or laser cutting, followed by sealing and packaging of a "stack" of such plates to achieve the desired total reactor volume.[24,15] This approach results in a near-linear cost of scale-up, [15] which has limited the market penetration of μreactors to the research and niche-chemical markets. Recently, the extension of mass-production techniques such as roll-embossing and laser cutting of individual metal plates has enabled the production of industrial-scale μreactors at competitive costs[25,26,7]

The planar μreactor design is a major achievement, it has demonstrated the potential of micro-channel technology, but it does not capture the true potential that micro-channels offer. It suffers inherent drawbacks in terms of thermal efficiency for heat-exchanger applications. In the planar μreactor design, the multitude of individual reaction channels are embedded in a single monolithic substrate, allowing significant thermal "cross-talk" between individual microchannels; more importantly, this means that each microchannel is provided with a direct conductive pathway to the outer surface of the micro-reactor. These drawbacks have been shown to play a significant role in reducing the effectiveness of micro-reactors.[27,28,29,30] Current solutions to this challenge involve the combustion of additional hydrogen and/or

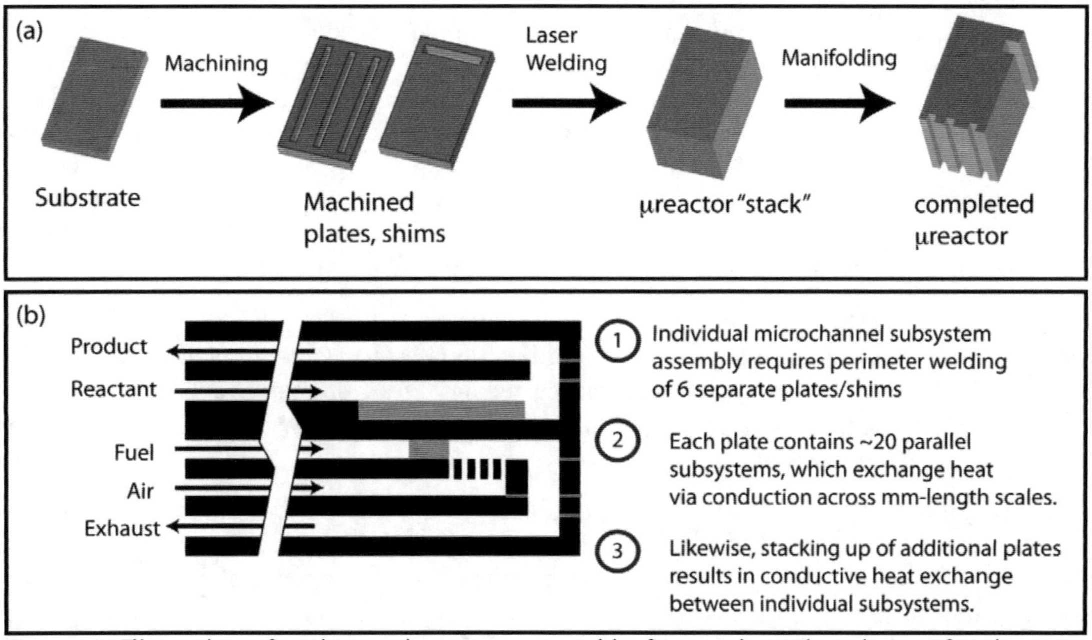


**Figure 1:** Illustration of a planar microreactor assembly for autothermal methane reforming to synthesis gas; (a) generalized process flow for microreactor construction, adapted from LeViness et al., 2011; (b) cross-section schematic a planar microchannel subsystem for integrated methane combustion and steam reforming, adapted from Tonkovich et al.

fuel to compensate for heat losses at the external surface of the micro-reactor system.[28] In addition to transverse or radial heat dispersion, solid-phase axial heat conduction further limits the thermal efficiency of planar microreactors. Analysis of microchannel heat-exchangers [31, 32] demonstrate the importance of maintaining significant axial thermal gradients within the solid phase to ensure maximum driving force for heat transfer between fluid volumes along the entire axial length of the exchanger. However, the high substrate-to-process volume ratios of existing planar systems result in significant axial dispersion of process heat by the solid phase, such that the lower limit of an isothermal slab is approached with commiserate losses in heat-exchanger efficiency.

| | **Planar System*** | **Tubular System**** |
|---|---|---|
| Subsystem Volume | 75 $cm^3$ | 17.9 $cm^3$ |
| Process Volume | 10 $cm^3$ | 9.3 $cm^3$ |
| Sealing Surface | 53.3 $cm^2$ | 21.4 $mm^2$ |
| Heat Transfer Surface | 256 $cm^2$ | 260 $cm^2$ |
| Surface Area-Process Volume Ratio | ~2,500 $m^{-1}$ | ~2,800 $m^{-1}$ |
| Sealing Area-Process Volume Ratio | ~5,000 $m^{-1}$ | ~4.6 $m^{-1}$ |
| Process Volume-System Volume Ratio | 0.14 | 0.52 |
| Volumetric Heat Transfer Flux | 65 $W/cm^3$ @ 4ms | 840 $W/cm^3$ @ 5ms |

*values estimated based upon dimensions and performance reported by Tonkovich et al.[53] for a single microchannel subsystem depicted in Figure 1.
**values estimated based upon nominal tube dimensions discussed in Section 3 and preliminary results for methane steam reforming.

## 3. Tube-in-Tube Microreactors (TTμR): A Novel Approach to Microreactor Design

P&E has developed a novel manufacturing process for realizing compact, high-efficiency radial microchannel systems with significant advantages over their planar counterparts. The heart of this technology is the use of individual tube-in-tube assemblies to realize thermally isolated μchannel subsystems (Figure 2). Individual tube-in-tube assemblies are constructed by coaxially mounting an inner stainless-steel microtube (nominally 1.47 mm OD) inside an outer stainless-steel μtube (nominally 2.03 mm ID). The outer tube is closed on one end, such that two radially-symmetric microchannels are formed by the inner diameter of the inserted tube and the annulus formed between the inner and outer tubes. Manufacturing of the tube-in-tube assembly is achieved via automated 200W continuous laser-welding system developed by P&E capable of achieving high-quality sealing at a rate of ~3 seconds/tube. Mechanical stabilization is provided at the sealed end of the outer-most tube via insertion in a "slip plate," which minimizes transverse deflection of individual tubes while allowing axial expansion with changing temperature. The P&E TTμR design offers several unique industrial advantages over existing planar μreactor, including:

- **Reduction in manufacturing costs:** By utilizing mass-produced tubular substrates and conventional (automated laser-welding) sealing technologies, dramatic reduction in cost of scale can be achieved, as compared to planar μreactors requiring customized manufacturing of individually-patterned plates.
- **Increased Durability, Scalability:** By minimizing the packaging and sealing contact area, device durability is increased by reducing the occurrence of gas leaks, seal failure; likewise, placement of seals is such that each seal can be inspected and repaired if necessary prior to final μreactor assembly. Finally, scale-up is no longer limited by sealing challenges (discussed in Section 2), as individual tubes are not sealed directly to each other.
- **Increased Resistance to Thermal Stresses:** By anchoring the tubular μreactor at one end and allowing the other to "float", the resulting system is capable of absorbing significant thermal gradients without the risk of failure due to thermal strain; likewise, because individual tubes are not sealed directly to each other, thermal expansion does not compromise seals.
- **Improved Power and System Density:** By minimizing substrate volumes through the use of the tube-in-tube architecture, breakthroughs in system density, portability construction costs are achieved.
- **Better Handling of High Pressure Operation:** the hoop stress advantage of, particularly, small OD cylinders allows larger pressure drops for a given wall thickness compared to planar technology. Alternatively, walls can be thinner for similar pressure differences providing better heat transport. The ability to operate at higher pressures may be particularly relevant when sequestering CO2.

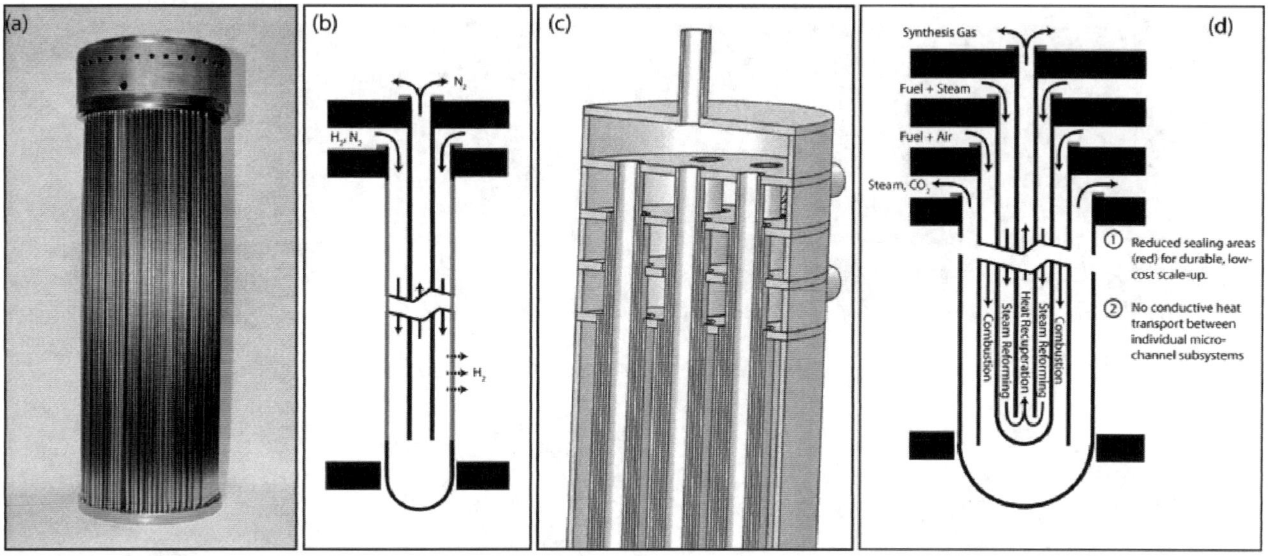


**Figure 2:** P&E tube-in-tube μchannel technology; (a) tube-in-tube Pd-μmembrane system, 6" x 12.5" containing 1400 individual μmembrane subsystems; (b) schematic of individual tube-in-tube μmembrane subsystem; (c) CAD drawing of proposed μreactor system showing packaging strategy; (d) schematic of planned μreactor system for coupling methane combustion with steam/autothermal reforming for compact production of hydrogen and/or synthesis gas.

## 4. Experimental

### 4.1. Manufacturing of TTμRs

A video will be shown of existing automated manufacturing methods currently employed in the production of commercial hydrogen Palladium membrane purifiers that is largely similar in the methods used to fabricate the novel tube-in-tube micro-channel devices. Hereunder, are various pictures showing tools and parts of the latter process.

Fig. 6, laser welding connecting tubes on one end only.

2800 tube-in-tube Pd-alloy micro-channels

Fig. 7, large parallel tubes assemblies

Figure 8 Automated laser welding positioning with computer vision software.

Figure 9 traditional hand weld attempt.

Fig. 10, laser weld result.

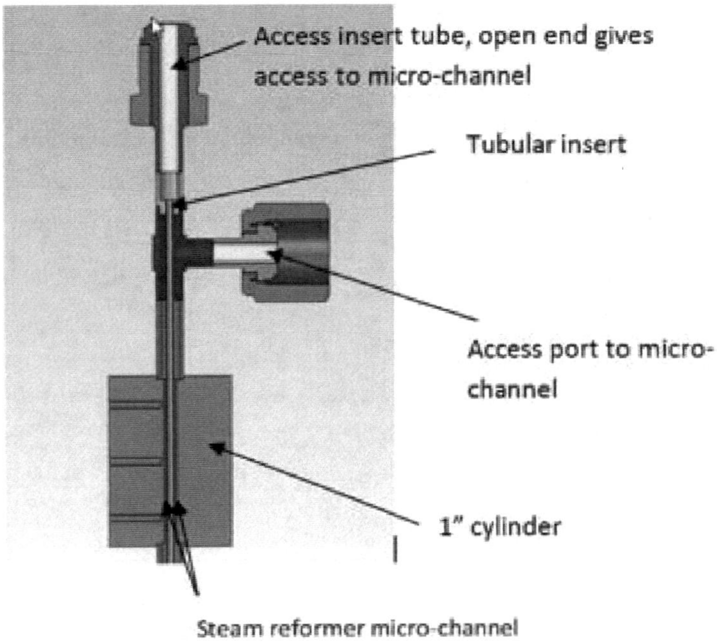


**Figure 3**: Experimental single cylindrical micro reactor systems developed by **P&E**, which allows the unique ability to precisely monitor heat flux and temperature in a single, isolated microchannel.

### 4.2. Prototype TTμR for Methane Steam Reforming

P&E's researchers have constructed a bench-scale TTμR system for evaluating heat transfer and reaction rates of individual micro-channel processes. This single TTμR system is comprised of a band heater attached to the outside of a 1" OD Aluchrom tube with bore diameter machined to match that of final tube-in-tube μreactor designs, see figure 3. The band heater allows accurate control of reactor temperature, to within +/- 0.1 K. The inner wall of the bore is coated with steam reforming catalyst and an insert tube is coaxially mounted within the bore to complete the TTμR geometry. Reactor temperatures are measured by thermocouples placed approx. 0.03" from the inner surface of the micro-channel. Heat flux from the band heater to the catalyst wall can then be accurately determined via geometry. Multiple cylindrical reactors are put in series with individual power control and samples can be extracted on either end of any of the reactors. This enables detailed and precise studies of reaction dynamics of the fuel, steam and catalyst in a test setup that has the essential internal features of TTμR based production systems. This apparatus provides, for the first time, operation of individual microchannels with precise monitoring and/or control of heat flux to the microchannel volume and inner-wall surface temperature while minimizing heat losses common to planar single-channel systems.

### 4.3. Two-Dimensional Simulations of P&E system using COMSOL

Detailed simulations of laboratory-scale single-tube μreactor described above were implemented using the COMSOL multiphysics programming environment. Gas velocity through individual μchannels was described using the Navier-Stokes equation assuming weakly compressible fluids in tandem with the continuity equation while mass transfer of individual species is described by Maxwell-Stefan multicomponent diffusion.[33] Fluid density (ρ) was calculated assuming ideal gas law and dynamic viscosity of the gas mixture is evaluated as a function of composition and temperature using the semiempirical formula developed by C.R.Wilke[34] with all values for Lennard-Jones parameters taken from Bird et al. (2006).[35] Fluid-phase heat transport was described by conduction and convection, assuming negligible radiative heat transport within the μchannel. Fluid-phase thermal properties were taken from physical property databases.[36] The resulting 2-dimensional, radially-symmetric computational fluid-dynamic (CFD) model is identical to models previously implemented by Vlachos' and Vacarro's research groups.[27,28,29,30,37]

Reaction, mass diffusion and heat conduction within the catalyst film was modeled assuming a combination of Darcy's Law (to describe convective transport in porous media) and Maxwell-Stefan diffusion and reaction. The resulting reaction-diffusion model has recently been shown to be capable of comparable accuracy to the more rigorous dusty-gas model of Jackson[38,39,40] for pore diameters > 500 μm.[41,42] Kinetics of methane pyrolysis and water-gas-shift provided in the literature for a $Ni/MgAl_2O_4$ catalyst by Xu and Froment[43] were employed. By imposing a constant temperature boundary condition at the catalyst-μreactor wall interface, i.e. the inner boundary of the catalyst layer, numerical solutions to the above system of partial differential equations predict extent of catalytic reaction as well as the total heat duty required to maintain the externally-imposed μreactor wall temperature.

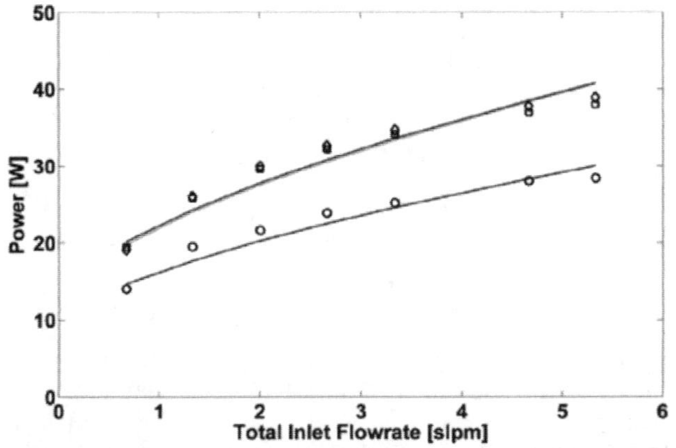


**Figure 4:** Comparison of preliminary simulation results (lines) of the net power entered into the reaction with experimental data points for single-tube microreactor provided by P&E; feed composition 25% $CH_4$:75%$H_2O$ at 10bar and (o) 600°C, (□) 650°C and (◇) 700°C.

## 5. Preliminary Results

Experimental data was initially obtained using a proprietary methane steam reforming catalyst washcoated within a single 1" μchannel system (described in Section 4.2) over a range of inlet gas superficial velocities of 25% methane, 75% steam in terms of the total heat supply added to the reactor to maintain a designated reactor-wall temperature (Figure 4). For the single TTμR described here, heat duties approaching 40W correspond to a volumetric heat flux of > 800W/cm$^3$, more than an order-of-magnitude improvement over competing planar designs.

Simulations using the model described above showed good agreement between COMSOL predictions and experimental data provided by

P&E relating heat duty vs. process flowrate for the 1" single-channel µreformer (Figure 3). Validation of the COMSOL design model allows theoretical exploration of different catalyst and/or reactor configurations for further improvement in process intensification. Additional experimental and simulation results will be presented in this talk.

## 6. Conclusion/outlook

Today, the advantages of micro-channel technology for next generation steam reforming have been demonstrated on a pilot scale. A radial microreactor is presented for efficient, scalable production of hydrogen and/or synthesis gas from natural gas at competitive costs. Modeling and experimental tools are introduced that provide, for the first time, operation of individual microchannels with precise monitoring and/or control of heat flux to the microchannel volume and the inner-wall surface temperature while minimizing heat losses common to planar single-channel systems. Simulation, in the absence of any refinement of kinetics or transport properties, showed good agreement between COMSOL predictions and experimental data relating heat duty vs. process flowrate. The presentation shows some of the tools of commercial production methods used for micro-channel hydrogen purifier manufacturing. These tools are very similar to tools needed for radial microreactor manufacturing and successful commercialization of this technology has the potential to achieve transformative breakthroughs in system cost, durability and thermal efficiency as compared to existing planar microreactor systems.

Future work is planned to provide the possibilities and limitations of the radial micro-channel reactor such as: the exceptionally low operational stresses, the possibility of manipulate heat- and mass-transport phenomena in catalytic-wall heat-exchanger microreactors to achieve transformational breakthroughs in catalyst utilization, process intensification and reactor stability.

**7. Acknowledgement** Funding support for the Development in the form of contract N00014-11-C-0194 was provided by the Office of Naval Research, Washington D.C., with technical oversight provided by the Naval Surface Warfare Center, Philadelphia PA.

## 8. References

[1] A. Kirubakaran, S. Jain, R.K. Nema, A review on fuel cell technologies and power electronic interface, **Renewable and Sustainable Energy Reviews**, 13 (2009) 2430-2440

[2] F. de Bruijn, The current status of fuel cell technology for mobile and stationary applications, **Green Chemistry**, 7 (2005) 132-150.

[3] A. Zagoria, R. Huycke, Refinery hydrogen management – the big picture, **Hydrocarbon processing**, 82 (2003) 41-46.

[4] N. Hallale, F. Liu, Refinery hydrogen management for clean fuels production **Advances in Environmental Research**, 6 (2001) 81-98

[5] D.C. Elliott, Historical Developments in Hydroprocessing Bio-oils, **Energy & Fuels**, 21 (2007) 1792-1815.

[6] M.E. Dry. The Fischer-Tropsch Process – Commercial Aspects. **Catalysis Today**, 6 (1990) 183-206.

[7] S. LeViness, A.L. Tonkovich, K. Jarosch, S. Fitzgerald, B. Yang, J. McDaniel. Improved Fischer-Tropsch Economics Enabled by Microchannel Technology. White Paper generated by Velocys (2011) http://www.oxfordcatalysts.com/press/wp/wp110224_microchannel_FT_white_paper_24Feb11.pdf

[8] M. Ball, M. Wietschel. The future of hydrogen – opportunities and challenges. **Int J Hydrogen Energy**, 34 (2009), 615-627.

[9] G. Kolios, B. Glockler, A. Gritsch, A. Morillo, G. Eigenberger. Heat-integrated reactor concepts for hydrogen production by methane steam reforming. **Fuel Cells**, 5 (2005), 52-65.

[10] G.D. Stefanidis, D.G. Vlachos. Intensification of Steam Reforming of Natural Gas: Choosing Combustible Fuel and Reforming Catalyst. **Chem. Eng. Sci.,** 65 (2010), 398-404.

[11] F. Galluci, M. Van Sint Annaland, J.A.M. Kuipers. Autothermal Reforming of Methane with Integrated $CO_2$ Capture in a novel fluidized bed membrane reactor. Part 1: Experimental Demonstration. **Topics in Catalysis**, 51 (2008), 133-145.

[12] A. Vigneault, S.S.E.H. Elnashaie, J.R. Grace. Simulation of a Compact Multichannel Membrane Reactor for the Production of Pure Hydrogen via Steam Methane Reforming. **Chem. Eng. Tech**., 35 (2012), 1520-1533.

[13] L. Barelli, G. Bidini, F. Gallorini, S. Servilli. Hydrogen Production through Sorption-Enhanced Steam Methane Reforming and membrane Technology: a Review. **Energy**, 33 (2008), 554-70.

[14] Balasubramanian B, Lopez Ortiz A, Kaytakoglu S, Harrison DP. Hydrogen from methane in a single-step process. **Chem Eng Sci** 54 (1999) 3543-3547.

---

[15] Jensen, K.F., "Microchemical Systems: Status, Challenges, and Opportunities," **AIChE Journal**, **45**(10), 2051-2054 (1999).

[16] Arana, L.R., S.B. Schaevitz, A.J. Franz, M.A. Schmidt and K.F. Jensen, "A Microfabricated Suspended-Tube Chemical Reactor for Thermally Efficient Fuel Processing," **J. Microelectromech. Syst**., 12(5), 600-612 (2003)

[17] Tonkovich, A., D. Kuhlmann, A. Rogers, J. McDaniel, S. Fitzgerald, R. Arora and T. Yuschak, "Microchannel Technology Scale-Up to Commercial Capacity," **Chem. Eng. Res. Des.,** 83(A6), 634-639 (2005).

[18] K. Shah, R.S. Besser. Key Issues in the Microchemical Systems-based Methanol Fuel Processor: Energy Density, Thermal Integration and Heat Loss Mechanisms. **J. Power Sources**, 2007 (166), 177-182.

[19] S.V. Karnik, M.K. Hatalis, M.V. Kothare. Towards a Palladium Micro-Membrane for the Water-Gas-Shift Reaction: Microfabrication Approach and Hydrogen Purification Results. **J. MEMS**, 12 (2003), 93-98.

[20] Wilhite, B.A., S.E. Weiss, J.Y. Ying, M.A. Schmidt and K.F. Jensen, "Demonstration of 23wt% Ag-Pd Micromembrane Employing 8:1 LaNi0.95Co0.05O3/Al2O3 Catalyst for High-Purity Hydrogen Generation," **Advanced Materials,** 18, 1701 (2006)

[21] E.R. Delsman, M.H.J.M. DeCroon, G.J. Kramer, P.D. Cobden, Ch. Hofmann, V. Cominos, J.C. Schouten. Experiments and Modeling of an Integrated Preferential Oxidation-heat Exchanger Microdevice. **Chem. Eng. J.,** 101 (2004), 123-129.

[22] A. Rouge, B. Spoetzl, K. Gebauer, R. Schenk, A. Renken. Microchannel Reactors for Fast Periodic Operation: The Catalytic Dehydrogenation of Isopropanol. **Chem. Eng. Sci.,** 56 (2001), 1419.

[23] R. Knitter, D. Gohring, P. Risthaus, J. Haubelt. Microfabrication of Ceramic Microreactors. **Microsyst. Technol.**, 7 (2001), 85

[24] K. Shah, X. Ouyang, R.S. Besser. Microreaction for Microfuel Processing: Challenges and Prospects. **Chem. Eng. Technol**., 28 (2005), 303.

[25] V .Hessel, B. Cortese, M.H.J.M. de Croon, Novel process windows – Concept, proposition and evaluation methodology, and intensified superheated processing. **Chem. Eng. Sci**., 66 (2011) 1426-1448.

[26] J.J. Brandner in "Micro Process Engineering – A Comprehensive Handbook, Volume 2", V. Hessel, A. Renken, J.C. Schouten, J. Yoshida (Eds.), WILEY-VCH Verlag GmbH & Co. KGaA, Weinheim, 2009. (Chapter 2 pp. 25-43).

[27] G.D. Stefanidis, D.G. Vlachos. High vs. Low Temperature Reforming for Hydrogen Production via Microtechnology. **Chem.Eng. Sci**., 64 (2009), 4856-4865.

[28] M.S. Mettler, G.D. Stefanidis, D.G. Vlachos. Enhancing Stability in Parallel Plate Microreactor Stacks for Syngas Production. **Chem. Eng.Sci.**, 66 (2011), 1051-1059.

[29] M.S. Mettler, G.D. Stefanidis, D.G. Vlachos. Scale-out of Microreactor Stacks for Portable and Distributed Processing: Coupling of Exothermic and Endothermic Processes for Syngas Production. **Ind. Eng. Chem. Res.,** 49 (2010), 10942-10955.

[30] G.D. Stefanidis, D.G. Vlachos. Intensification of Steam Reforming of Natural Gas: Choosing Combustible Fuel and Reforming Catalyst. **Chem. Eng. Sci**., 65 (2010), 398-404.

[31] A. Moreno, K.M. Murphy, B.A. Wilhite. Parametric Study of Solid-Phase Axial Heat Conduction in Thermally Integrated Microchannel Networks. **Ind. Eng. Chem. Res**., 47 (2008), 9040-9054.

[32] T. Steif, O.-U. Langer, K. Schubert. Numerical Investigations of Optimal Heat Conductivity in Micro Heat Exchangers. **Chem. Eng. Technol**. 22 (1999), 297.

[33] C.F. Curtiss, R.B. Bird, Multicomponent Diffusion, **Ind. Eng. Chem. Res.,** 38 (1999) 2515-2522.

[34] C.R. Wilke, A Viscosity Equation for Gas Mixtures, **The Journal of Chemical Physics**, 18 (1950) 517-519.

[35] B. Bird, W.E. Stewart, E. N.Lightfoot, Transport Phenomena, revised Second ed., 2006

[36] R.H. Perry, D.W. Green. Perry's Chemical Engineers' Handbook (7th Ed). McGraw-Hill, (1997).

[37] Vaccaro et al., Modelling of a Catalytic Micro-Reactor Coupling Endothermic Methane Reforming and Combustion, published Online: 10/04/2012, citation Information: International Journal of Chemical Reactor Engineering. Volume 8, Issue 1, Pages –, ISSN (Online) 1542-6580, DOI: 10.2202/1542-6580.2121, March 2010.

[38] R. Jackson, Mass Transport and Reaction in Porous Catalysts, Elsevier, Amsterdam, 1977.

[39] E.A. Mason, A.P. Malinauskas, Transport in Porous Media: The Dusty Gas Model, Elsevier, New York, 1983.

[40] J.W. Veldsink, R.M.J. van Damme, G.F. Versteeg, W.P.M. van Swaaij. The use of the dusty-gas model for the description of mass transport with chemical reaction in porous media. **Chem. Eng. J.,** 57 (1995), 115-125.

[41] R. Suwanwarangkul, E. Croiset, M.W. Fowler, P.L. Douglas, E.Entchev, M.A. Douglas. Performance comparison of Fick's, dusty-gas and Stefan-Maxwell models to predict the concentration overpotential of a SOFC anode. **J. Power. Sources**, 122 (2003), 9-18.

[42] J. Solsvik, H.A. Jakobsen. Multicomponent mass diffusion in porous pellets: effects of flux models on the pellet level and effects on the reactor level. Application to methanol synthesis. **Can. J. Chem. Eng.,** articles ASAP (2012).

[43] J. Xu, G.F. Froment, Methane Steam Reforming, Methanation and Water-Gas-Shift: I. Intrinsic Kinetics. **AIChE J., 35** (1989), 88-96.

# (304f) Kinetic Modeling of Methanol to Olefins (MTO) On SAPO-34 Catalyst

**Guozhen Qi**, Siqing Zhong and Hongyuan Zong, Sinopec, Shanghai, China

Methanol to olefins (MTO) coupled with transformation of coal or natural gas to methanol gives an interesting and promising way to produce ethylene and propylene. The kinetics of methanol conversion to olefins on SAPO-34 catalyst was studied in an isothermal multiple fixed bed reactors in parallel. A reaction kinetic model including five lumps was proposed on the basis of Hydrocarbon Pool parallel reaction and parallel deactivation of catalyst. The influence of water and coke deposition was considered in the model. The simulation in fixed bed reactor using the obtained general kinetic equations was performed and also compared with the experimental value. The result showed a good match between the estimated and experimental values. The average activity of catalyst showed a corresponding evolution with the increasing coke formation. Some water could be effective to reduce the deactivation of catalyst and methanol conversion did not drop instantaneously but instead it remained constant for some time before deactivation breakthrough. An empirical model for the coke formation was obtained and made the coke content as a function of time on stream, temperature, WHSV, water/methanol in feed. The obtained model showed a very good linearity correlation.

A simulation using the bubble fluidized bed reactor model coupled with the obtained kinetic model to predict the performance of MTO was performed. The values calculated by the model were identical to the experimental ones. Besides, the probability application of riser reactor in MTO process was also investigated.

# (304g) Numerical Investigation of Effect of Hollow Cross-Over Disk On Flow and Heat Transfer in Tube of Cracking Furnace

**Yuan Zong**, Chemcial Engineering, East China University of Science and Technology, Shanghai, China, Xiaogang Yang, Glyndŵr University , Wrexham, United Kingdom and Ling Zhao, East China University of Science and Technology, Shanghai, China

This paper presents the influence of main parameters of a novel hollow cross-over disk (HCD) on flow, heat transfer enhancement and flow resistance in tube of ethylene cracking furnace. The location of HCD in the tube, geometric size and shape of HCD are studied. Global analysis of thermal performance and local analysis of vorticity flux, turbulence mixing and friction factor are performed based on computational fluid dynamics (CFD). The numerical results indicated the overall Nusselt number of tube decreased when HCDs' location was far from the inlet and the space between HCD pair decreased. While the total pressure drop of the tube had no significant change with the variance of HCDs' location. The effects of the geometrical size of the HCD on heat transfer enhancement were analyzed from the view point of the field synergy principle and it was found that the increase of heat transfer enhancement was always accompanied with the decrease of field synergy angle between the velocity and temperature gradient when the geometrical parameters of HCD were changed. This confirms that the basic mechanism of heat transfer enhancement by HCD can be well described by the field synergy principle. This present results show that heat transfer enhancement can be achieved without much flow resistance using strategies of geometrical optimization and appropriate placement for HCDs, which is benefit to energy consumption decreasing and coke formation inhabitation for tubular cracking furnace.

# (306a) Evaluation of Process Capability and Robustness for a Lyophilized Product

**Michael M. Choi**, GBPRD, Teva, West Chester, PA

Evaluation of Process Capability and Robustness for a Lyophilized Product

Michael Choi, Daniel Kim, Marazban Sarkari, Ronald Simko, David Unger, Stelios Tsinontides

Teva Global Branded Products Research and Development

**Abstract**

Process capability for the manufacturing of lyophilized commercial batches at a new contract manufacturing organization (CMO) site was evaluated against the current CMO site using statistical process control (SPC) calculations. Key process parameters, material attributes, and quality attributes were examined using univariate and multivariate control charts, Box plots, probability plots, and capability ratios. Within[i] and between[ii] variations were evaluated to arrive at the type of control charts used in the analysis. Outliers were verified using Box and probability plots in conjunction with control charts. Simultaneous control charts were used to assess the impact of the critical process parameters (CPP) on critical quality attributes (CQA). Overlaid control charts were used to compare CMO performances side by side. The analysis was used to determine the adequacy of the controls and monitors at the new site, success of tech transfer, and impact of statistical outliers on critical quality attributes.

---

[i] variation within group of samples

[ii] variation between groups of samples

# (306b) Understanding Thermodynamics of Aqueous Film Coating Process Using Data Logger

Ganeshkumar A. Subramanian, Drug Product Science and Technology, Bristol Myers Squibb, New Brunswick, NJ

Understanding Thermodynamics of Aqueous Film Coating Process Using Data Logger

**Authors: G. Subramanian, P. Pandey, S. Gour, D. Bindra, S. Badawy**

Abstract:

Pharmaceutical film coating process is typically monitored by inlet and exhaust condition of temperature and humidity with measurement devices located outside the coating pan. Current configurations of coating pans do not allow continuous measurement of the temperature and humidity within the tablet bed. Vast differences in heat and mass transfer efficiencies across different coating pan sizes can result in differences in drying and film forming dynamics across various coating pans at the same outlet temperature and humidity. Similarly, different local temperature and humidity conditions (and hence film forming dynamics) can be expected for coating operations carried out at same exhaust temperature, if such temperature is achieved by different means such as changing inlet air volume, air temperature, suspension spray rate or suspension solids concentration. Considerable insight into the dynamics of the coating process can be gained by using data loggers that can be incorporated in the tablet bed and placed such that they can collect temperature and humidity conditions in the tablet bed continuously. A design of experiments (DOE) with coating process performed at various macro-environmental conditions was carried out to determine their effects on the local-environmental thermodynamics in the tablet bed. The data collected from the data loggers enabled correlation between the changes to the macro-environment with changes in the local environment temperature and humidity in the tablet bed. Data from the data logger was used to develop a simple thermodynamic model based on heat and mass balance at different coating conditions across different pan sizes.

# (306c) Process Robustness Aided by Retrospective Multivariate Data Analysis of Factory Data - a Case Study

**Sankar Raghavan**, Technical Operations, Merck & Company, Riverside, PA, Jeffrey Foster, Manuf. Tech. Support, Merck & Company , Riverside, PA, Dale Kopas, American Quality Associates and Ed Warner, Merck & Company

Process robustness aided by retrospective multivariate data analysis of factory data – a case study

Sankar Raghavan, Cherokee Pharmaceuticals L.L.C. a wholly owned subsidiary of Merck & Company, Sankar.Raghavan@merck.com , 570-271-2106

Jeff Foster, Cherokee Pharmaceuticals L.L.C. a wholly owned subsidiary of Merck & Company, Jeffrey.Foster@merck.com, 570-271-2173

Dale Kopas, dakopas@gmail.com, 609-417-1707

Ed Warner, Merck & Company, edwarner4@gmail.com, 908-670-1902

Presenter: Sankar Raghavan

Keywords: Multistep process, Noisy data, Multivariate analysis, Continuous improvement

Purpose: Illustrate the power of combining process knowledge and advanced statistical analysis in addressing intricate process issues

Abstract

Quality by Design is a powerful tool that can be applied to continuously improve performance and robustness of processes. However, with complex processes, it can be effort intensive and costly to do particularly when the process has multiple steps and the scale-up and the interaction effects are not fully understood. Since most pharmaceutical processes are batch, factory data with its normal variability provides a vehicle to generate such data without incurring excessive cost. Such a case study is presented – The example involves a manufacturing process for an antibiotic drug substance that involved four reactions and several down stream purification steps. Multiple factors influenced each of these steps. Identification of key factors that affected the process proved difficult as univariate and bivariate approaches such as SQC/SPC control charts and simple linear regression models explained only a small fraction of the process variability. Multiple interactions added to the complexity, and the presence of a high degree of co-linearity between process variables in the X-matrix lead to problems with matrix inversion in traditional multivariate regression models such as stepwise regression. To correct the situation, data mining was attempted by advanced multivariate methods, specifically Random Forest and Partial Least Squares (PLS) to supplement process knowledge.

The paper describes the analysis of inherently noisy factory data for an entire campaign consisting of over 200 individual batches and more than 100 variables by a combination of advanced statistical methods and physical reasoning based on expert knowledge of the process. The method was effective in explaining the broad factory trends. Key outcomes were:

1. The methodology helped rank factors in terms of their importance. This made it possible to focus on a subset of few factors.
2. Implementation of lessons derived from the studies has resulted in a more robust process with improvements in purity, yield and productivity.

3. The work pointed to areas of opportunity to improve the process.

The method aligns well with the F&DA initiative to continuously improve robustness of processes through the life cycle of a process. The methodology should be broadly applicable for the continuous improvement of complex processes.

# (306d) Application of Weber Number-Based Regime Map for Scaling and Control of Pharmaceutical Coating Sprays

Sip C Liew[1], **Ariel Muliadi**[1] and Paul Sojka[2], (1)Chemical Engineering, Purdue University, West Lafayette, IN, (2)Mechanical Engineering, Purdue University, West Lafayette, IN

A Weber-number based regime map for describing the patternation of twin-fluid pharmaceutical coating sprays is proposed. The map was developed according to the dynamics of the spray evolution at the point where the two shaping air flows meet (the 'junction' point)—depending on the spray drop size, drop velocity, shaping air velocity, and surface tension at the junction point, the shaping air can either redistribute the drops, cause additional (secondary) drop breakup, or pinch the spray. A Gaussian spray results when the spray evolution process is driven mainly by drop redistribution; a homogeneous pattern is observed when there exists balance between the redistribution and secondary breakup effects; a dumbbell-shaped spray results when the pinching effect dominates spray evolution. The We-based regime map was constructed by plotting the junction point We against a dimensionless spray parameter calculated from the product of two flow ratios: the shaping-to-atomizing air flow rates ratio and the liquid supply rate-to-maximum nozzle liquid flow rate capacity ratio. Experimental results show a linear relationship between the junction point We and this dimensionless spray parameter. More importantly, different regions of the We-dimensionless spray parameter plot correspond to the three different spray patternations. Several scale-up scenarios we have explored demonstrate the We-based regime map reliability in achieving homogenous spray patterns. These examples also support the We-based regime map applicability in providing quantitative solutions for the design and optimization of sprays for pharmaceutical coating applications.

# (306e) Characterizing Heterogeneity of Environmental Conditions in Various Bioreactor Scales Used for Cell Cultivation

**Miroslav Soos**, Benjamin Neunstoecklin and Massimo Morbidelli, Institute of Chemical and Bioengineering, ETH, Zurich, Switzerland

Large scale mammalian cell culture up to 20000L is developing more and more to an industrial standard. Process up scaling to those volumes is a fundamental necessity to guarantee process performance, product amount and product quality. Therefore, it is of outmost importance to be aware of scale dependent variables influencing the process. Reactor geometry, shear stress distribution, dissolved O2 and CO2 distribution and gas mass transfer rates can be very different in between scales, resulting in non-desired variations of process performance.

We studied scale dependent parameters (e.g. shear stress, kLa, DO, pCO2, and mixing time) from small to largest scale bioreactors, covering a volume range from 3L up to 15000L, using CFD simulation. At each scale, CFD simulations were validated against relevant scale dependent parameters (e.g. kLa, DO distribution, mixing time) measured experimentally. Comparisons of the computational and experimental values were in good agreement supporting the used approach. Such CFD simulations were consequently used to characterize heterogeneity of the environmental conditions (i.e. shear stress, DO, mixing time etc.) and to understand limitations of actual lab scale bioreactors currently used for process development. We will discuss those limitations and propose more suitable downscale approaches to fit large scale conditions already during process development.

# (321a) Microchannel Rapid Temperature Swing Adsorption of CO2 for Methane Production

**Kriston Brooks**[1], Scot Rassat[2] and Paul Humble[2], (1)Energy and Environmental Directorate, Pacific Northwest National Laboratory, Richland, WA, (2)Pacific Northwest National Laboratory, Richland, WA

A rapid thermal swing adsorption system combined with Sabatier reactor was developed to demonstrate the applicability of a NASA microchannel Mars *In Situ* Propellant Production (ISPP) system. A thermal swing adsorption pump was designed to collect and compress $CO_2$ by rapidly temperature cycling eight cells filled with Molesieve 13X. Recuperation of energy between cells was used to minimize heat and cooling requirements. The collected $CO_2$ can be combined hydrogen from electrolyzed water to produce methane fuel using a microchannel Sabatier reactor.

Tests were performed to compress $CO_2$ by 10 fold using a feed stream of 10% $CO_2$ and 90% $N_2$ at 1, 2, and 4 minute cycle times. With the initial testing, a compression ratio of 5.3 was achieved with 71 to 79% thermal recuperation efficiency. Subsequent modeling work has refined the design and the number of cells required to improve the overall system efficiency and simplify the process. These results indicate similar thermal efficiencies can be achieved using a system with only four cells rather than eight. Modeling results further demonstrate design changes required to obtain improved efficiencies and compression ratios. Results of both the original experimental work and the more recent modeling efforts will be presented.

# (321b) Solubilities of CO2 in Aqueous Solutions of Ionic Liquid and Monoethanolamine

**Fan Xu**[1], Haifeng Dong[2], Xiangping Zhang[2], Hongshuai Gao[3], Zhanli Wang[1], Suojiang Zhang[2] and Baozeng Ren[4], (1)Institute of Process Engineering, Beijing, China, (2)Institute of Process Engineering, Chinese Academy of Sciences, Beijing, China, (3)State Key Laboratory of Biochemical Engineering, Key Laboratory of Green Process and Engineering, Institute of Process Engineering, Chinese Academy of Sciences, Beijing, China, (4)Zhengzhou University, Zhengzhou, China

With the deterioration of climate change, $CO_2$ capture and sequestration have been paid much attention to reduce greenhouse gas emissions. Currently aqueous solutions of alkanolamines or their mixtures are commonly used in many industrial processes to capture $CO_2$. However, drawbacks such as solvent loss and high regeneration costs due to the high water content drive researchers for new superior technologies [1]. Ionic liquids (ILs), due to their wide liquid range, low point, tunable properties, negligible vapor pressure, high $CO_2$ solubility and reasonable thermal stability, are emerging as promising candidates to capture $CO_2$, but it is difficult to realize industrialization owing to its high viscous and high cost. Recently, the idea of mixing ILs and alkanolamines has been received great attention from the industries, since its advantages and capabilities in reducing the problem caused by the usage of conventional alkanolamines solutions are promising [2-4]. In this work, a low viscous IL ([$C_2$OHmim][$N(CN)_2$]) has been synthesized and the solubilities of $CO_2$ in aqueous mixture of IL-MEA were measured by instruments with high accuracy at temperature 313.15K and 333.15K, over $CO_2$ partial pressure ranging from 100 to 1000kPa and IL concentration varying from 5% to 30%. Then the corrected PVT equation has been used to calculate the absorption of $CO_2$. The results indicate that 0.583 mol $CO_2$/mol MEA could be reached under the condition of 40°C and 100.66 kPa when appropriate amount of IL has been added, which only dropped by 1.7% when compared with 30% MEA aqueous at the same condition. The results sufficiently demonstrated that the addition of the IL can reduce regeneration energy consumption to 27% with little affecting their absorption performance.

**References**

[1] Chinn, D.; Vu, D. Q.; Driver, M. S.; Boudreau, L. C. $CO_2$ removal from gas using ionic liquid absorbents. *US Patent*, 7,527,775 B2, 2009

[2] Camper, D.; Bara, J. E.; Gin, D. L.; Noble, R. D. Room-temperature ionic liquid – amine solutions: Tunable solvents for efficient and reversible capture of $CO_2$. *Ind. Eng. Chem. Res*., 2008, 47, 8496-8498.

[3] Zhao, Y.; Zhang, X.; Zeng, S.; Zhou, Q.; Dong, H.; Tian, X.; Zhang, S., Density,

Viscosity, and Performances of Carbon Dioxide Capture in 16 Absorbents of Amine+ Ionic Liquid+ $H_2O$, Ionic Liquid+ $H_2O$, and Amine+ $H_2O$ Systems. *J. Chem. Eng., Data*, 2010, 55, 3513-3519.

[4] Ahmady, A.; Hashim, M. A.; Aroua, M. K., Absorption of carbon dioxide in the aqueous mixtures of methyldiethanolamine with three types of imidazolium-based ionic liquids. *Fluid Phase Equilibria*, 2011, 309, 76-82.

# (321c) A Nove Amino-Functionalized Ionic Liquid for CO2 Capture

**Jinzhi Zhang**, Haifeng Dong and Xiangping Zhang, Institute of Process Engineering, Chinese Academy of Sciences, Beijing, China

Ionic liquids (ILs), especially amino-functionalized ionic liquids, have been found to be effective for $CO_2$ capture because of their unique properties such as negligible vapor pressure, high thermal stability and good absorption capacity[1-2]. As reported [3-5], ILs of more amine groups could afford higher $CO_2$ solubility. To explore deeply into this viewpoint, a new ionic liquid is synthesized and used in lab scale of $CO_2$capture experiments to identify the relationship between the structure characteristics and capture performance in terms of the absorption rate and capacity.

In present work, a novel amino-functionalized ionic liquid was synthesized and characterized by $^1$H NMR, $^{13}$C NMR, IR and MS. The physicochemical properties including glass transition temperature and thermal decomposition temperature were tested. Experiment result shows that this ionic liquid is capable of rapid capture 0.174g $CO_2$ / g IL at 303.15K and atmospheric pressure, which is the theoretical maximum for absorption of $CO_2$ chemically. The interactions of the ionic liquid with $CO_2$ were further investigated by comparison of both the FTIR and $^{13}$C NMR spectra of the gas-untreated and gas-treated materials to support the experimental and the theoretical results.

Moreover, the synthesized ionic liquid could be recycled under vacuum at 100°C.

**References**

[1] J. L. Anthony, E. J. Maginn, J. F. Brennecke, *J. Phys. Chem. B* **2002**, *106*, 7315-7320.

[2] L. A. Blanchard, D. Hancu, E. J. Beckman, J. F. Brennecke, *Nature* **1999**, *399*, 28-29.

[3] Z. Xue, Z. Zhang, J. Han, Y. Chen, T. Mu, *International Journal of Greenhouse Gas Control* **2011**, *5*, 628-633.

[4] J. M. Zhang, S. J. Zhang, K. Dong, Y. Q. Zhang, Y. Q. Shen, X. M. Lv, *Chem.-Eur. J.* **2006**, *12*, 4021-4026.

[5] Y. Zhang, S. Zhang, X. Lu, Q. Zhou, W. Fan, X. Zhang, *Chemistry - A European Journal* **2009**, *15*, 3003-3011.

# (321d) Assessing Anhydrous Tertiary Alkanolamines for Gas Purifications

**David J. Heldebrant**[1], Paul M. Mathias[2], Kash Afshar[2], Satish Reddy[2], Arnold R. Smith[3], Rubin J. McDougal[4], Louis V. Jasperson[4], Phillip K. Koech[5], Mark Bearden[6] and Charles J. Freeman[1], (1)PNNL, Richland, WA, (2)Fluor Corporation, Aliso Viejo, CA, (3)Fluor Corporation, Aliso Viejo, WA, (4)Wiltec Research Company, Provo, UT, (5)Pacific Northwest National Laboratory, Richland, WA, (6)Energy & Environmental, Pacific Northwest National Lab, Richalnd, WA

**Assessing Anhydrous Tertiary Alkanolamines for Gas Purifications**

Paul M. Mathias, Kash Afshar, Satish Reddy, Arnold R. Smith, Rubin J. McDougal
Louis V. Jasperson, Phillip K. Koech, Mark D. Bearden, Charles J. Freeman, David J. Heldebrant

Anhydrous tertiary alkanolamines chemically react with $CO_2$ and $H_2S$, with greater selectivity for the latter. This is in direct contrast to aqueous amine-based solvent systems with higher selectivity for $CO_2$ due to stronger chemical fixation than for $H_2S$. Anhydrous tertiary alkanolamines exhibit pressure-induced chemical fixation of $CO_2$ to form zwitterionic ammonium alkylcarbonate ionic liquids, while the same tertiary alkanolamines react with $H_2S$ at atmospheric pressures to form hydrosulfide ionic liquids. This difference in capture pressure implies that anhydrous alkanolamines could be chemically selective for $H_2S$ over $CO_2$ under specific conditions. We present here the first vapor-liquid equilibrium (VLE) data of anhydrous ethyldiethanolamine (EDEA) with $CH_4$, $C_3H_8$, $H_2S$ and $CO_2$ at 10 and 50 °C measured by the PTx and PTxy methods. The data are modeled in Aspen Plus® using an electrolyte NRTL model. Data trends and phenomena are discussed for each gas. We also compare anhydrous EDEA's simulated performance for $CO_2$ and $H_2S$ separations to other solvents such as Fluor Solvent, Econosolv, and aqueous methyldiethanolamine (MDEA) for representative gas-processing applications.

# (321e) Development of a Novel Carbonate Absorption Process with Crystallization-Enabled High Pressure Stripping for Post-Combustion CO2 Capture: Kinetic Study of Bicarbonate Salt Crystallization

**Qing Ye**, University of Illinois at Urbana-Champaign, Champaign, IL

Development of a Novel Carbonate Absorption Process with Crystallization-Enabled High Pressure Stripping for Post-Combustion $CO_2$ Capture: Kinetic Study of Bicarbonate Salt Crystallization

Qing Ye[1,2], Manoranjan Sahu[2], Yongqi Lu[2] and Xinlei Wang[1]

1 Agricultural and Biological Engineering, University of Illinois at Urbana-Champaign

2 Illinois State Geological Survey, University of Illinois at Urbana-Champaign

A novel absorption process with high pressure stripping is being developed at our group for the post-combustion $CO_2$ capture application. The process employs a carbonate salt aqueous solution, such as potassium carbonate (PC), as a solvent. The $CO_2$ absorption operates at elevated temperature and the bicarbonate in the $CO_2$ rich solution from the absorber is crystalized in a separate cooling-crystallizer. Slurry of bicarbonate crystals is used in the stripper for $CO_2$ stripping at high pressure. Compared to conventional mono-ethanol-amine (MEA) processes, the carbonate-based process consumes less stripping heat and $CO_2$ compression work. A preliminary techno-economic analysis reveals that the energy use of the novel process is 40-50% less than that of its MEA counterpart.

Crystallization of the bicarbonate from the $CO_2$-rich PC solution is a key step in the carbonate-based process. Kinetic studies of the bicarbonate crystallization were performed under continuous mix suspension, mixed product removal (MSMPR) mode using an automated calorimetric reactor system. Parametric tests were conducted to investigate the effect of the PC concentration, $CO_2$ loading, addition of absorption promoter, temperature, residence time, stirring speed, seeding, etc. on the kinetics of the crystallization and morphology of crystal products. Crystal solids were characterized by particle size distribution (PSD), turbidity, SEM, and XRD analysis techniques. Kinetic data, such as the nucleation rate constant, size-dependent growth rate constant, and agglomeration/breakage factor, were obtained by fitting the differential PSD data under various operating conditions.

This presentation will provide a summary of the results from the crystallization kinetic study and discuss their implications for the crystallizer design.

**Doped lithium orthosilicate for absorption of $CO_2$ at steam atmosphere with low $CO_2$ partial pressure**

**Abstract:** Calcium and lithium containing oxides are regarded as options for the capture of $CO_2$ in sorption-enhanced steam reforming of methane (SEMSR) system. Among the latter, lithium orthosilicate (Li4SiO4) is considered as one of the most potential materials due to its large capacity, high selectivity and good cycle performance. Improved properties can be obtained by appropriately doping different elements into $Li_4SiO_4$. In this study, the best amount of doping elements (K, Fe) into $Li_4SiO_4$ were calculated by Factsage 5.5. Therefore, different lithium orthosilicate based absorbents, potassium doped and iron doped lithium orthosilicate according to the calculated results from Factsage 5.5 were prepared and characterized by XRD, SEM. For the study of practical applicability in SESMR, the absorption ability was investigated by TGA under low $CO_2$ partial pressure (5%~20%) influenced by different steam atmosphere (5%~30%). In addition, the cyclic absorption/desorption performance of different absorbents were also investigated. The results show that potassium and iron can enhance the absorption capacity at low $CO_2$ partial pressure and cyclic ability respectively and the best absorbent with suitable amount of potassium and iron was acquired.

**Key words: $CO_2$ acceptors; Lithium orthosilicate; doping method**

## 1 Introduction

$CO_2$ acceptors are widely used in industrial process such as sorption enhanced methane steam reforming, flue gas decarburization, energy storage system and etc. Thus, it is important to obtain a solid $CO_2$ acceptor which has high $CO_2$ absorption capacity at moderate temperature with good selectivity for $CO_2$ and good regeneration ability.

Thus far, several materials, such as hydrotalcite-like materials, CaO-based sorbents, and lithium-containing materials[1], have been proposed for $CO_2$ capture. Among these materials, lithium orthosilicate ($Li_4SiO_4$) is considered to be one of the most potential materials. $Li_4SiO_4$ has been reported to adsorb $CO_2$ more than 30 times faster than $Li_2ZrO_3$ and is lighter and cheaper than $Li_2ZrO_3$.

$$Li_4SiO_4 + CO_2(g) = Li_2CO_3 + Li_2SiO_3$$

$Li_4SiO_4$ has high $CO_2$ absorption capacity. However, the temperature in the real SESRM process is between 500□~600□. Moreover, the concentration of $CO_2$ is as low as about 10%. Under such condition, the $CO_2$ absorption ability of $Li_4SiO_4$ will decrease dramatically. In order to improve the defect, exotic elements can be doped into $Li_4SiO_4$. C. Gauer [2]etc. added elements such as Al, Fe and K into $Li_4SiO_4$, concluding that for the practical use of $Li_4SiO_4$, doping vacancies through doping Fe can not only increase the sorption ability but also the cycle performance. M. Seggiani etc[3]. doped K element using simple addition method and eutectic method into Li4SiO4 and compared the two methods drawing the conclusion that simple addition method can well promote the sorption ability while has a bad cycle performance. Mayra Y. Veliz-Enriquez etc[4]. prepared $Li_{2-x}K_xZrO_3$ with different quantities of $K_2CO_3$ and then concluded that the best x is 0.2 using solid solution theory and experiment.

In the real SEMSR process, not only can the temperature reach the best absorption temperature(700□) and can the concentration of $CO_2$ be high enough for sorption, but also the steam exist to influence the sorption ability of $Li_4SiO_4$. Hughes, Manovic, Reijers, Sun, Yan, Zeman etc[5-10]. investigated the effect of steam on absorption of CaO for $CO_2$ holding the opinion that steam can facilitate the sorption ability of CaO for $CO_2$. Esther Ochoa-Fernández[11] studied the effect of steam on pure $Li_4SiO_4$ for $CO_2$ absorption but the reason was not indicated.

In this article, $Li_4SiO_4$ with different quantities of K were prepared through solid synthesis using eutectic method. The best amount of $K_2CO_3$ was calculated by Factsage software and the results were proved by characterization method such as XRD, SEM and TG. The sorption ability of different samples for $CO_2$ was investigated under the same condition as in the real SEMSR process. Finally, the effect of sorption condition on sorption ability such as temperature, pressure of $CO_2$ and the amount of steam were analyzed.

## 2 Experimental part

### 2.1 Characterization of samples

Solid-state synthesis of analytical grade lithium carbonate and highly dispersed silicon dioxide ($SiO_2$) was performed by adjusting a molar ratio of 2.05:1 to obtain $Li_4SiO_4$ powder. After homogenization in a vibrating disc mill for a minute the initial mixture was calcined in air at 750□ for 6h.

For simple addition doping, the molar ratio of starting materials was adjusted to $Li_2CO_3:SiO_2:K_2CO_3=(2.05+A):1:(2A)$.

### 2.2 Characterization of samples

The sample was identified by X-ray diffraction (XRD). A diffractometer (RIGAKU D/MAX 2550 VB/PC, Japan) coupled to a copper-anode X-ray tube was used. The K$\alpha$ wavelength (1.54056nm) was selected with a diffracted beam monochromator. Li4SiO4 was identified by the corresponding Joint Committee on Powder Diffraction Standards (JCPDS) in virtue of MDI Jade 5.0 software. The surface structure and particle size of the sample was characterized by scanning electron microscopy (SEM) (JEOL JSM-6360LV, Japan).

A WRT-3P scale was used to analyze the temperature ranges and kinetic behaviors of the CO2 sorption process.

## 3 results and discussions

### 3.1 Theoretic effects of doped K element on $Li_4SiO_4$

Doping substance which is able to form eutectic compound with $Li_2CO_3$ can improve the $CO_2$ sorption ability of $Li_4SiO_4$. There are two common alkali metal salt including $K_2CO_3$ and $Na_2CO_3$ that can form eutectic compound with $Li_2CO3$. Fig.1 and Fig.2 show the binary phase diagram of $Li_2CO_3/K_2CO_3$ and $Li_2CO_3/Na_2CO_3$. When the amount of $K_2CO_3$ is 0.4, the melting point of the eutectic compound is 500□. While the amount of $Na_2CO_3$ is up to 0.5, the melting point is 500□. $K_2CO_3$ and $Na_2CO_3$ can both decrease the melting point dramatically but $K_2CO_3$ is better for its smaller amount needed.

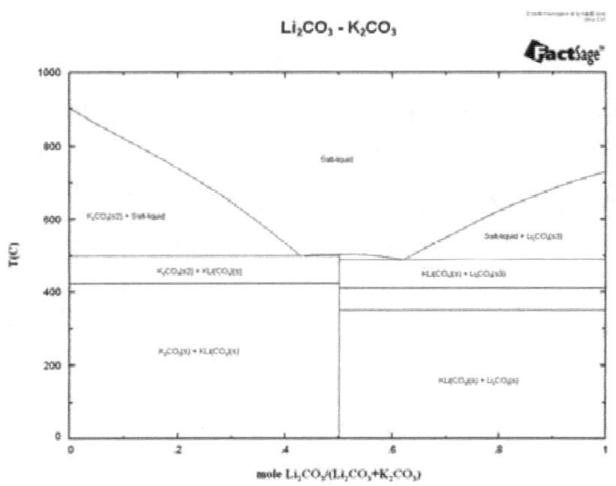


Fig.1 $Li_2CO_3$-$K_2CO_3$ binary phase diagram

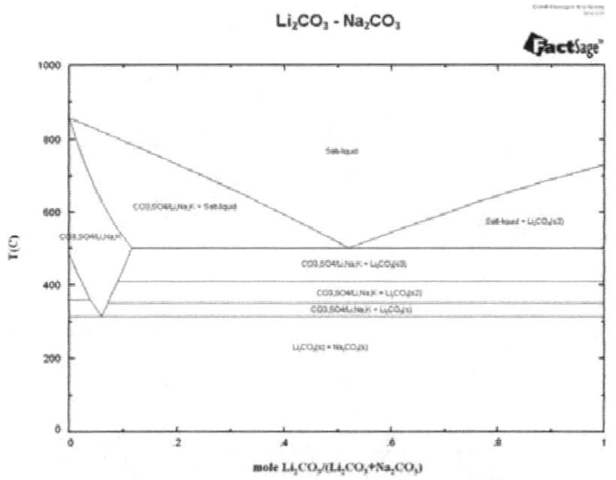


Fig.2 $Na_2CO_3$-$K_2CO_3$ binary phase diagram

Fig.3 shows the calculated equilibrium composition using Factsage software of starting materials with different amount of $K_2CO_3$

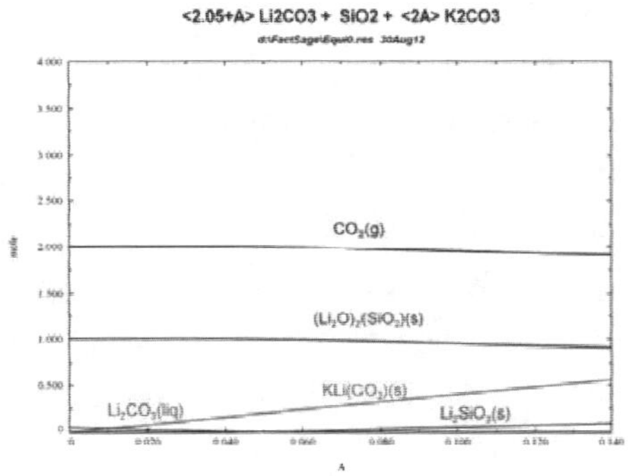


Fig.3 Calculated equilibrium composition of $Li_2CO_3$+$SiO_2$+$K_2CO_3$ calcined at 750□ for 6h

The value of A in the experiment of C. GAUER is 0.1. With the increase of A, which means the increase of amount of $K_2CO_3$, $Li_2CO_3$ that remains in the product will disappear gradually. While A increases continually, $Li_2SiO_3$ appears as a by-product, which has no promotion for the sorption of $CO_2$. A new substance, $KLi(CO_3)$, which is the eutectic compound formed by $K_2CO_3$ and $Li_2CO_3$, can be found with the increase of $K_2CO_3$. The melting point decrease more as the amount of $K_2CO_3$ increase thus the sorption ability for $CO_2$ is improved more. In summary, the value of A should be maintained between 0.04~0.06 so the absorbent can have more $KLi(CO_3)$

while less impurity.

**3.2 influence of amount of $K_2CO_3$ to sorption ability of K-$Li_4SiO_4$**

Table.1 shows the value of A in different samples. The molar ratio of starting materials is as follows: $Li_2CO_3$:$SiO_2$:$K_2CO_3$=(2.05+A):1:(2A)

Table.1 value of A in different samples

| sample | 1 | 2 | 3 | 4 | 5 | 6 |
|---|---|---|---|---|---|---|
| A | 0 | 0.06 | 0.08 | 0.10 | 0.12 | 0.14 |

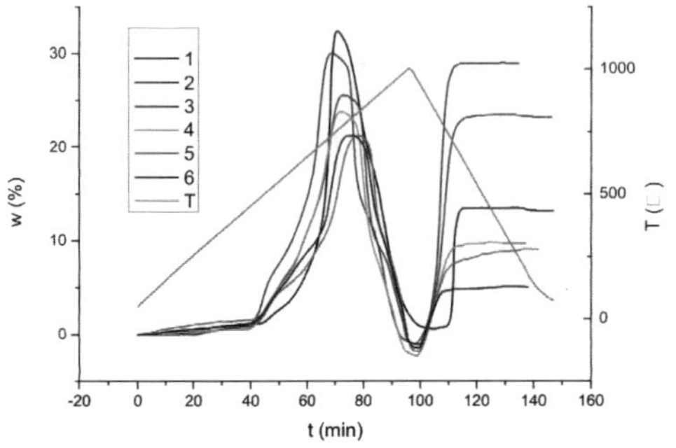


Fig.4 thermo analytical curve of different samples

The thermo analytical curve in Figure 4 was obtained by heating the samples in the pure $CO_2$ atmosphere with a heating rate of 10□/min from 0°C to 1000℃ and then with a cooling rate of 20℃/min from 1000℃ to room temperature. The samples which have $K_2CO_3$(No.2~No.6) has a clearly improvement of $CO_2$ sorption ability at 50min when the temperature is about 500□ compared with pure $Li_4SiO_4$(No.1), which indicates that the eutectic compound can indeed improve the $CO_2$ sorption ability. No.2 sample has the most significant increase at 500□. Nevertheless, the sorption ability decreases as the amount of $K_2CO_3$ increases. This can be attributed to the increase of the $Li_2SiO_3$, which has no positive effect for $CO_2$ sorption. The results are well consistent with the results calculated by Factsage before.

During the temperature decrease process, all samples absorb $CO_2$ for the second time. This process is rather short for the cooling rate is as twice as much as the heating rate. Even though, No.2 sample can have a weight gain of 30%, which is obviously better than other samples, within such a short time.

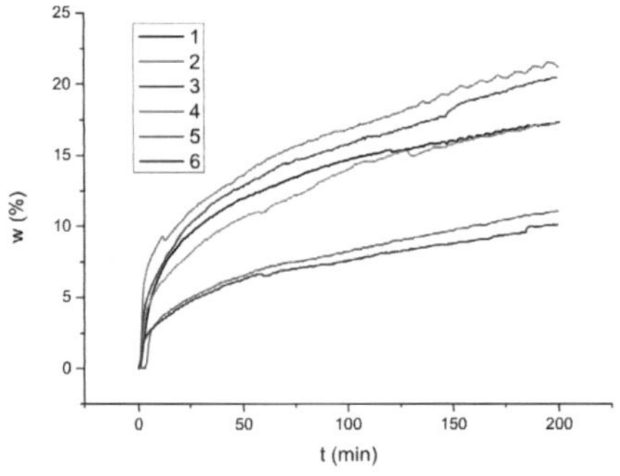


Fig.5 Isothermal curves of different samples at 575□ and 100% $CO_2$ atmosphere

In the real SESRM process, the low the temperature is, the less energy the reaction consumes. Fig. 5 shows the isothermal analysis curves of different samples at 575□ and 100% $CO_2$ atmosphere. No.2 and No.3 samples have better sorption performance than No.1 sample and No.2 has a bigger increase. Compare No.2 sample with No.1 sample, the rate of rapid reaction stage increases. This can be attributed to the formation of eutectic compound, which restrains the formation of the double shell because liquid phase eutectic compound can facilitate the diffusion of $Li^+$ and $O^{2-}$ and thus higher conversion is needed to form the solid double shell. With the increase of $K_2CO_3$, the sorption ability, on the contrary, decreases. The reason is similar with the reason for the results of the thermal analytical curves.

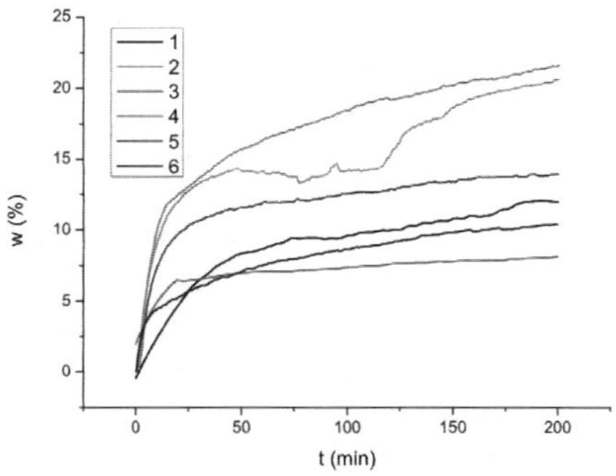


Fig.6 Isothermal curves of different samples at 575□ and 10% $CO_2$ atmosphere

Fig.6 shows the isothermal curves of different samples at 575□ and 10% $CO_2$ atmosphere. The rate of the rapid reaction stage of No.1 sample is significantly low than any other samples because of the low $CO_2$ concentration. While for the doped with K samples, No.2~No.6 samples, the rate of the rapid reaction stage increases apparently. This can be attributed to the formation of eutectic compound which can increase the amount of $Li^+$ and $O^{2-}$ on the surface and thus facilitating the following reaction.

$$2Li^+ + O^{2-} + CO_2 \underset{k_4}{\overset{k_3}{\rightleftharpoons}} Li_2CO_3 \qquad (3)$$

For No.2 sample, the weight gain can easily reach to 13% within 20min and to 22% within 200min, which is clearly higher than that of No.1 sample. As the amount of K increases, the sorption ability decreases. The reason is also the same as the reason for the results of thermal analytical curves.

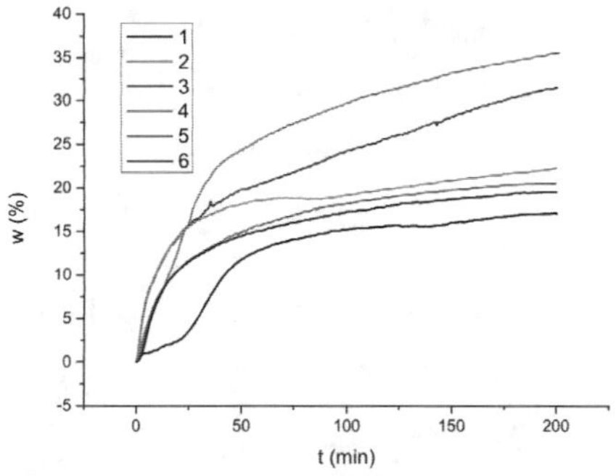


Fig.7 Isothermal curves of different samples at 575□, 10% $CO_2$ and 20% steam atmosphere

Fig.7 shows the isothermal curves of different samples at 575□, 10% $CO_2$ and 20% steam atmosphere. For No.1 sample, the weight gain within 200min increases from 13% to 15%, while fro No.2 sample, the weight gain increases from 22% to 35%. For the other samples, the sorption capacity also increases more than No.1 sample.

### 3.3 Characterization of samples

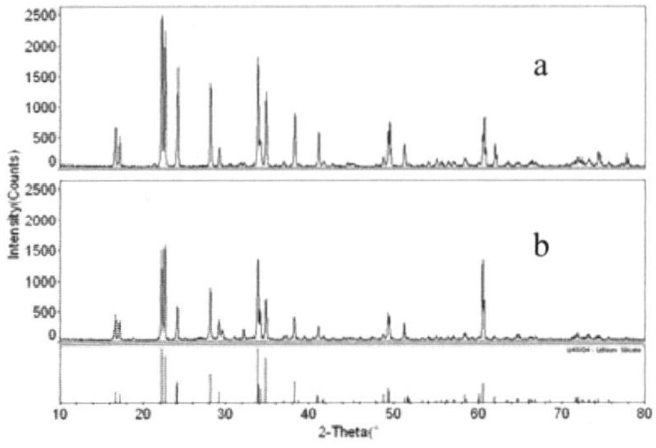


a-pure $Li_4SiO_4$; b-No.2 sample

Fig.8 XRD patterns of different samples

Fig.8 shows the XRD patterns of pure Li4SiO4 and No.2 sample. They were all fitted to JCPDS file 37-1472, and only the highly crystalline structure of $Li_4SiO_4$ was presented. This can be explained by the solid solution theory. Small quantities of $K_2CO_3$ can be dissolved into the $Li_4SiO_4$ solvent as a solid solute, so no substances including K can be found. This also explains why the sorption ability decreases while the amount of K increases.

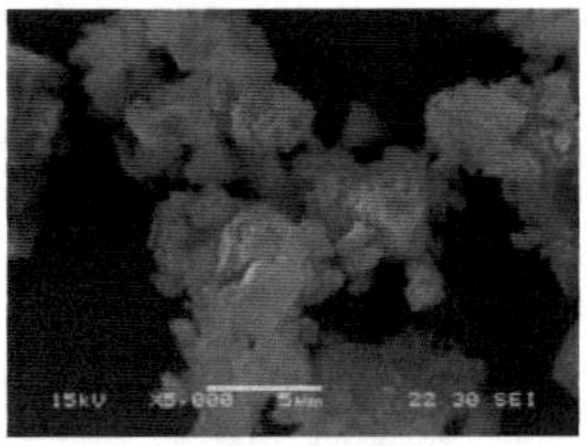

a-pure $Li_4SiO_4$

b-No.2 sample

Fig.9 SEM images of different samples

Fig.9 shows the SEM images of different samples. Pure $Li_4SiO_4$ and No.2 sample both have smooth surface and are fin powders. Pure $Li_4SiO_4$ has a smaller diameter of 10μm, while No.2 sample has an agglomeration phenomenon with a diameter of 30μm and has small particle which is $Li_4SiO_4$ with K on its surface. Thus, even though No.2 sample has bigger particle which is not good for chemical absorption, the sorption ability is much better than that of pure $Li_4SiO_4$. The results show that the formation of eutectic substances can improve the sorption ability significantly.

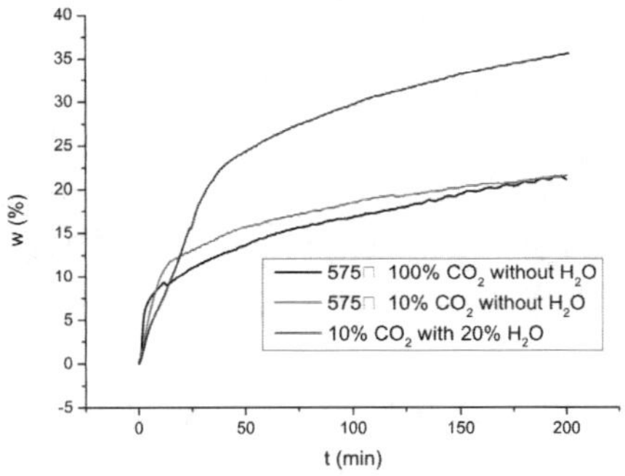


Fig.10 Isothermal curves of No.2 sample under different conditions

Fig.10 shows the Isothermal curves of No.2 sample under different conditions. For No.2 sample, the sorption ability maintains while the concentration of $CO_2$ decreases. While the water steam occurs, the sorption ability even improves with a weight gain of 20% with 30min. The improvement mainly displays in the extension of rapid reaction stage. When there is no steam, the rapid reaction stage ends when the weight gain is about 13% while 23% when there is steam. This can be attributed to the inhibition of the formation of double shell and the fact that steam can facilitate the mobility of $Li^+$ and $O^{2-}$.

## 4 Conclusion

(1) Doping K element into $Li_4SiO_4$ can form eutectic substance with the $Li_2CO_3$ produced by $Li_4SiO_4$ after $CO_2$ sorption thus facilitating the $CO_2$ sorption ability of $Li_4SiO_4$.

(2)It was comfirmed by calculation using Factsage and experiments results that the optimum doping ratio was $Li_2CO_3$:$SiO_2$:$K_2CO_3$=2.11:1:0.12. Then the rapid reaction stage of the adsorbent will be extended, and the maximum absorption capacity will be improved obviously.

(3)Under the same condition as in real SESRM process, the ability of $CO_2$ acceptors obtained maintains when the concentration of $CO_2$ decreases and the sorption ability even improves when steam occurs. The weight gain is more than 20% within 30min at 575℃, 10%$CO_2$ and 20% steam atmosphere.

## 5 References


[1] Li Li, Yuan Wenhui, Wei Chaohai. Adsorbents and adsorptive process for carbon dioxide at high temperature[J]. Chemical Industry and Engineering Progress, 2006, 25(8): 918-922

[2] Gauer C, Heschel W. Doped lithium orthosilicate for absorption of carbon dioxide[J]. Journal of Materials Science, 2006,41(8): 2405-2409

[3] M. Seggiani, M. Puccini, S. Vitolo. High-temperature and low concentration $CO_2$ sorption on $Li_4SiO_4$ based
sorbents: Study of the used silica and doping method effects[J]. International Journal of Greenhouse Gas Control 5 (2011): 741–748

[4] Mayra Y, Veliz-Enriquez, Gonzalo Gonzalez, Heriberto Pfeiffer. Synthesis and $CO_2$ capture evaluation of $Li_{2-x}K_xZrO_3$ solid solutions and crystal structure of a new lithium-potassium zirconate phase[J]. Journal of Solid State Chemistry,2007, 180: 2485–2492

[5] Robin W. Hughes, Dennis Lu, Edward J. Anthony, and Yinghai Wu. Improved long-term conversion of limestone-derived sorbents for in situ capture of $CO_2$ in a fuidized bed combustor[J]. Ind. Eng. Chem. Res, 2004, 43, 5529-5539

[6] Vasilije Manovic, Dennis Lu, Edward J. Anthony. Steam hydration of sorbents from a dual fluidized bed $CO_2$ looping cycle reactor[J]. Fuel, 87 (2008) 3344-3352

[7] Hendricus Th. J. Reijers, Sacha E. A. Valster-Schiermeier, Paul D. Cobden, and Ruud W. van den Brink. Hydrotalcite as $CO_2$ sorbent for sorption-enhanced steam reforming of methane[J]. Ind. Eng. Chem. Res. 2006, 45, 2522-2530

[8] Ping Sun, John R. Grace, C. Jim Lim, and Edward J. Anthony. Investigation of attempts to improve cyclic $CO_2$ capture by sorbent hydration and modification[J]. Ind. Eng. Chem. Res. 2008, 47, 2024-2032

[9] Aiyu Yan, Bin Liu, Yonglai Dong, Zhijian Tian, Dezheng Wang, Mojie Cheng. A temperature programmed desorption investigation on the interaction of $Ba_{0.5}Sr_{0.5}Co_{0.8}Fe_{0.2}O_{3\_\delta}$ perovskite oxides with CO2 in the absence and presence of $H_2O$ and $O_2$[J]. Applied Catalysis B: Environmental 80 (2008) 24–31

[10] Frank Zeman. Effect of steam hydration on performance of limesorbent for CO2 capture[J]. International Journal of Greenhoue Gas control, 2, ( 2008 ) 20 –209

[11] Esther Ochoa-Fernández, Tiejun Zhao, Ph.D., Magnus Rønning, De Chen. Effects of steam addition on the properties of high temperature ceramic $CO_2$ acceptors[J]. Journal of Environmental Engineering, 2009, 6: 397-403

# (325a) Data Integration in Pilot Plants for Goal Optimization

**Hsiang Yee Lai**, Chemistry and Chemical Engineering Division, Southwest Research Institute, San Antonio, TX

Southwest Research Institute has years of experience building, commissioning, and operating pilot and demonstration plants to support the fuels and energy industry.

Testing results at pilot plant scale provide insight for design and operation of larger scale units. Pilot plant designs vary from single to multiple unit operations. Specifically, Southwest Research Institute has operated multiple unit operations such as hydrotreaters, distillation columns, and fixed bed and circulating fluidized bed reactors.

The results and data interpretation of pilot plants are vital to optimization of new and existing processes, demonstration, marketing, and economics. Southwest Research Institute has developed a method to minimize the time required to integrate data across multiple platforms, bringing together analytical and operating data. In turn, process development goals are quickly achieved and project schedules are shortened.

This presentation will discuss numerous unit operations and how using data integration from multiple sources can optimize process goals in pilot plants.

# (325b) Pilot-Scale Testing of an in Tank Process to Remove Radionuclides From Waste Water

**Michael Poirier**, Savannah River National Laboratory, Aiken, SC

Pilot-Scale Testing of an in Tank Process to Remove Radionuclides From Waste Water

The Savannah river Site is developing a new waste treatment process to remove radionuclides from radioactive waste. The process adds monosodium titanate to a waste tank and mixes the tank with rotating mixer pumps. The monosodium titanate sorbs strontium and select actinides. The waste is filtered to remove the monosodium titanate and entrained sludge. The filtrate is treated with an ion exchange process to remove cesium. The decontaminated liquid is grouted. The concentrated radionuclides are eventually vitrified.

The authors conducted pilot-scale testing of the sorption and solids removal parts of the process. The pilot-scale vessel was constructed of acrylic and was nominally eight feet in diameter. The vessel contained two to four rotating mixer pumps. Simulated sludge, monosodium totanate, and crystalline silico titanate were added to the vessel to simulate expected waste tank conditions. The testing determine the pump requirements to mobilize the sludge, monosodium titanate, and crystalline silicotitanate, as well as the contact time needed for the monosodium titanate to sorb the soluble strontium.

The presentation will discuss the testing, the scaling of the test equipment, the test results, and the scaling of the test results.

# (325c) Benchmarking of a New Absorption Mini-Plant

**Chinmay Kale**[1], Philip Lutze[2] and Andrzej Górak[2,3], (1)Laboratory of Fluid Separation, Department of Biochemical and Chemical Engineering, TU Dortmund University, Dortmund, Germany, (2)Department of Biochemical and Chemical Engineering, Laboratory of Fluid Separations, TU Dortmund University, D-44227 Dortmund, Germany, (3)TU Dortmund University, Department of Biochemical and Chemical Engineering, Laboratory of Fluid Separations, D-44227 Dortmund, Germany

**Benchmarking of a new absorption mini-plant**

Chinmay Kale, Philip Lutze, Andrzej Górak

TU Dortmund University, Department of Biochemical and Chemical Engineering, Laboratory of Fluid Separations, Emil-Figge-Strasse 70, D-44227 Dortmund, Germany; phone: +49 (0) 231 755-3034, e-mail: chinmay.kale@bci.tu-dortmund.de

**Keywords:**Mass transfer kinetics, hydraulics, desorption, reactive absorption

**Suggested theme:** Process Development Division: Pilot Plants - Best Practices In Pilot Plants (12b01)

Flue gases from coal fired power plants and oil refineries are major contributors to $CO_2$ emissions and global warming. Post combustion $CO_2$ capture by reactive absorption using aqueous amines is seen as one of the most attractive solutions to reduce $CO_2$ emissions. Reaction kinetics and thermodynamics of this process has been extensively studied over the last decade. Before this technology can be successfully applied on industrial scale, a detailed analysis of mass transfer kinetics is necessary. In this regard, the behaviour of the process in columns of different dimensions is of specific interest.

For this purpose, a new absorption mini-plant has been installed at Laboratory of Fluid Separations, TU Dortmund university, Germany. This plant has three absorption columns having internal diameters DN25, DN50 and DN100 and variable heights from 1 m to 4 m. It also comprises of an upstream gas humidifier and a downstream batch distillation column for absorption solvent regeneration and uses common auxiliary equipment for the operation e.g. pumps, flow meters etc. The plant is equipped with a sophisticated control system which allows pressure drop, temperature and concentration (gas and liquid) measurements over the height of the columns. Different diameters, packings and column heights allow user to study the influence of column internals and hydrodynamics on absorption / desorption processes. The experimental results can be used to validate the rigorous absorption model.

Benchmarking of this mini-plant was carried out in three steps – a) hydraulic measurements to identify the operating range, to measure pressure drop and liquid hold-up, b) measurement of liquid side NTU and HTU by carrying out desorption experiments of $CO_2$ from $H_2O$ into air and c) reactive absorption of $CO_2$ using monoethanolamine (MEA). Results of hydraulic and desorption experiments were compared with the results from open literature. Results of the reactive absorption were compared with results from other pilot absorption columns for the same gas as well as liquid loading and same compositions of the input streams. A generic non-equilibrium stage model was validated with the experimental results of reactive absorption $CO_2$.

After comparison of the model and experimental results without fitting any transfer parameters, it was found that the absorption mini-plant produced predictive experimental results with acceptable accuracy. Therefore, the mini-plant will be further used for detail mass transfer kinetics and scale-up related studies.

# (325d) Development of a Short-Cut Model for Three-Phase Liquid Separation Dynamics for a Hydroformylation Mini-Plant

David Müller, Erik Esche, Michael Müller and Günter Wozny, Chair of Process Dynamics and Operation, Berlin Institute of Technology, Berlin, Germany

**Development of a Short-Cut Model for Three-phase Liquid Separation Dynamics for a Hydroformylation Mini-Plant**

David Müller*, Erik Esche**, Michael Müller***, Günter Wozny****

Chair of Process Dynamics and Operation, Berlin Institute of Technology, Sekr. KWT-9, Str. des 17. Juni 135, D-10623 Berlin, Germany

Corresponding author: *david.mueller@tu-berlin.de,

**erik.esche@tu-berlin.de, ***m.mueller@tu-berlin.de,****guenter.wozny@tu-berlin.de

Within the framework of the Collaborative Research Centre SFB/TR 63 InPROMPT, "Integrated chemical processes in liquid multiphase systems", a novel process concept for the hydroformylation of long chain alkenes in micro emulsions is investigated and developed at Berlin Institute of Technology (Technische Universität Berlin), Germany. In industry, hydroformylation is an important application in the field of homogenous catalysis and has been established as a standard process for the production of short-chained aldehydes from alkenes. The application for higher alkenes (longer than $C_{12}$) on the other hand, has not yet been established.

The investigated novel process concept sees opportunities for the continuous hydroformylation of long-chained alkenes in the creation of a micro emulsion system. Through the application of a surfactant this micro emulsion system can be formed and the hydrophilic rhodium-ligand-complex catalyst required for the reaction can be brought into contact with the alkene. The reaction itself is initiated by injecting syngas ($H_2$&CO) into the micro emulsion system formed in a continuously stirred tank reactor. After the reaction, due to the phase separation into an aqueous (catalyst-rich), an emulsion (mixed), and an organic (product-rich) phase, the valuable rhodium catalyst is separated and recycled. To investigate and optimize the described process concept, a mini-plant is built at Berlin Institute of Technology.

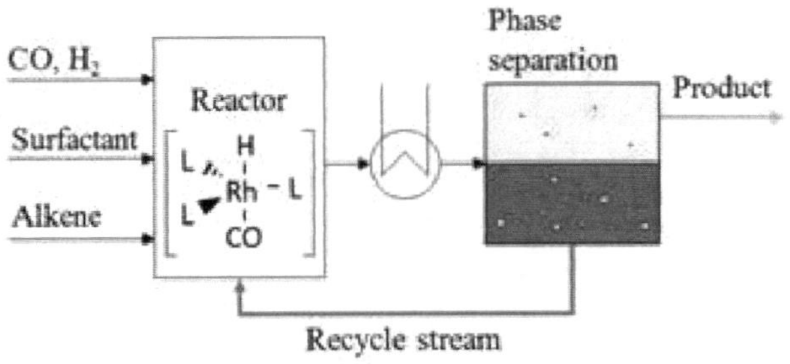


Figure 1: Novel process concept for the hydroformylation of long-chained alkdehydes.

The crucial aspects of the concept with regards to technical and economic feasibility are the separation steps to recycle the rare and expensive catalyst. Since there is barely any thermodynamic data on the system available, the temperature and concentration sensitity proves to be a challenging issue. The goal of this contribution is to develop a model of the three-phase liquid separation dynamics for the temperature control of the decanter. Thus, the quality of the phase separation can be adjusted accordingly.

The solution approach is divided into several steps. Firstly, experiments in graduated test tubes for various concentrations and temperatures are performed and the height of each of the phases is measured over time. The idea is that the optimal separation time at a certain temperature indirectly represents the necessary length in a horizontal flow decanter, which can be adjusted by manipulating the superficial velocity of the inlet stream. In the second step, the previously obtained results are used to derive a polynomial function to estimate the heights of each of the three phases $h_j$: $h_j = f(t, T, c_i)$. These are dependent on time t, the temperature level T, and the concentrations of alkenes, aldehydes, surfactant, and catalyst. Furthermore, the switch from 3-phase to 2-phase systems is included through the intersection of two polynomials describing the phase interfaces. Thirdly, an experimental set-up with a glass replication of the actual decanter in the mini-plant is used to test the validity of the determined results from the test tubes for the decanter.

As a next step, adjustments to the derived equations are made to incorporate the pressure dependency of the system. Afterwards, the results are implemented in the mini-plant allowing for an effective temperature control. Hence, through the determined model the optimal operation conditions can be maintained in the mini-plant and a step towards economic feasibility is taken.

**Acknowledgment**

This work is part of the Collaborative Research Centre "Integrated Chemical Processes in Liquid Multiphase Systems" coordinated by the Technische Universität Berlin. Financial support by the German Research Foundation (Deutsche Forschungsgemeinschaft, DFG) is gratefully acknowledged (TRR 63).

# (325e) Detection and Diagnosis of Blockage in Externally Parallelized Microreactors

**Osamu Tonomura**, Kyoto university, Kyoto, Japan and Shinji Hasebe, Kyoto University, Kyoto, Japan

Microreactors engage the attention of researchers and engineers in the pharmaceutical and chemical industries as well as universities. When the production capacity of micro chemical plants is increased by numbering-up approach, it is important to realize the uniform flow distribution among the parallelized microreactors. In addition, a blocked microreactor needs to be identified as early as possible to achieve the stable long-term operation of micro chemical plants. However, it is not practical to install the sensors in all the microreactors from the viewpoint of cost or space. In this research, a system that can detect and diagnose a blocked microreactor by using just two flow sensors is developed. The effectiveness of the developed system is demonstrated by numerical and experimental investigations, and it is clarified that the developed method has high robustness to changes in the fluid properties and the microreactor characteristics such as pressure drop. The developed system will be applicable to various types of micro chemical plants with parallelized microreactors.

# (372a) Integration of Pervaporation in the Methyl Acetate Synthesis

**Thomas Winkler**, Susanne Lux and Matthäus Siebenhofer, Institute of Chemical Engineering and Environmental Technology, Graz University of Technology, Graz, Austria

**Integration of Pervaporation in the Methyl Acetate Synthesis**

*T. Winkler, S. Lux, M. Siebenhofer*

*Graz University of Technology, Institute of Chemical Engineering and Environmental Technology, Graz/Austria*

In recent years, there has been a significant rise in the number of pervaporation processes for industrial applications. The field of interest for the application of this unit operation has a span from dehydration processes to equilibrium limited reactions. Former applications conducted by pervaporation already represent the state of the art in the dehydration of organic solvents. The main benefits of pervaporation are based on low operation temperatures, partial evaporation and the nature of the driving force. The outlined advantages may result in energy savings compared to conventional unit operations when dealing with thermodynamic limitations.

In this research activity we focus on the separation of azeotropic mixtures. The methyl acetate synthesis serves as a model reaction, since it forms two low boiling azeotropes with methanol and with water. The reverse reaction, the hydrolysis of methyl acetate, as well as transesterification reactions are facing the same issues. The azeotropic mixture of methyl acetate and methanol represents the bottleneck in all these reactions. The major focus of our investigations is to overcome this limitation by applying a pervaporation process. We conducted a screening of various commercially available hydrophilic and organophilic membranes. Hydrophilic membranes enhance the permeate flux of polar components. The active layer of the investigated hydrophilic membranes consists of crosslinked PVA. An increase in the degree of crosslinking leads to a significant improvement in selectivity. The maximum permeate flux of the investigated PVA-membranes ranges from 0.9 to 9.4 $kg \cdot m^{-2} \cdot h^{-1}$. In the case of organophilic membranes a rise in the permeate flux up to values of 72 $kg \cdot m^{-2} \cdot h^{-1}$ was observed. Organophilic membranes with an active layer consisting of POMS show a better separation performance than PDMS based layers although the difference in permeate flux is not that pronounced. A model for the permeate flux based on an adapted solution-diffusion model was proposed.

# (372b) From Racemate to Single Enantiomer with 100% YIELD

Markus Fuereder, Christian Femmer, Sven Panke and Matthias Bechtold, Department of Biosystems Science and Engineering, ETH Zurich, Basel, Switzerland

Integration of SMB enantioseparation and mild enzymatic racemization allows for obtaining the target enantiomer from the racemic mixture in theoretically 100% yield- compared to 50% yield with a stand-alone chiral SMB [1].

Next, such a process gives access to both enantiomers with one SMB installation and one enzyme compared to the three enzymes required in dynamic kinetic resolution approaches for the same flexibility.

In this work we present the design and operation of an integrated process realized by coupling of an SMB equipped with Chirobiotic TAG (Sigma Aldrich, USA) columns [2], a nanofiltration concentration deviceand an enzyme membrane reactor containing an amino acid racemase for the production of enantiopure amino acids, specifically D-methionine (Fig.1).

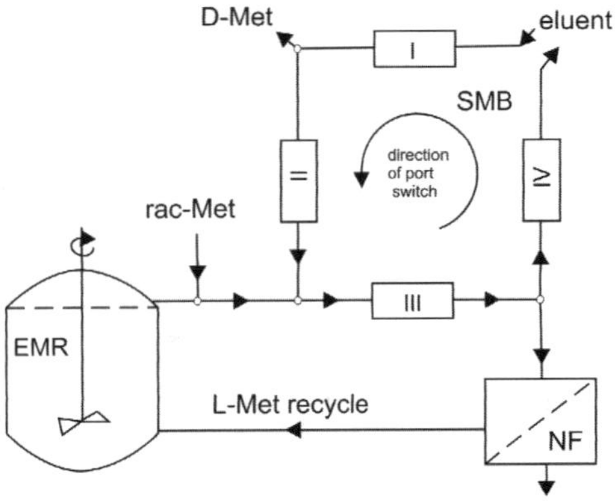


However, due to the direct coupling the same solvent needs to be applied in all unit operations. Hence, the design and optimization of a fully integrated production system require a detailed characterization and accurate parameterization of all involved units with respect to potentially inter-operable solvent compositions, specifically reaction kinetics and stability, adsorption isotherms and rejection of the nanofiltration membrane. Based on these parameters model-based design is applied in order to identify appropriate operating points for aset of specifications such as product purity. The usefulness of the obtained design is evaluated by experimental runs of the fully integrated process. This constitutes the first report on the experimental implementation of integration of chiral SMB and enzymatic racemization.

References

1) M. Bechtold, S. Makart, M. Heinemann, S. Panke, Journal of Biotechnology 124 (2006) 146

2) M.Fuereder, S.Panke, M.Bechtold, Journal of Chromatography 1236 (2012), 123

# (372c) Reactive Distillation for Synthesis of Ethyl Levulinate From Biobased Resources

**Aspi K. Kolah**, Lars Peereboom, Kelly O'Connell, Carl T. Lira and Dennis J. Miller, Chemical Engineering and Materials Science, Michigan State University, East Lansing, MI

In order for USA to reduce its dependence on foreign oil, it is necessary to produce liquid transportation fuels from lignocellulose biomass. Levulinic acid can be synthesized from waste lignocellulosic biomass by a thermo-chemical route. Levulinic acid can be converted into its ethyl and butyl esters by reaction with the corresponding alcohol, ethanol or butanol. Ethyl levulinate and butyl levulinate have been identified as promising diesel fuel blending agents arising from their high oxygen content. When burned in diesel engines, they give rise to very low soot and NOx emissions. Ethyl levulinate is also used for synthesis of levulinic ketals by reaction with glycerol; these ketals are used as green plasticizers and cleaning solvents.

Esterification of levulinic acid is an equilibrium limited chemical reaction. In order to improve conversion over the conventional batch reactor followed by distillation, the versatile process intensification technique reactive distillation, where simultaneous chemical reaction and distillation take place in a single piece of equipment, is used for levulinic acid esterification.

In this paper, we present results of recent experimental studies at the Michigan State University Reactive Distillation Facility to synthesize ethyl levulinate from the heterogeneous ion exchange resin catalyzed reaction of levulinic acid with ethanol. In developing the process concept, we have studied the batch kinetics and equilibrium for esterification of levulinic acid with ethanol using homogeneous and heterogeneous catalysts. A pseudo-homogeneous model has been used to represent the batch kinetic data using MATLAB. Vapor-liquid equilibrium studies for important binary mixtures has been experimentally generated and co-related using the NRTL-HOC model.

Continuous reactive distillation experiments were conducted in the MSU reactive distillation facility. The catalytic zone of the reactive distillation column is packed with Katapak-S filled with Amberlyst-15; the column was run at atmospheric pressure. High conversion of levulinic acid (~98%) was observed. A process concept has been developed that allows ethyl levulinate containing less than 0.1% of lactones which form as side products.

Experimental studies are complemented by process analysis using Aspen Plus simulation software. By implementing the kinetic model and phase equilibrium data generated in our laboratories, we are able to accurately model pilot-scale column behavior and simulate commercial-scale facilities.

# (372d) Experimental Investigation and Model Validation of a Heterogeneously Catalyzed Reactive Distillation Process to Intensify the Synthesis of n-Butyl Acrylate From Acrylic Acid and n-Butanol

**Alexander Niesbach**, Ron Fuhrmeister, Jan Daniels, Benjamin Schröter, Philip Lutze and Andrzej Górak, TU Dortmund University, Department of Biochemical and Chemical Engineering, Laboratory of Fluid Separations, D-44227 Dortmund, Germany

**Experimental investigation and model validation of a heterogeneously catalyzed reactive distillation process to intensify the synthesis of n-butyl acrylate from acrylic acid and n-butanol**

**Alexander Niesbach[*], Ron Fuhrmeister, Jan Daniels, Benjamin Schröter, Philip Lutze, Andrzej Górak**

Laboratory of Fluid Separations, TU Dortmund University, Germany

([*]Corresponding Author's E-mail: alexander.niesbach@bci.tu-dortmund.de)

**ABSTRACT**

The development of innovative apparatuses and techniques has led to process intensification and achieving ecologic and economic improvements, such as decreasing energy consumption and increasing process efficiency. One of those is reactive distillation (RD) which exploits the synergy of the combining reaction and separation at the same place and time. This combination allows to overcome limitations occurring for the reaction and separation alone such as azeotropes or chemical equilibria. Hence, RD may lead to increased process yields. The concept of RD has been widely studied for low carbon numbers esterifications.

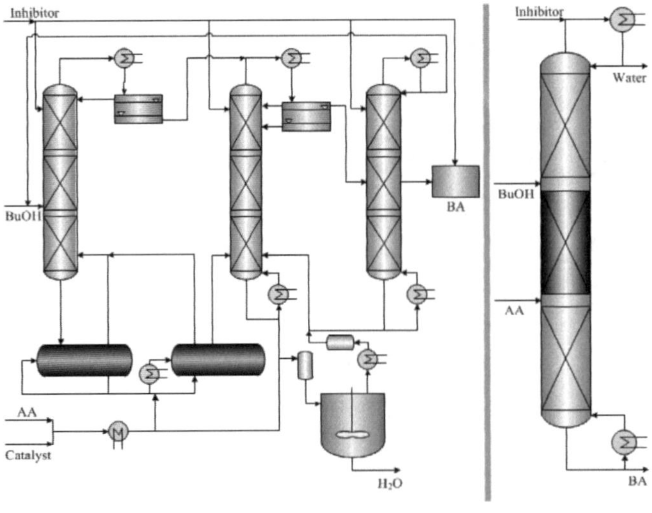


Figure 1: Comparison of conventional and intensified process for the production of *n*-butyl acrylate.

In this work, the heterogeneously catalyzed synthesis of *n*-butyl acrylate (BA) from acrylic acid (AA) and *n*-butanol (BuOH) in a reactive distillation column is investigated experimentally and theoretically. The reaction is catalysed by a strong-acid ion exchange resin. Due to the limitation of the esterification by the chemical equilibrium, process intensification using an RD column may increase the yield and lead to a reduced side-product formation compared to the base case design, see Fig.1.

Additionally, through the use of RD as an integrated process for this reaction system, the number of apparatuses can be reduced significantly compared to the conventional process.

Hence, savings in operational and investment costs are expected. Further savings can be achieved by replacing the conventional homogeneous catalyst by a heterogeneous catalyst as the separation and recycle of the homogeneous catalyst is not necessary and the effort for waste water treatment is reduced significantly.

The main barrier for the application of an RD column is the polymerisation tendency of both acrylic acid and *n*-butyl acrylate which increases rapidly with increasing temperature. To prevent those components from polymerising, inhibitors need to be added to the process. To ensure a safe operation of a pilot-scale reactive distillation column during the experimental investigation, a detailed study was performed during this work to determine the influence of temperature, inhibitor concentration and the atmosphere on the resulting inhibition period in lab-scale batch experiments. Based on these experiments, a concept for adding polymerisation inhibitors to the column was developed and implemented. Prior to the experimental investigation, the inhibitor dilution was studied and a concept for the startup and the shutdown of the column was developed.

Afterwards, pilot plant experiments were conducted in a glass column with an inner diameter of 50 mm and an effective packing height of 5.7 m. For this purpose a fractional factorial design of experiments has been developed and a set of experiments was performed to validate the model in a wide range of the main operational parameters. Temperature and concentration profiles were measured along the pilot scale RD column. An online data-reconciliation was used to monitor the steady-state condition.

The experimental results were compared to a nonequilibrium-stage model implemented in the simulation environment Aspen Custom Modeler® (Klöker et al, 2005). This model considers multicomponent mass and heat transfer rates and hydrodynamics. Thermodynamic and physical properties are taken into account by a link to the software Aspen Properties®. Liquid phase activity coefficients were calculated using the UNIQUAC model. The Hayden O'Connell equation of state was used to account for non-idealities in the vapour phase. A comparison of experimental and simulation results is shown in Figure 2.

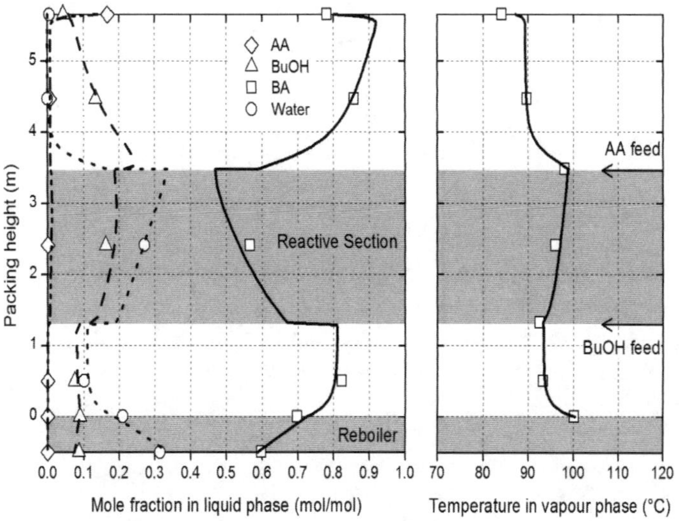

Figure 2: Experimental (symbols) and simulation (lines) results of a column profile used for the model validation.

This comparison was done for all experimental results to validate the model in a wide range of the operational parameters. The validated model will be used for further optimization studies to investigate the optimal process for the synthesis of *n*-butyl acrylate from acrylic acid and *n*-butanol using reactive distillation.

**Acknowledgement:** The research leading to these results has received funding from the European Community's Seventh Framework Programme (FP7/2007-2013) under grant agreement n° 228867, $F^3$-Factory

**References**

1. Klöker, M., E.Y. Kenig, A. Hoffmann, P. Kreis and A. Górak, *Chemical Engineering and Processing*, **44**, 617-629 (2005).

# (372e) Off-Cut Recycling in Batch Reactive Distillation Processes

**Yu-Lung Kao** and Jeffrey D. Ward, Department of Chemical Engineering, National Taiwan University, Taipei, Taiwan

Batch reactive distillation (BREAD) is an integrated process which combines the advantages of reactive distillation and the flexibility of batch processes. BREAD process performance can be improved by withdrawing an off-cut (slop cut) during the batch. In non-reactive batch distillation processes, it has been shown that off-cut withdrawal can reduce total batch time compared to operation without off-cut withdrawal, especially when the separation is difficult (Christensen and Jorgensen, 1987). However, very few authors have considered the collection of off-cut for BREAD processes (Venimadhaven et al., 1999; Wajge and Reklaitis, 1998), and we are not aware of any that consider the recycling of off-cut for BREAD processes.

Recently, we have also shown that for BREAD processes off-cut withdrawal improves the process in terms of the batch capacity (CAP) (Kao 2010; Kao and Ward, 2010). Usually, the off-cut that is collected is not discarded directly, instead it must be processed to recover valuable materials or treated to remove hazardous pollutants.

In this work, we consider an alternative which is to recycle off-cut to the next batch as a part of initial feed. This transforms the operation problem from single batch problem into a batch-to-batch problem. A canonical set of quaternary reversible reaction $A + B \leftrightarrow C + D$ is used to investigate the design of batch campaigns with off-cut recycle. The pseudo-steady state concept proposed by Luyben (1988) for non- reactive batch distillation is applied to BREAD processes to simplify the optimization problem. A pseudo-steady state for a batch campaign means that for a fixed operating recipe the amount and composition of off-cut produced in the previous batch (which is included in the feed to the present batch) is identical to the amount and composition of off-cut produced in the present batch.

The optimal operating policy based on maximizing CAP is found for a process operating at pseudo-steady state, and the result is compared with the optimal operating policy for single batch operation when the off-cut is not recycled. The results show that the optimal CAP when off-cut is recycled is similar to the case when off-cut is collected but not recycled, and is significantly more than the case when off-cut is not used. Therefore, recycling off-cut may not only save the trouble processing the worthless off-cut, but also make the process more economical.

**References**

Christensen FM, Jorgensen SB. Optimal-Control of Binary Batch Distillation with Recycled Waste Cut. Chem Eng J Bioch Eng. Feb 1987;34(2):57-64.

Kao YL. Effect of Relative Volatility Ranking On the Design and Operation of Batch Reactive Distillation Systems. Master Thesis. National Taiwan University, Taiwan. July, 2010.

Kao YL, Ward JD. Effect of Relative Volatility Ranking On the Design and Operation of Batch Reactive Distillation Systems. AIChE Annual Meeting. Salt Lake City, USA. Nov. 7-12, 2010.

Luyben WL. Multicomponent Batch Distillation .1. Ternary-Systems with Slop Recycle. Ind Eng Chem Res. Apr 1988;27(4):642-647.

Venimadhavan G, Malone MF, Doherty MF. A novel distillate policy for batch reactive distillation with application to the production of butyl acetate. Ind Eng Chem Res. Mar 1999;38(3):714-722.

Wajge RM, Reklaitis GV. An Optimal Campaign Structure for Multicomponent Batch Distillation with Reversible Reaction. Ind Eng Chem Res. May 1998;37(5): 1910-1916.

# Optimal Design of Catalytic Distillation Column for the Hydrolysis of Methyl Acetate

***Huidong Zheng**, Yanru Zeng, **Suying Zhao**[*] and Jiaojiao Bai, College of Chemistry and Chemical Engineering, Fuzhou University, Fuzhou, China*

**Abstract:**

Optimal design is a three-step process, including feasibility analysis, synthesis analysis and design step, The final results could then be considered as an initial value for the further design.The process could screen several kinds of column configuration, determine the corresponding optimal configuration and operation parameters. The method proposed in this work provided important reference for the optimal design of catalytic distillation column for the hydrolysis of methyl acetate.

**Keywords:** Optimal design; Catalytic distillation column; Hydrolysis of methyl acetate.

## Introduction

Large amount of methyl acetate (MA) is produced as byproduct in the production of purified terephthalic acid (PTA) and polyvinyl alcohol (PVA) , while a ratio of 1.68 tons of MA per ton of PVA[1]. As containing methanol (MeOH) and some other impurities, the purity of MA is not high enough for industrial applications. In industry the mixture is usually hydrolyzed to produce methanol (MeOH) and acetic acid (HAc) or esterified to produce purity MA both by catalytic distillation (CD). In view of the hydrolysis product acetic acid and methanol in PVA and PTA industry can be used as the raw materials, and at the same time the high economic benefits would be made as a product to sell, so this paper discussed the hydrolysis process of methyl acetate.

Catalytic distillation (CD) is a special reactive distillation (RD) process with solid catalysts. The combination of the separation process with catalytic reactions in a CD column shows up many advantages[2-3], particularly for reversible reactions with small equilibrium constant such as the reaction of MA hydrolysis.

In recent years the number of papers in the CD process has grown rapidly for the MA process design[4-6], steady-state behavior description[7], dynamics[8] and control[9-10]. For the systems studied, related papers and patents in hydrolysis[11] are much less than that of esterification process[3,12-13]. The hydrolysis was first investigated by Fuchigami[14], the reactive column used was packed with cation exchange resin molded with polyethylene powder. It was reported that a near complete conversion of MA was achieved with a high feed ratio of $H_2O$/MA and the total energy consumption of the optimum process was estimated to be only about half of the conventional hydrolysis process. However, the catalyst used in this research was not readily available on an industrial scale. Kim and Roh[15], followed the work of Fuchigami, with the azeotrope feed of MA/MeOH. Almost the same conversion can be obtained in a high feed ratio using the total reflux configuration.

Han et al.[11] reported a research on a modified sieve-type distillation tray, where liquid–solid catalytic reaction zone and a vapor–liquid separation zone on the tray coexisted. And fixed-bed pre-reactor was installed before the CD column. The highest total conversion was only about 50%. Wang et al.[16], following the configuration of Han et al. with more experimental date, changed the operating condition by increasing the feed ratio of $H_2O$/MA and the reflux ratio. The conversion of MA was improved to 72%. In the simulation, Lee[17] proposed a hydrolysis process consisting of one CD column with a reactive reflux drum and two separation columns, while the column is operated with total reflux. With an excess of water, almost complete conversion of MeAc could be obtained and trace amount of MeAc was detected in the column base. The control of an reactive distillation column was investigated by Muhammad[18]. Three control structures were proposed and evaluated for both ideal and real reaction systems. The design and control of methyl acetate hydrolysis using reactive distillation was studied by Yu-Der Lin et al.[19]. The proposed flowsheet was similar with Lee, but a systematic design procedure was used to generate the flowsheet based on the total annual cost (TAC). A plantwide control structure was developed followed by process identification and controller tuning.

In the last two year, Xin Gao et al.[2] developed an equilibrium stage model for the simulation of the process of MA hydrolysis. In the model, the influences of the reactive kinetics, residence-time, liquid holdup, and separation efficiency of the catalytic packing were considered. The model was in good agreement with experimental data. Optimal effect parameters and design factors of new technology on energy consumption and conversion were obtained. This new process had a higher conversion of MA compared with the results in previous research and was found to be energy efficient.

The design methods for reactive distillation processing developed over the last decades are divided into three categories ： those based on graphical/topologycal considerations, those based on optimization techniques and those derived from heuristic/evolutionary considerations[20] .The graphical methods are called so, because the decisions are made on the basis of graphical information. This information is often generated using models. These approaches rely on the behavior of residue curves or distillation lines.

The second group of design methods is composed of those methodologies, which are supported by the power of computational subroutines and take into account important elements of an overall design strategy (e.g. types of phase equilibrium, number of components, occurrence or not of chemical reaction on a given stage).A third trend of conceptual design approaches may be identified from the evolution of RD as a derivation from conventional distillation processes. Residue curve map techniques and fixed-point algorithms are very useful tools for preliminary screening and design of RD systems.

In addition, MA is regarded as our main research object in the team, including the measurement of fundamental thermodynamic data[21], dynamics and catalysts action researches[22-24], and the development of new processes with experiment and simulation[25-29].The vapor-liquid phase equilibrium data[21] of MA-MeOH-H2O-HAc was measured at atmospheric pressure, and the activity coefficients of liquid compositions was calculated. In the field of catalyst research and kinetics, the macro-reaction kinetics on the hydrolysis of the methyl acetate catalyzed by cation-exchange resins was studied at the atmospheric pressure and in a range of temperature. The kinetics model under the experimental conditions was established[22]. Furthermore, A new process composed of a fixed-bed and a distillation column with a side withdraw was proposed with simulation and experiments[25].

Although many works have been done in our team about this subject, China's domestic study is still rarely published on the RD design, and the technology is immature. Therefore, this article presented a relative systematic method on the optimal design of catalytic distillation column for the hydrolysis of methyl acetate.

**Feasibility analysis**

In this paper, the process of the hydrolysis of methyl acetate is discussed. The corresponding reaction equation is as followed:

$$CH_3COOCH_3(MA)+H_2O \leftrightarrow CH_3COOH(HAc)+CH_3OH(MeOH)$$

The main tools used in the step contained the reactive residue curve maps (rRCM) method and reactive extractive curve maps (rExCM) method. The RCM technique has traditionally been considered a powerful tool for flow sheet development and preliminary design of conventional multicomponent separation processes. It does not only represent a good approximation of actual equilibrium behavior, but it also allows one to perform feasibility analysis of separation processes where non-ideal and azeotropic mixtures are involved[30]. The rExCM concept[31] enables to formulate a necessary condition for the feasibility of a reactive double-feed unit which is very similar to the necessary condition formulated for single-feed RD columns and based on the analysis of reactive residue curve maps.

***Residue Curve Maps without reaction***

Figure 1 showed the residue curve maps for ternary mixture of methyl acetate hydrolysis reaction system at 1 atm. In the system two binary azeotropes (points B and C, respectively )were displayed in the figure1, plot B represented the azeotrope of methyl acetate-methanol, with the azeotropic temperature 326.62 K (53.47□) , composed of the 66.6% molar of MA and 33.4% of MeOH (mole fraction in the following), likewise plot C represented the azeotrope of methyl acetate-water, with the azeotropic temperature 329.72 K (56.57 ℃) , composed of the 91.7% molar MA and 8.3% of $H_2O$. From the figure 1, all the residual curves pointed from the low-boiling-point mixture to the high-boiling-point one. However showed in figure 1b, the residual curves pointed to methyl acetate when the feed stream contained a majority of methyl acetate and little water. The line BC was named distillation boundary, in the either side of the boundary, the end of curve was different. As methyl acetate-methanol azeotrope was the lowest-boiling-point azeotrope, it was not easy to separation them in the column, and also it made the reactive distillation design difficult in some degree.

Further analysis was implemented to make sure whether the mentioned azeotropes disappeared or the new azeotropes appeared when the four components coexisted. A residue curve map of the quaternary system was displayed in figure 2 and no new azeotropes were found, the mentioned azeotropes B and C still existed. The analysis of the singular point suggested that all the curves started from the point B, which named point B a unstable node, and all the curves ended to the vertexes representing water, methyl acetate and acetic acid, which were named stable node, the vertex methanol and point C were neither the starting point of the curves nor the ending point, namely for saddle point. In order to observe the residue curve map of the quaternary system clearly, the three-dimensional residue curve map was transformed into two-dimensional one through the coordinate transformation, and the transformation method was seen in the literature [32]. In Figure 2b most of the curves ended to the vertexes representing acetic acid and water. Because of the existence of four components and two azeotropes in the hydrolysis of methyl acetate system, it was difficult in separating them. So the reaction was always employed, and the residue curve map with reaction could be studied in the following.

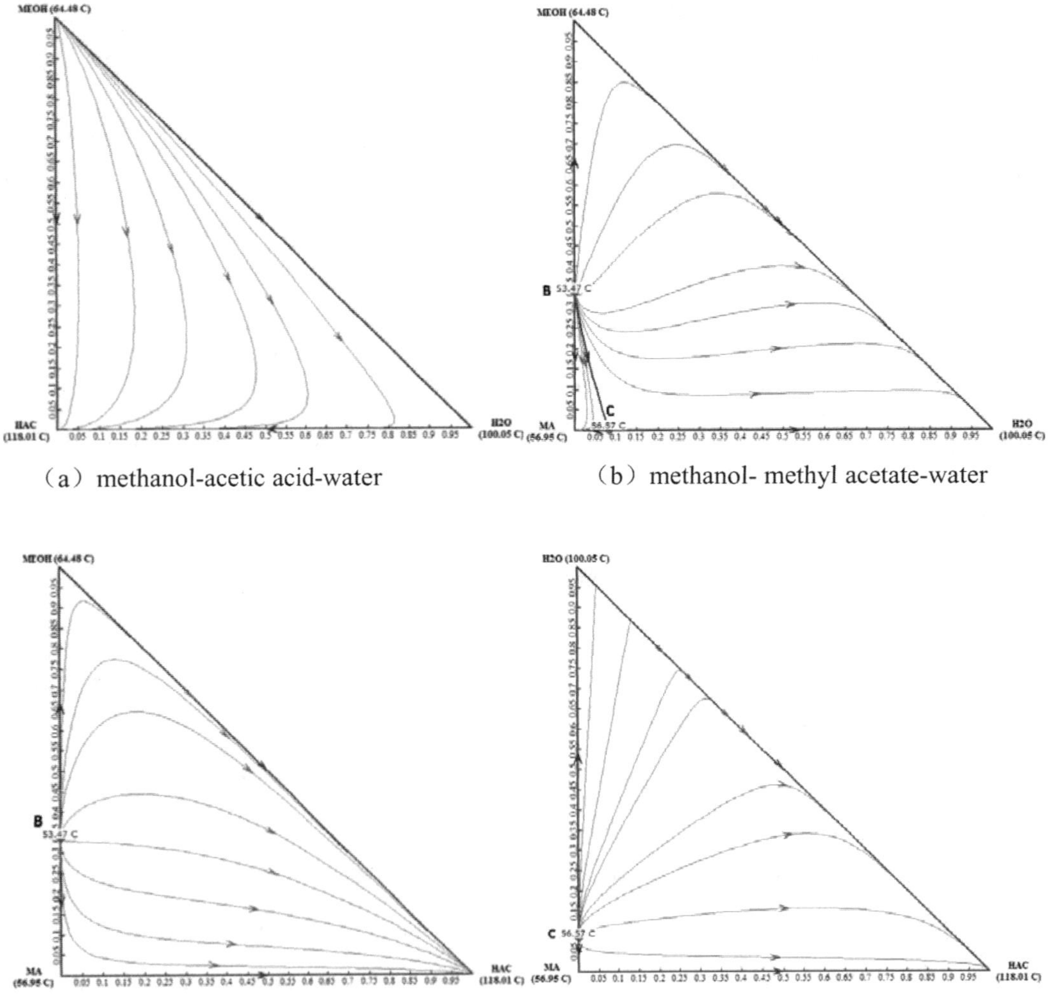


（a）methanol-acetic acid-water

（b）methanol- methyl acetate-water

（c）methanol- methyl acetate - acetic acid

（d）water- methyl acetate - acetic acid

Figure 1. The residue curve maps of ternary system at 1atm

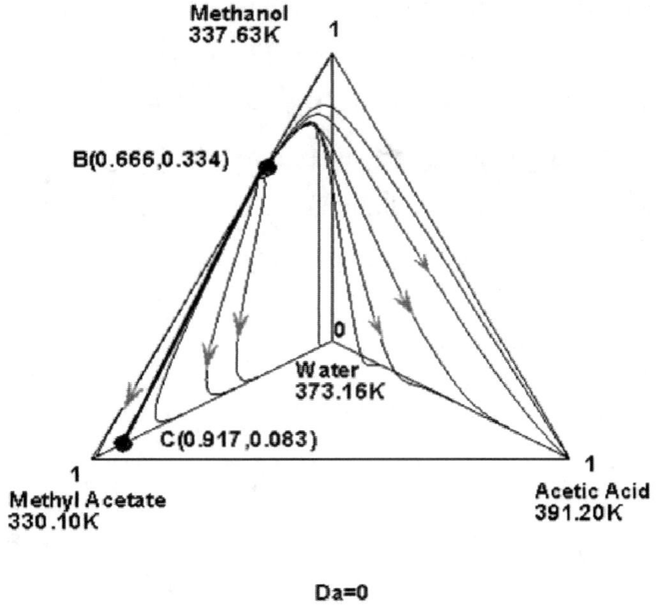


Figure 2（a）.The residue curve maps of quaternary system at 1atm

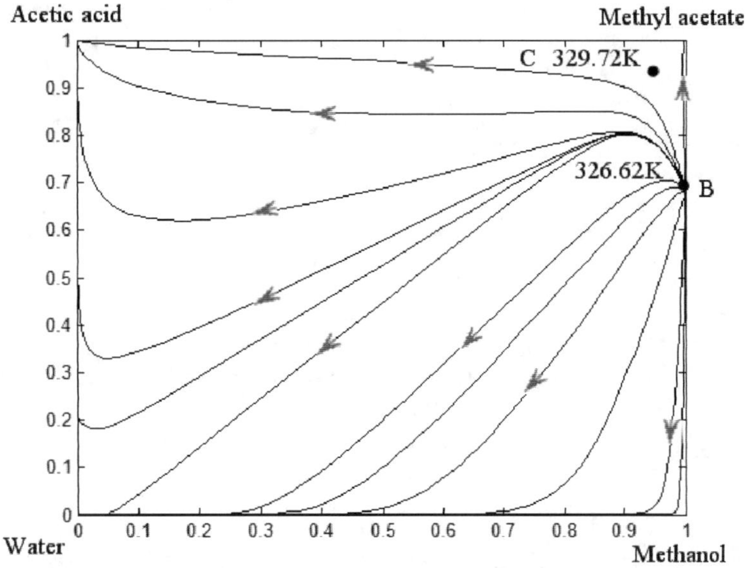


Figure 2（b）.The residue curve maps of quaternary system at 1atm via coordinate transformation

### *Reactive Residue Curve Maps*

The residue curve map with reaction of methyl acetate hydrolysis was showed with different Da (Da reflects the impaction of reaction to the residue curve maps) at 1 atm in figure 3, and the impaction of the reaction to the azeotropes was researched with different Da. The study suggested that the residue curve was not change with the increasing of Da when the reaction balanced, also it mean that the phase equilibrium force and the reaction equilibrium force achieved mutual balance. In order to understand how the reaction influenced the residual curves more clearly, some cases with the same initial concentration and different Da values was studied, the curves showing in figure 4. From the diagram, the curves changing more when the reaction joined in the system, the distillation boundary BC was disappeared and when Da =1 and Da = 10 the curve overlapped together, that was to say the system had reached the force balance in Da = 1.

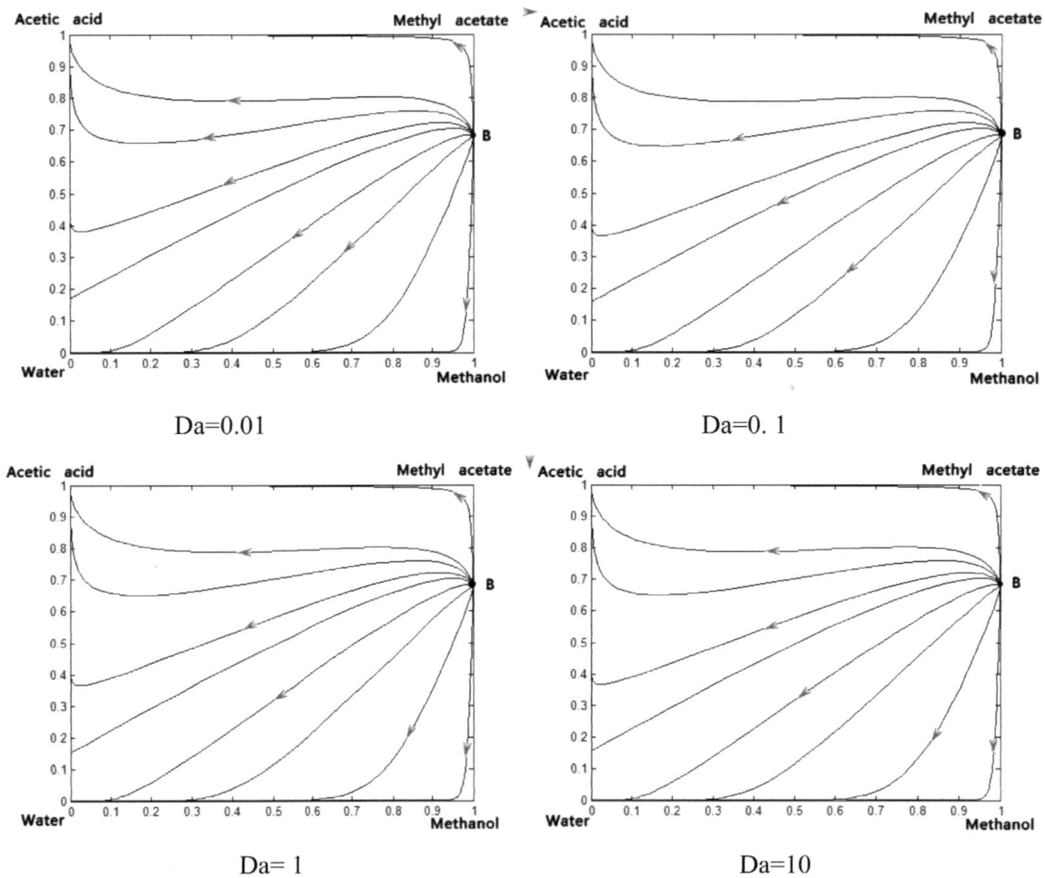

Figure 3. The reaction residue curve maps with different Da

At the same time, in the hydrolysis process of methyl acetate, the MEOH and the HAC was regarded as the final production. However, according to the rule of the rRCM analysis, if one or both products are saddles, the concentration profile can be shifted in the direction of the saddle using a column with two feed stages. Because of the MeOH confirmed to be a saddle, the CD column would be required to be set up with two feed stage in order to obtain the pure component. Some methods were developed to assess the feasibility or to design double-feed reactive distillation columns. In this paper, the ExCM method was employed to analysis the system. The specific application will be investigated in our later researches.

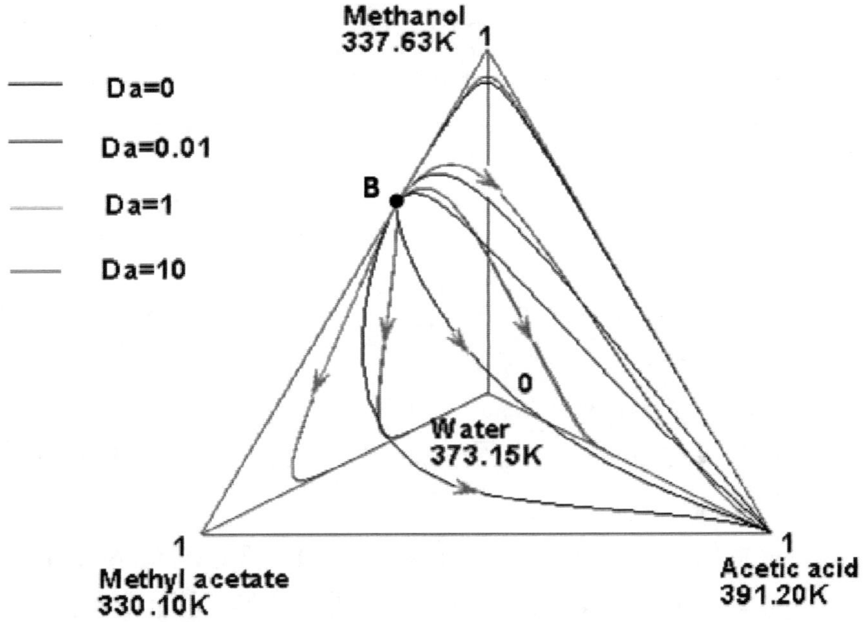


Figure 4. The reaction residue curve maps with different Da value

**Synthesis Analysis**

Synthesis, which confirms the results obtained through the first step based on a more rigorous analysis and determines the column configuration able to achieve the goals formerly specified; the composition profiles are deduced from the mass balance equations stage by stage under neglection of all the thermal effects in the synthesis step which is based on boundary value design method. On the other hand, the boundary value design method is also applied to analyze the effect of CD column configuration under the different synergies between reaction and separation. Sequentially, combined with the first step, the general rule of synergies in the whole unit is obtained. The CD column configuration parameters corresponding to specified conditions such as total number of theoretical stages, location and size of rectifying/stripping section, location and size of reactive section, location of the feed plate(s) are optimized under different feed compositions in the second step.

According to the feasibility analysis, a double-feed reactive distillation was needed to achieve the pure MeOH. Except for that, the data required for the step and inherited from the feasibility analysis step are: the distillate and bottom compositions;

the basic configuration of the column: requirement or not of a pure separation section, finite or infinite reflux or reboiling ratios. The feed composition is determined by the mixtures directly produced from the industry to deal with.

Now the computation of the composition profiles are preceding combined the reaction kinetics, mass balance and the phase equilibrium. The plate's number in each region such as the pure separation and the reaction section will be determined with the profiles. The location of feed plate and the minimum reflux ratio can be obtained in the same way.

**Design step**

The previous steps lead to a column configuration and a reflux ratio that permits to reach the initial product specifications. But these results have been obtained neglecting all thermal phenomena (heat of reaction, thermal loss, heat of mixture, etc.). The objective of the simulation step is to recalculate the operating parameters without modifying the column configuration by taking into account the heat balances. The steady-state simulation with MESH model in Aspen Plus was employed to optimize the operation parameters such as reflux ratio, heat duty and distillation/bottom flow.

The heat of reaction for the hydrolysis of methyl acetate can be considered as negligible (Q=20kJ/mol). The heat of vaporization of the different components are very close. The assumptions formulated in the synthesis step are nearly satisfied, and the actual reflux ratio determined through the design step, according to the reflux ratio determined through the synthesis step. Hence, the design step could only be carried out after the synthesis analysis.

**Conclusion**

The proposed method was characterized by relying on a minimal set of information concerning the physicochemical properties of the system and the reaction, introducing a set of methods ranged from putting forward various solutions to the optimization, presenting a novel design and analysis process for the hydrolysis of MA. The systematical and universal design method was also proposed upon the study of the mechanism and its regulation method for the optimal design of CD column in order to minimize the equipment and operation cost. The results obtained through this method leads to a reliable initialization point for a further optimization of the operating conditions and the process design.

Although several successful sets of CD column for MA hydrolysis are built, there still remain some problems, such as low hydrolysis conversion, high energy consumption and complex column configuration. The method proposed in this work can provide important reference and direction for the further improvement of MA hydrolysis and also other CD processes.

However, the limits exited in the method also have been pointed out. Although the RD processes are really promising, the interactions existing between reaction and separation functions lead to nonlinearities which strongly complicate the control of such processes. For that reason, studies concerning the design of RD processes including controllability are urgent affair.

On the other hand, the procedure should be extended to kinetically controlled reaction. A pure numerical analysis should also be substituted to the graphical analysis to be able to investigate reactive systems whatever the number of components and reactions.

## References


[1] Dirk-Faitakis, C.B.; An, W.Z.; Li, T.B.; Chuang, K.T. Catalytic distillation for simultaneous hydrolysis of methyl acetate and etherification of Methanol. *Chem. Eng. Process.* 2009, 48, 1080-1087.

[2] Xin Gao, Xingang Li, Hong Li. Hydrolysis of methyl acetate via catalytic distillation: simulation and design of new technological process. *Chem. Eng. Process.* 2010, 49,1267-1276

[3] Bernd Bessling, Jan-Martin Löning, Andre Ohligschläger. Investigations on the synthesis of methyl acetate in a heterogeneous reactive distillation process. *Chem. Eng. Technol.* 1998;21(5):393-400.

[4] R.S. Russ, F. Chen, M. F. Malone, M. F. Doherty. Computer-aided tools for the design of reactive distillationsystems.*Comput.Chem.Eng.*1999: s955-s962.

[5] Hao-Yeh Lee, Yi-Chen Lee, I-Lung Chien, Hsiao-Ping Huang. Design and control of a heat-integrated reactive distillation system for the hydrolysis of methyl acetate. *Ind. Eng. Chem. Res.* 2010, 49, 7398-7411.

[6] Akram Avami, WolfgangMarquardt, YadollahSaboohi, KorbinianKraemer. Shortcut design of reactive distillation columns. *Chem. Eng.Sci.* 2012,71, 166–177

[7] Robert S.Huss,Fengrong Chen,Michael F.Malone, Michael F.Doherty. Reactive distillation for methyl acetate production.*Comput.Chem. Eng.* 2003, 27,1855-1866

[8] S. Gr.unera, K.-D. Mohla, A. Kienleb, E. D. Gillesa,b, G. Fernholzc, M. Friedrich. Nonlinear control of a reactive distillation column. *Control Eng. Pract.* 3003, 11, 915-925.

[9] Marten Volker, Christian Sonntag, Sebastian Engell. Control of integrated processes: A case study on reactive distillation in a medium-scale pilot plant. *Control Eng. Pract.* 2007, 15, 863-881.

[10] M. V. P. Kumar and N. Kaistha. Temperature based inferential control of a methyl acetate reactive distillation column. *Chem. Eng. Res. Des.* 2007,85, *1268-1280.*

[11] S.J. Han, Y. Jin, Z.Q. Yu. Application of a fluidized reaction-distillation column for hydrolysis of methyl acetate. *Chem. Eng. J.* 1997, 6, 227-230.

[12] Victor H.Agreda, Lee R.Partin, both of Kingsport Tenn. Reactive distillation process for the production of methyl acetate: US Patent, 4425595, 1984-03-06

[13] Victor H.Agreda, Ronnie D.Lilly, both of Kingsport Tenn.Preparation of ultra high puriyt methtl acetate: US Patent, 4939294, 1990-07-03.

[14] Y. Fuchigami, Hydrolysis of methyl acetate in distillation column packed with reactive

packing of ion exchange resin. *J. Chem. Eng. Jpn.* 1990, 23, 354–358.

[15] Kim, K.J., Roh, H.D. Reactive distillation process and equipment for the production of acetic acid and methanol from methyl acetate hydrolysis: US Patent ,5770770, 1998.

[16] Wang, J., Ge, X., Wang, Z., Jin, Y. . Experimental studies on the catalytic distillation for hydrolysis of methyl acetate. *Chem. Eng. Tech.* 2001,24 (2), 155.

[17] Lee, M.M. . Method and apparatus for hydrolyzing methyl acetate. US Patent 20020183549A1. 2002.

[18] MuhammadA. Al-Arfaj, William L. Luyben. Comparative control study of ideal and methyl acetate reactive distillation. *Chem. Eng. Sci. 2002, 57, 5039*

[19] Yu-Der Lin, Jun-Hong Chen, Jian-Kai Cheng. Process alternatives for methyl acetate conversion using reactive distillation. 1. Hydrolysis. *Chem. Eng. Sci. 2008, 63, 1668*

[20] Almeida-Rivera, C.P. Designing reactive distillation processes with improved efficiency. *TU Delft, Delft University of Technology*, 2005.

[21] Liang'en Wang, Suying Zhao. Vapor-liquid equilibrium of the methyl acetate - alcohol - water - acetic acid quaternary reaction system. *J. Chem.Ind.Eng.* 1995, 46(1), 57.

[22] Zhishan Zhao, Liang'en Wang, Suying Zhao. Kinetics study on methyl acetate hydrolysis catalyzed by cation exchange resins. *Chem. Eng.* 1996, 24(5), 28-32.

[23] Yanxiang Wu, Zhishan Zhao, Liang'en Wang, Suying Zhao. Kinetics of hydrolysis of methyl acetate and the effectiveness factor of catalyst capsule. *Eng. Chem. Metal.* 1999, 20(3), 241-246

[24] Suying Zhao, Liang'en Wang, Zhishan Zhao, Yi Yan. A method for detecting the effectiveness factor of the catalyst capsule without full immersion type. *J. Fuzhou University* . 1999, 27(4), 103-107

[25] Suying Zhao, Jingzhao Huang, Liang'en Wang, Guoqiang Huang. Coupled reactor/column process for hydrolysis of methyl acetate. *Chin. J. Chem. Eng.* 2010, 18(5), 1-6

[26] Suying Zhao, Jinyin Zhou, Baichuan Yang, Liang'en Wang. An integrated process of catalytic distillation and fixed bed for methyl acetate hydrolysis. Chem. *Ind. Eng. prog.*2011,30(4),725-728

[27] Huidong Zheng, Suying Zhao,Yanru Zeng, Liang'en Wang, Binghui Xu. Catalytic distillation coupled with a side reactor for methyl acetate hydrolysis. *Chem. Ind. Eng. prog.* 2012, 31(9),1-5

[28] Ting Qiu, Liang'en Wang, Suying Zhao, Huidong Zheng. Simulation of the catalytic distillation process for hydrolysis of azeotrope of methyl acetate and methanol. *Chem. Ind. Eng. prog.* 2003,22,66-70

[29] Zhishan Zhao, Ge Kuang, Liang'en Wang, Suying Zhao, Jiaqi Liu. Improvement of BP algorithm on network training and application of ann in catalytic distillation column simulation. *Chem. Eng.* 1998, 26(6), 44-46

[30] C. P. Almeida-Rivera, P. L. J. Swinkels, J. Grievink. Designing reactive distillation processes: present and future. *Comput. Chem. Eng*. 2004, 28, 2002.

[31] Raphaele Thery-Hetreux, Xuan Mi. Meyer, Michel Meyer and Xavier Joulia. Feasibility analysis, synthesis, and design of reactive distillation processes: a focus on double-feed processes. *AIChE J.*.2011,0,1-11.

[32] Weixin Su. Study of the production of ethyl acrylate by reactive distillation processes. *Taiwan: National Taiwan University of science and technology*. 2006.

# (372g) Process Improvements Through Heat Integration in Distillation Schemes: Design and Retrofit

Faisal Baksh, Chemical Techonology Development, SABIC, Riyadh, Saudi Arabia

Process Integration through modification and innovation in unit operations is a valuable way of affecting Process Intensification in the chemical process industry. Simultaneous design and optimization of reaction, separation, heat exchange and utilities systems is well recognized as an effective approach to Process Integration and Process Intensification. In this work we have constructed and compared distillation sequences with and without complex columns together with heat integration of the exchanger network.

Results are presented for the optimization of a distillation train for SABIC's 'α-SABLIN' technology, which is commercial and implemented at one of SABIC's affiliated companies, Jubail UNITED, as a 150,000 kton/year plant. α-SABLIN produces Liner Alpha Olefins (LAO) from ethylene.

Genetic Algorithm (stochastic optimization) approach, is used to construct and optimize distillation sequences based on shortcut equations. Complex column arrangements like side strippers, dividing wall columns and advanced operations like vapor recompression, heat pumps are also incorporated in the optimization process. Each scheme is subjected to heat integration through pinch analysis and compared for capital and operating costs. Results for the best schemes are cross-checked through rigorous simulations, reanalyzed and a detailed costing is performed. Results show that **maximum operating cost savings of 22% and maximum capital savings of 17% are possible** by the use of heat-integrated complex column schemes. Furthermore, a column retrofit methodology is proposed, which incorporates design heuristics based on Column Grand Composite Curves (CGCC) and column hydrodynamic curves into the optimization algorithm. This enables the user to perform automatic retrofit of distillation columns for energy savings and/or capacity increase.

# Concentrate Enhanced Reverse Osmosis: A New Process for RO Concentrate and Brackish Water Treatment

*Anthony Tarquin, PhD.; Guillermo Delgado, PhD Candidate*

*University of Texas at El Paso*

Advances in reverse osmosis (RO) technology have led to a significant reduction in the production cost of water desalination. Brackish water desalination, however, is often challenging because of the high costs associated with residual RO concentrate management, especially at inland locations. The city of El Paso owns and operates the largest inland desalination plant in the world [1]. The Kay Bailey Hutchison (KBH) desalination plant treats brackish water from the Hueco Bolson and produces, when operated at full capacity, 27.5 million gallons of water every day. The KBH desalination plant uses reverse osmosis (RO) membranes to treat the brackish water. This results in a production of 3 million gallons of RO concentrate every day. The concentrate is disposed via injection wells into the ground [2]. El Paso Water Utilities (EPWU), the Texas Commission of environmental Quality (TCEQ), and the U.S. Bureau of Reclamation have supported several research projects that can reduce the volume of this concentrate, increase the production of fresh water, and minimize the cost of disposal of the concentrate. One project in particular [3] started as pilot Seawater Reverse Osmosis system (SWRO) to evaluate the performance of seawater RO membranes to treat the KBH concentrate. However, the excellent results obtained thanks to the design and not only the membrane performance made this research to evolve to a new treatment process. The Concentrate Enhanced Recovery Reverse Osmosis (CERRO©) process (UTEP Provisional patent application number 61/233,761) is a new batch treatment process to reduce the volume of concentrate produced during RO treatment.

## Concentrate Enhanced Recovery Reverse Osmosis Process

The CERRO process consists of one or several seawater reverse osmosis membranes in parallel working in a batch configuration system. It is the batch configuration that allows the process to reach higher recoveries than conventional RO systems. The water is pretreated in the feed tank, treated by the seawater RO membrane, the permeate is recovered, and the concentrate is recirculated to the feed tank. A detailed description of the process is shown in the next sections.

## Pretreatment

There is a tank where the feed water is pretreated by the addition of sulfuric acid and antiscalants. The pH of the feed water needs to be between 3.5 and 4.5 in order to eliminate the presence of alkalinity. This has to be done to avoid the precipitation of calcium carbonate which can foul the membranes. Antiscalants are added to the water to maximize the treatment time and to avoid damaging the membranes due precipitation of calcium sulfate and/or polymerization of silica.

## Treatment

After pretreatment is done, the feed water is treated by the seawater RO membranes. Several membranes can be used in the process, but they have to be working in parallel. Membranes connected in series are not needed in the CERRO process. The permeate produced during treatment can be recollected. The concentrate is send back to the feed tank. This is done in order to increase the recovery of the system since seawater RO membranes have very low recovery rates. The operating pressure of the system is from 500 psi's to 900 psi's. The system can be operated at a constant pressure or at a constant permeate flow. At a constant pressure, the permeate flow will drop during the process because the osmotic pressure of the water will increase as it gets more concentrated. On the other hand, the operating pressure of the system will rise if the system is operated at a constant flow. Figure 1 shows the permeate flow of a test made using the CERRO process at a constant pressure of 700 psi's.

One important characteristic of the CERRO process is that the membranes can be cleaned after a batch. A fast flush with permeate or feed water between batches increases the life of the membranes and reduces the risk of fouling and scaling the membranes.

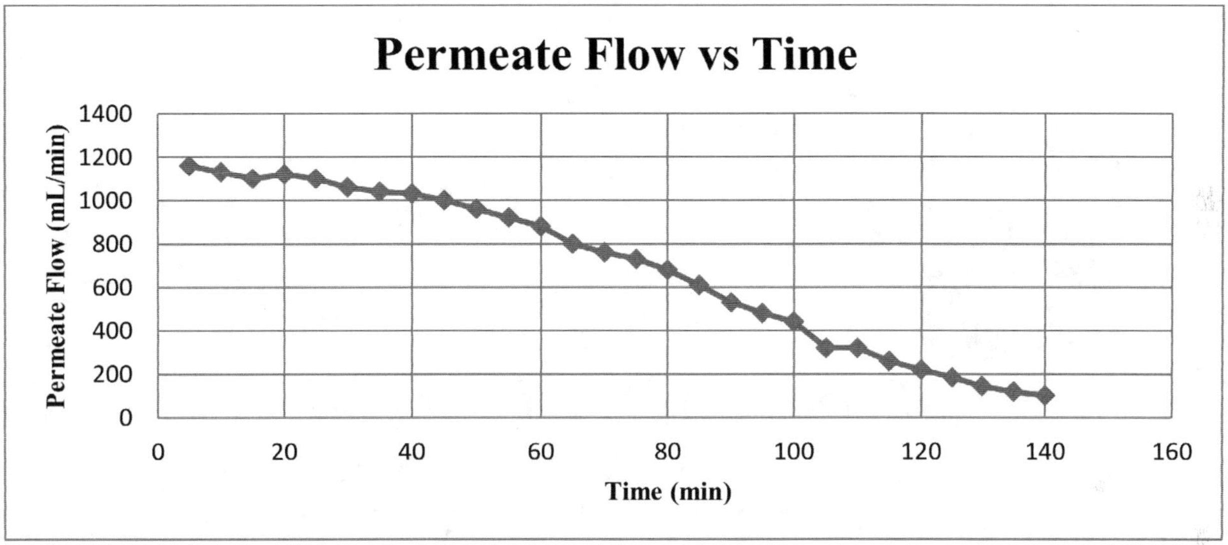


**Figure 1: Permeate Flow of the CERRO Process operated at 700 psi's**

One limitation for any membrane treatment system is the precipitation of sparingly soluble salts (i.e. $CaSO_4$) [4]. Usually, the presence of this type of salts decreases the recovery of fresh water due precipitation in the system during treatment. Several tests using the CERRO process have shown that the system can work under conditions of supersaturation. This happens because the treatment time of the system can be faster than the induction time of precipitation. This characteristic makes the CERRO process an excellent option for treating water with high concentrations of calcium sulfate and/or silica as will be shown in the case studies. Figure 2 shows a schematic of the design of the CERRO process.

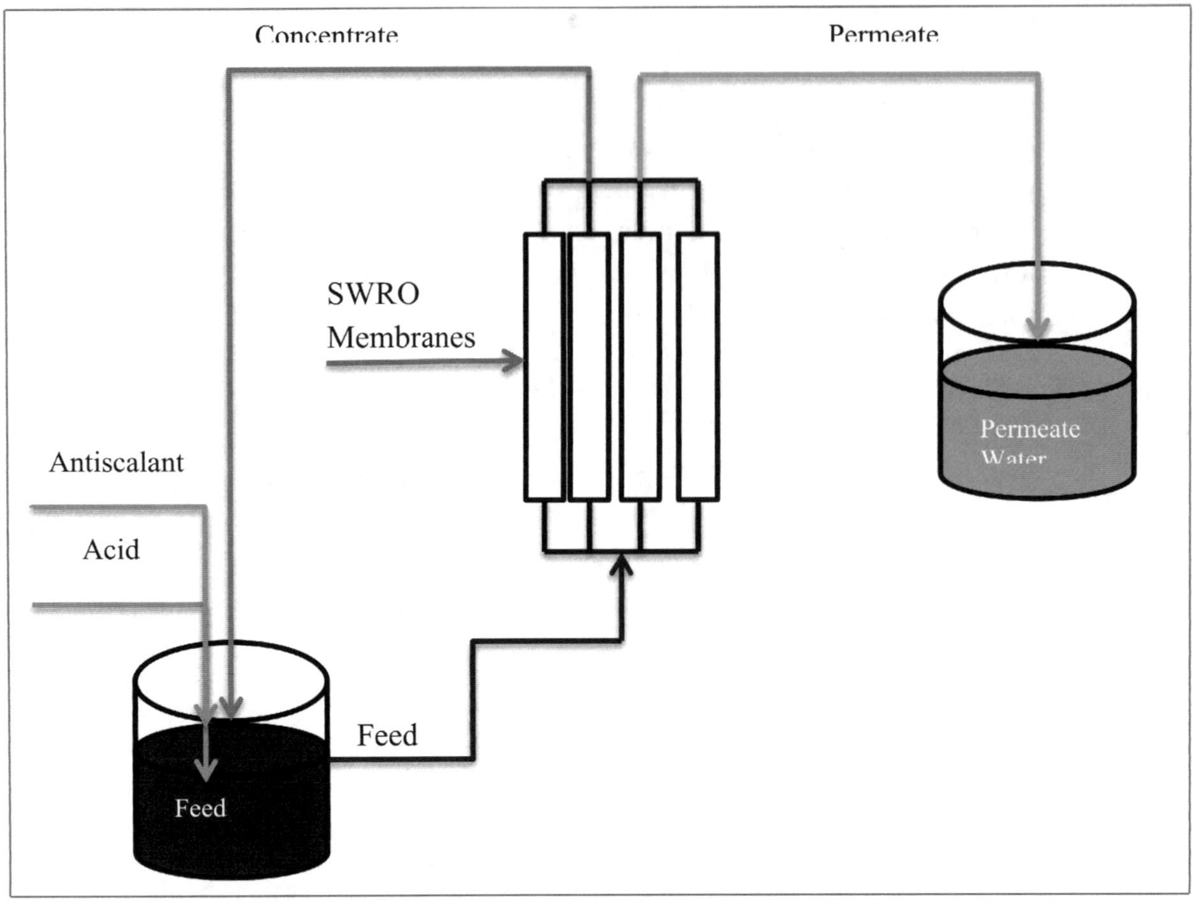


**Figure 2: Schematic of the CERRO Process**

As mentioned before, the CERRO process was designed as an effort to reduce the volume of RO concentrate generated in the KBH desalination plant. Pilot testing using this system has shown that 70% to 90% of the RO concentrate can be recovered using the CERRO process. However, other research projects have shown that the CERRO process can be used in brackish water containing high concentrations of calcium sulfate. Conventional RO systems will have low recovery rates due calcium sulfate precipitation. However, the CERRO process have shown an exceptional performance recovering water from these type of source waters, working under supersaturation conditions, and, in some cases, even when precipitation occurs during treatment.

Three case studies where the CERRO process was used are presented in the next sections.

**Case Study 1: KBH Concentrate Volume Reduction**

The KBH Desalination Plant, located in El Paso, Texas, is the largest inland desalination plant in the United States. This plant produces a concentrate volume of two million gallons per day when operated at full capacity. Table 1 shows the characteristics of the KBH concentrate [3].

**Table 1: Characteristics of the KBH Concentrate**

| | Concentration | | |
|---|---|---|---|
| **Parameter** | **Min** | **Average** | **Max** |
| Cl | 4840 | 5089 | 5540 |
| $SO_4$ | 1050 | 1111 | 1200 |
| Alk-T | 400 | 427 | 445 |
| Ba | 0.042 | 0.31 | 0.48 |
| EC (uS/cm) | 16600 | 18122 | 20200 |
| Fe | 0.03 | 0.07 | 0.12 |
| T Hard | 2050 | 2291 | 2430 |
| Mn | 0.11 | 0.17 | 0.22 |
| ortho-P | 0.11 | 0.19 | 0.49 |
| pH (pH units) | 7.6 | 7.8 | 8 |
| Ca | 281 | 608 | 937 |
| K | 43 | 113 | 759 |
| Mg | 85.7 | 161 | 183 |
| Na | 1730 | 2810 | 3260 |
| Sr | 8.74 | 17.1 | 30.1 |
| $SiO_2$ | 26.9 | 131 | 173 |
| TDS | 10300 | 10722 | 11200 |
| Temp (°C) | 23.3 | 25.4 | 26.3 |
| Turb (NTU) | 0.06 | 0.3 | 1.94 |

(Concentrations in mg/L except as noted in table)

Note that the average concentration of silica is 130 mg/L. There is a high risk of membrane fouling due silica polymerization if this concentrate is treated in a conventional RO system. During the testing using the CERRO process, silica concentrations of 800 mg/L and above were reached. This is because the treatment time was faster than the polymerization time of the silica. Figure 3 shows the silica concentration at different times during one experiment.

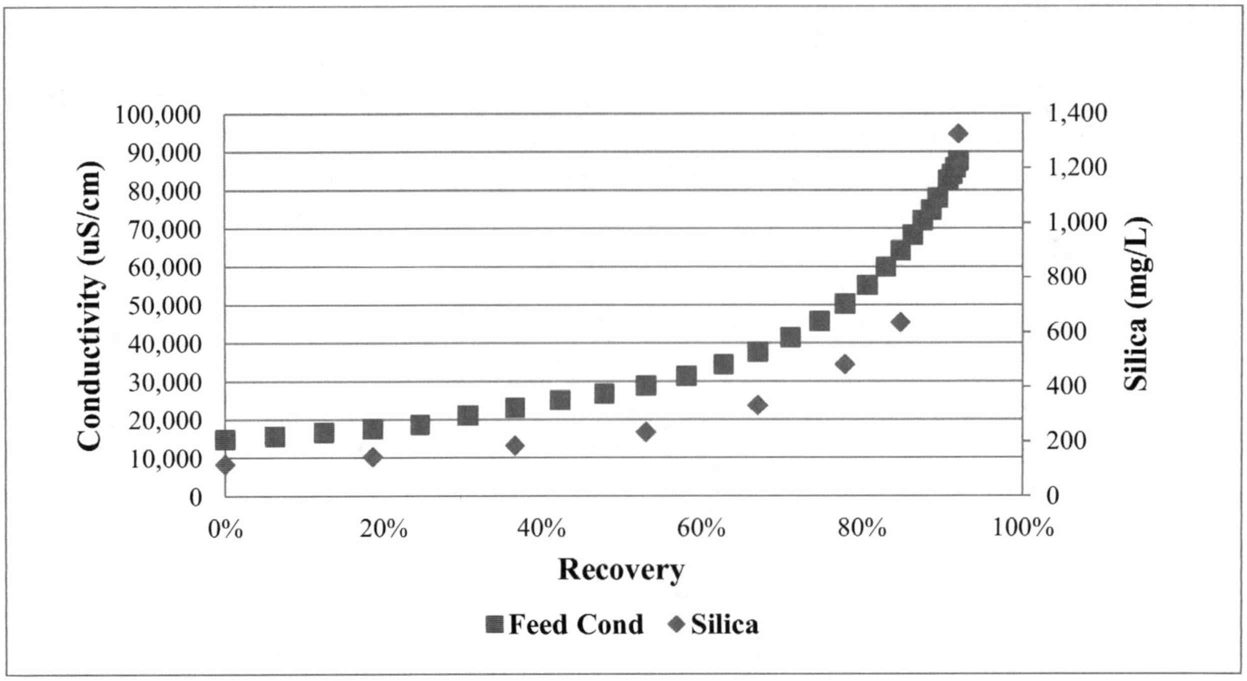


**Figure 3: Silica concentration at Different Times during Testing**

During this test, the recovery of the system was 92%. The silica concentration was 1,325 mg/L (average of a triplicate). Precipitation occurred after the system was shut down and the final concentrate was removed from the feed tank. After a quick flush with permeate water, the SWRO membrane did not showed any signs of being fouled. Several tests have been made treating KBH concentrate using the CERRO process. Table 2 shows the results of some tests.

**Table 2: Results of Several Tests in the KBH Plant**

| | CERRO Final Concentrate | | | | | | | | |
|---|---|---|---|---|---|---|---|---|---|
| **Test (date)** | **Ca** | **Mg** | **SO4** | **SiO2** | **Cl** | **TDS** | **Cond (uS/(cm)** | **Recovery** | **Notes** |
| 8/17/2010 | 3,500 | 851 | 18,500 | 597 | 28,250 | 71,347 | 82,600 | 84% | Spiked with $Na_2SO_4$ |
| 10/1/2010 | 4,100 | 1,215 | 13,540 | 183 | 35,900 | 65,660 | 93,500 | 90% | |
| 10/8/2010 | 3,300 | 1,337 | 12,400 | 178 | 32,500 | 67,347 | 81,800 | 88% | |
| 11/12/2010 | 600 | 851 | 17,480 | 368 | 21,500 | 60,100 | 69,800 | 84% | Spiked with $Na_2SO_4$ |
| 7/9/2010 | 4,000 | 972 | 11,080 | 247 | 34,500 | 69,652 | 87,500 | 82% | Spiked with NaCl |
| 6/10/2010 | 4,080 | 886 | 12,900 | 780 | 27,750 | 54,540 | 79,800 | 90% | |
| 12/21/2010 | 3,800 | 3,645 | 11,720 | 382 | 39,000 | 76,154 | 86,500 | 91% | Spiked with $MgCl_2$ |

(Concentrations in mg/L except as noted in table)

In some of the experiments, the KBH concentrate was spiked with different salts (i.e. NaCl, $Na_2SO_4$, $MgCl_2$). This was made in order to change the chemistry of the KBH concentrate to simulate a different water source. As can be seen in table 2, the change in the chemistry of the KBH concentrate did not affect the performance of the CERRO process. More specifically, in the test made in 11/12/2010, it can be seen that the calcium concentration is relatively low. This is because calcium sulfate precipitated during the test. Even though there was precipitation inside the membrane, a fast rinse with permeate after the test avoided any damage to the membrane.

### Case Study 2: Treatment of Alamogordo, New Mexico Brackish Water

Alamogordo groundwater is naturally contaminated with gypsum. Analysis made in the groundwater showed that 60% of the total dissolved solids are calcium and sulfate ions. The Brackish Groundwater National Desalination Research Facility (BGNDRF) is a research facility where desalination systems are tested using this specific groundwater. The CERRO process was tested with this brackish water having recoveries from 60% to 90%. Table 3 shows the characteristics of Alamogordo groundwater.

**Table 3: Characteristics of Alamogordo Brackish Water**

| | Cations (mg/L) | | | Anions (mg/L) | | | | |
|---|---|---|---|---|---|---|---|---|
| **Sample ID** | **Ca** | **Mg** | **Na** | **Cl** | **F** | **NO3** | **SO4** | **TDS (mg/L)** |
| **Alamogordo BW** | 421 | 254 | 676 | 533 | 2.43 | 28 | 3,001 | 5,942 |

The presence of high concentrations of calcium and sulfate ions makes the treatment of this groundwater very difficult. The CERRO process was used in order to evaluate the maximum recovery that can be achieved using reverse osmosis. Table 4 shows the results of a test made using the CERRO process. In this experiment, the ground water was treated in a RO pilot unit until a 50% recovery was reached. The concentrate of this unit was treated in the CERRO process until precipitation occurred. The CERRO process achieved a 36% recovery which gave a total recovery of 68% of the ground water. No antiscalants were used in this test.

**Table 4: Final Results of a Test Where no Antiscalants Were Used**

| | Cations (mg/L) | | | Anions (mg/L) | | | | |
|---|---|---|---|---|---|---|---|---|
| **Sample ID** | **Ca** | **Mg** | **Na** | **Cl** | **F** | **NO3** | **SO4** | **TDS (mg/L)** |
| RO Concentrate | 1,020 | 563 | 1,320 | 1,209 | 2.76 | 53.55 | 6,053 | 10,950 |
| CERRO Final Feed | 530 | 846 | 1,994 | 1,817 | 1.93 | 77.59 | 7,599 | 13,531 |
| CERRO Permeate | 5.26 | 2.89 | 10.98 | 35.68 | 0.33 | 5.21 | 26.05 | >100 |

The high content of calcium sulfate in the groundwater is a limiting factor for the treatment process. However, it was found that calcium sulfate is more soluble when sodium chloride is present in the water

[5]. One batch of groundwater was treated using the CERRO process. However, the concentration of sodium chloride was raised to 10,000 mg/L. This was made in order to increase the solubility of calcium sulfate and increase the final recovery. Figure 4 shows two tests made using the CERRO process. One Batch was made with the addition of sodium chloride and the other batch was used as a control with no changes in the characteristics of the groundwater.

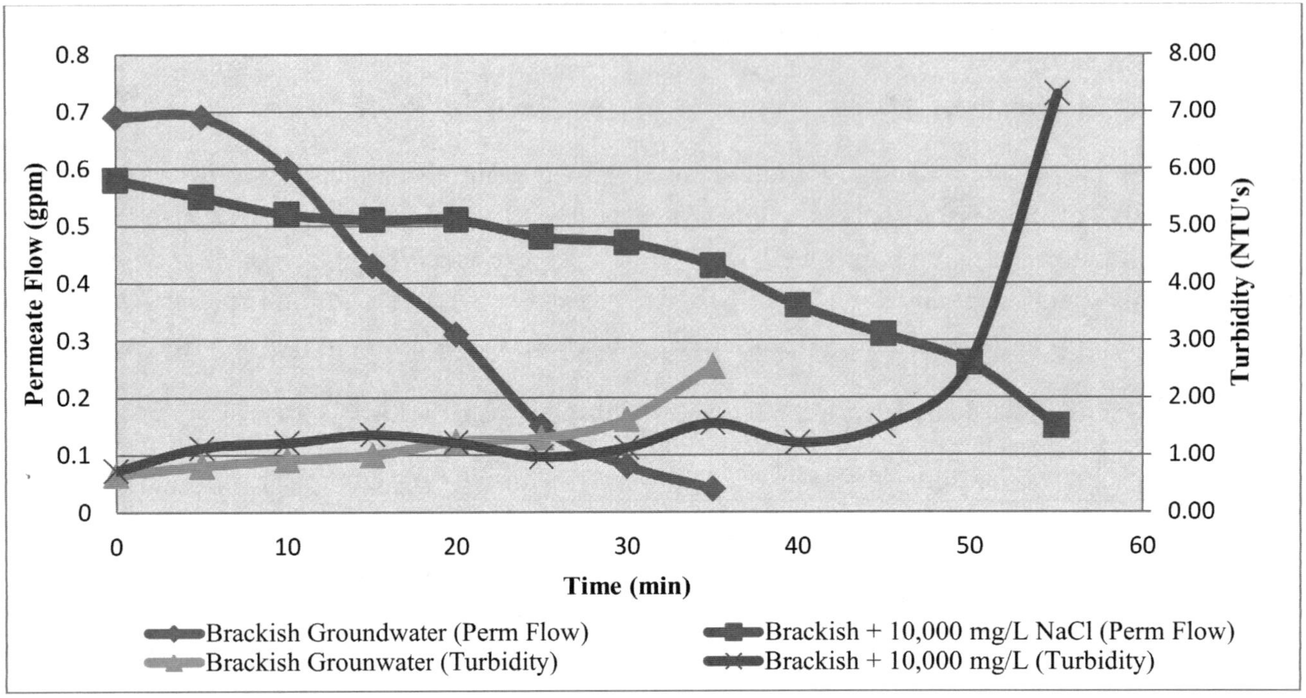


**Figure 4: Permeate Flow and Turbidity of Two Batches Using the Cerro Process**

The recovery in the experiment used as a control was 68% before precipitation. The recovery in the batch with the addition of sodium chloride was 87%. The addition of sodium chloride in the water clearly increased the recovery of the system. During this research, it was found that the main limitation for the CERRO process was the high concentrations of calcium sulfate. However, the system showed an excellent performance at supersaturation concentrations.

**Case Study 3: Treatment of Alto, New Mexico Brackish Water**

Brackish groundwater found in Alto, New Mexico was treated using the CERRO process. This water has high concentrations of calcium sulfate. Test done using the CERRO process showed that 96% recovery can be achieved. Table 5 shows the characteristics of the groundwater.

**Table 5: Characteristics of Alto Groundwater**

| | Concentration (mg/L) | | | | | |
|---|---|---|---|---|---|---|
| | **Ca** | **Na** | **Cl** | **PO4** | **SO4** | **TDS** |
| **Alto Brackish Water** | 237 | 105 | 165 | <1 | 887 | 1,779 |

Several tests were made using the CERRO process with this groundwater as a feed. The system reached a recovery higher that 90% in every test. It was found that precipitation occurred when 94% to 96% recovery was reached. Even though the system reached recoveries above 90%, the production of drinking water from this water source resulted to be cost prohibitive due to the high cost related to the management of the concentrate. In order to be a suitable project, a recovery of 98% or higher is necessary. A research project is being developed where the calcium sulfate is removed from the concentrate by the addition of lime. The excess in calcium forces the precipitation of calcium sulfate. After the calcium sulfate is removed, the concentrate will be treated in the CERRO process a second time increasing the final recovery.

The CERRO process is a novel process where brackish water and RO concentrate can be treated. The system has shown an excellent resistance to membrane fouling because of two main reasons. First, the compact design of the system allows a fast operation time which allows the system to work under supersaturation conditions. Second, the capacity of the system to work in a batch configuration permits that the membranes can be cleaned continuously. This characteristic of the system is particularly helpful to protect the membranes even under supersaturation conditions. Finally, the main limitation of the system is the precipitation of calcium sulfate. Future projects will intent to find a suitable method to remove the calcium sulfate from the concentrate, treat the concentrate a second time and increase the maximum recovery of the system.

Bibliography

*1. EPWU. Setting the Stage for the Future. . 2007;2012.*

*2. Delgado G. Treatment of RO concentrate using VSEP technology. . 2009.*

*3. Tarquin AJ, Fahy MP, Balliew JE. Concentrate Volume Reduction Research in El Paso, Texas. ASCE Conf Proc. 2010;371:358.*

*4. Gloede M, Melin T. Physical aspects of membrane scaling. Desalination. 2008;224:71-75.*

*5. Messinger A, Anna Hahn D. A Dictionary of Chemical Solubilities: Inorganic. In: , 1921:951.*

Demonstration of Zero Discharge Desalination Technology for Concentrate Management

Tom Davis[1], Brad Biagini[2], Malynda A. Cappelle[1], Bernie Mack[2], and Emily Gilbert[2]

[1]*Center for Inland Desalination Systems, University of Texas at El Paso*
[2]*Veolia Water Solutions & Technologies*

Zero Discharge Desalination (ZDD) technology utilizes electrodialysis metathesis (EDM) to remove troublesome ions from membrane desalination concentrate so that the solution can be further processed to increase the yield of fresh water from brackish groundwater that is difficult to treat by RO or NF alone. ZDD technology is particularly applicable to treatment of groundwater containing appreciable concentrations of scaling salts such as gypsum ($CaSO_4$), calcium phosphate, barium sulfate, etc. In contrast to conventional electrodialysis, which has two membranes and two solution compartments in a repeating unit, EDM utilizes four ion-exchange membranes, two depleting compartments, and two concentrating compartments. The arrangement of membranes and transport of ions in one EDM repeating unit are illustrated in Figure 1. More than 100 of these repeating units are installed between a single pair of electrodes in a commercial-size EDM stack.

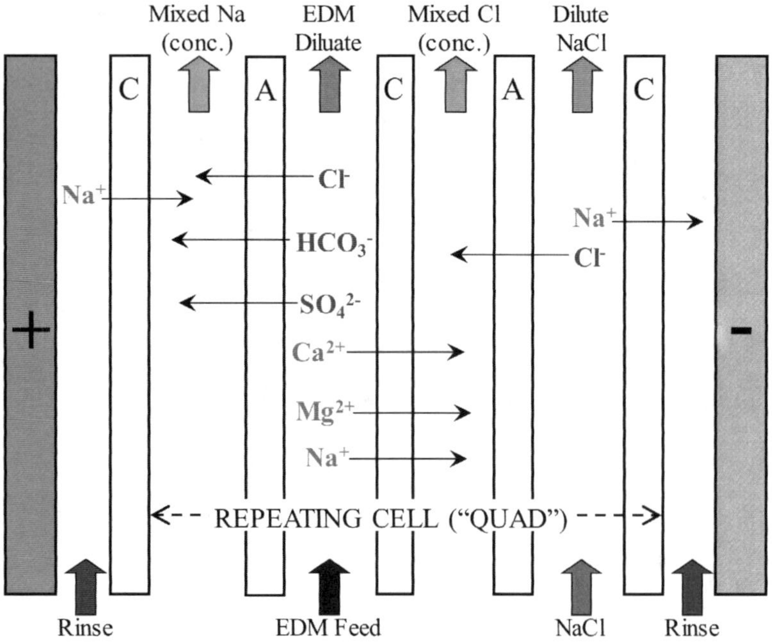

Figure 1. Membrane Arrangement and Ion Transport in EDM.

The RO concentrate flows through one of the depleting compartments, and the diluate can be returned to the system as RO feed or blended with RO permeate depending on project-specific water quality requirements. A solution of NaCl flows through the second depleting compartment. Application of an electric potential to the electrodes of the EDM stack causes the anions from the RO concentrate to migrate through anion-exchange membranes where they meet $Na^+$ ions from the NaCl compartment to form a concentrated salt stream rich in $Na_2SO_4$. Similarly, the cations from the RO concentrate meet $Cl^-$ ions from the NaCl stream to form another concentrated salt stream rich in $CaCl_2$. Because $Na_2SO_4$ and $CaCl_2$ are highly soluble salts, the concentrations of these two streams can achieve very high levels with minimal scale potential. In fact, when $CaSO_4$ is the dominant salt in the groundwater, all of the salts removed from the groundwater can be carried away in the two EDM concentrate streams, each containing only about 1% of the water that was in the RO feed. Thus the ZDD process has the potential to recover 98% of the water from groundwater compared to typical yields of about 75% when the same groundwater is treated by RO alone.

The high recovery obtainable by the ZDD process is particularly beneficial in arid regions where groundwater is scarce and concentrate disposal is problematic. Typical disposal methods – dumping in a sewer, deep-well injection, evaporation ponds, or hauling by truck – are impractical, too expensive, or damaging to the environment in many locations. At the very least, the ZDD technology can reduce the costs and or size of conventional disposal methods leading to lower operational or capital costs. The ZDD technology has the advantage that the salts are in a highly concentrated solution. In this concentrated state, the salts can be separated and recovered as marketable products. In one example of the ZDD process illustrated in Figure 2, the salt-depleted stream from the EDM is returned to the RO feed, and the two EDM concentrate streams are combined to precipitate $CaSO_4$, which has potential markets for soil augmentation or the production of gypsum wallboard. The supernatant from the precipitation contains NaCl that can be recovered and reused in the EDM stack. Continuing laboratory research is directed toward recovery of useful products from the EDM concentrate streams so that even more water can be recovered to further approach Zero Discharge Desalination.

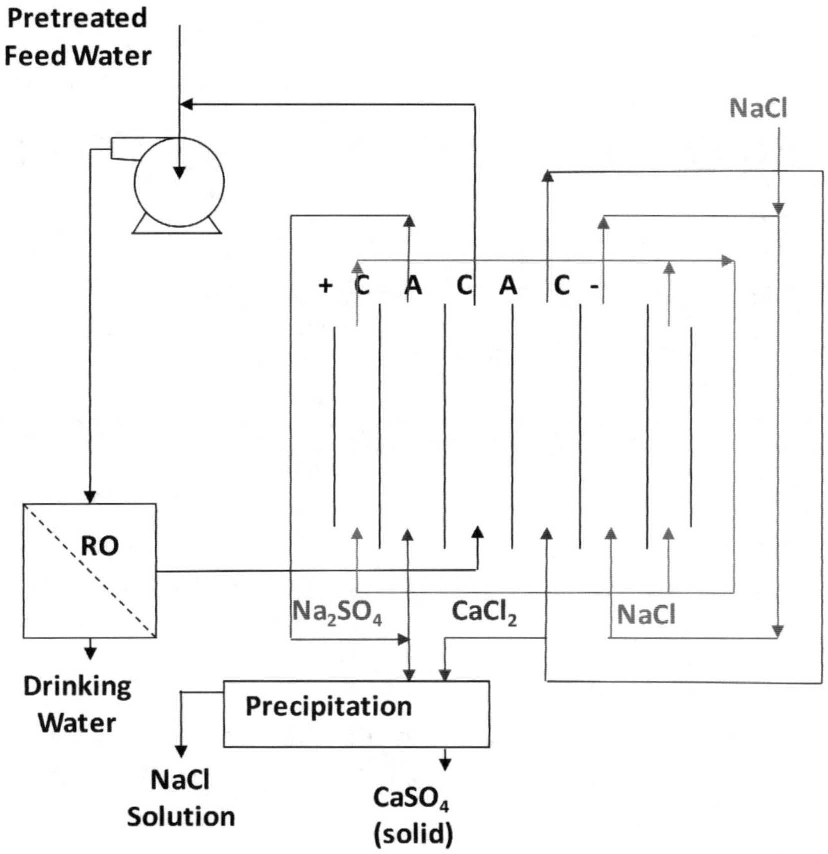


Figure 2. Flow Diagram of ZDD Process

The ZDD technology has been evaluated in progressively larger pilot studies at the Brackish Groundwater National Desalination Research Facility (BGNDRF) in Alamogordo, New Mexico. From the first pilot with 1 gpm capacity to the latest with 20 gpm capacity, refinements to the process have been incorporated. Major efforts have been directed toward dealing with silica that occurs in the groundwater of that region. In the 1-gpm pilot, precipitation of silica was observed in the stream that circulated through the EDM diluate compartments and the RO concentrate. Ions were removed from that stream by the EDM, and water was removed by the RO. However, because silica is not ionized appreciably at neutral pH, the silica accumulated in that circulating stream until it became supersaturated and precipitated. Precipitation of the silica could be prevented by purging some of that stream, but the purge stream represented a loss of water from the process. The water loss was reduced by treating the purge stream to remove silica and then returning the liquid to the process. Although the silica removal was successful, it added complexity and chemical cost to the ZDD process. The ultimate solution was to replace the RO membranes with nanofiltration (NF) membranes. Certain NF membranes are highly permeable to silica, and the presence of silica is not harmful to drinking water. With the selected NF membranes install 20-gpm pilot plant, and the silica problem was completely solved. Elimination of the silica purge allowed 98% of the brackish water to be produced as fresh water.

Table 1. Concentrations (mg/L) of solutes in the ZDD process streams.

| | | NF Streams | | | EDM Streams | | | |
|---|---|---|---|---|---|---|---|---|
| **CONSTITUENT** | **Source** | **Feed** | **Perm** | **Conc** | **Feed** | **Diluate** | **Mixed Cl** | **Mixed Na** |
| **ANIONS** | | | | | | | | |
| F | 1.7 | 1.4 | 0.9 | 2.5 | 1.8 | 0.8 | ND | 61.3 |
| Cl | 290 | 210 | 260 | 160 | 140 | 100 | 89,000 | 11,766 |
| $SO_4$ | 1,300 | 1,300 | 9.50 | 3,600 | 2,600 | 1,300 | 1,600 | 90,764 |
| **CATIONS & METALS** | | | | | | | | |
| Ca | 250 | 220 | 19 | NT | 380 | 160 | 17,000 | 716 |
| Mg | 94 | 94 | 7.50 | NT | 190 | 86 | 7,400 | 310 |
| Na | 370 | 350 | 150 | NT | 580 | 350 | 21,000 | 57,078 |
| **OTHER CONSTITUENTS** | | | | | | | | |
| $SiO_2$ | 24.0 | 27.0 | 23.0 | NT | 33.0 | 33.0 | 5.7 | 11.4 |
| pH | 7.4 | 7.0 | 7.0 | 7.0 | 6.9 | 6.6 | 1.8 | 7.9 |
| $HCO_3$ | 205 | 110 | 70 | 135 | 128 | 43 | NT | 4,447 |
| Conductivity | 2,857 | 2,724 | 872 | 5,021 | 4,271 | 2,404 | 124,800 | 109,000 |
| TDS (at 180 °C) | 2,300 | 2,180 | 471 | 4,980 | 3,840 | 1,940 | 133,000 | 146,834 |

The concentrations of ions in the process streams of the 20-gpm pilot plant are shown in Table 1. The concentrations of calcium and sulfate ions in the source water and the EDM concentrate streams are highlighted. The concentrations of both the calcium and the sulfate increased by a factor of almost 70. Also noteworthy is the fact that the NF membrane was so permeable to silica that the silica level in the feed and diluate of the EDM were only slightly elevated.

# (378c) Water Purification with Closed Circuit Reverse Osmosis

**Richard Stover**, Desalitech, Ltd., Newton, MA

The demand for clean water for industrial, agricultural and domestic use is increasing around the world. Reverse osmosis (RO) is an important method for water purification and desalination. Advances in membrane performance have increased permeate recovery, improved reliability and reduced energy consumption, however, broader application of RO has been limited, in part, by its energy requirements, costs, risks of fouling and scaling and operational challenges posed by source water changes. An improved technology that further increases permeate recovery, lowers costs, lowers energy consumption and increases the reliability and flexibility of RO would help advance the beneficial and efficient uses of locally-produced, sustainable water sources.

Closed circuit (CC) RO is being applied for water purification and desalination in commercial installations in several international locations. CC RO uses standard equipment and membranes in a batch-like configuration that operates at a lower average feed pressure than conventional RO, thereby reducing energy consumption. The process exposes membranes to varying feed compositions as the system concentrates and dilutes and provides high cross flow across short membrane arrays. These conditions improve membrane performance by improving flux distribution and enhancing cross flow. The process also has the potential to prevent fouling and scaling and disrupting bio-film formation and scale precipitation which, in turn, may reduce chemical use and membrane clean-in-place frequencies. The permeate recovery percentage can be adjusted at the control panel without modification of system hardware, giving operators the means to adjust system performance as necessary to cope with source water changes. A recovery range of 15 to over 97% has been demonstrated in single-stage units, and energy consumption reduced by 35%.

The objective of this study is to evaluate CC RO as an alternative water treatment process to conventional RO for achieving high recovery and brine minimization. The author presents the technical principles of the CC RO processes and operating data from running plants. Process design and performance models are developed and validated with the purpose of adapting CC RO for new applications. This paper will help stakeholders better understand this emerging technology and the potential general applicability of CC RO processes in desalination and water reuse.

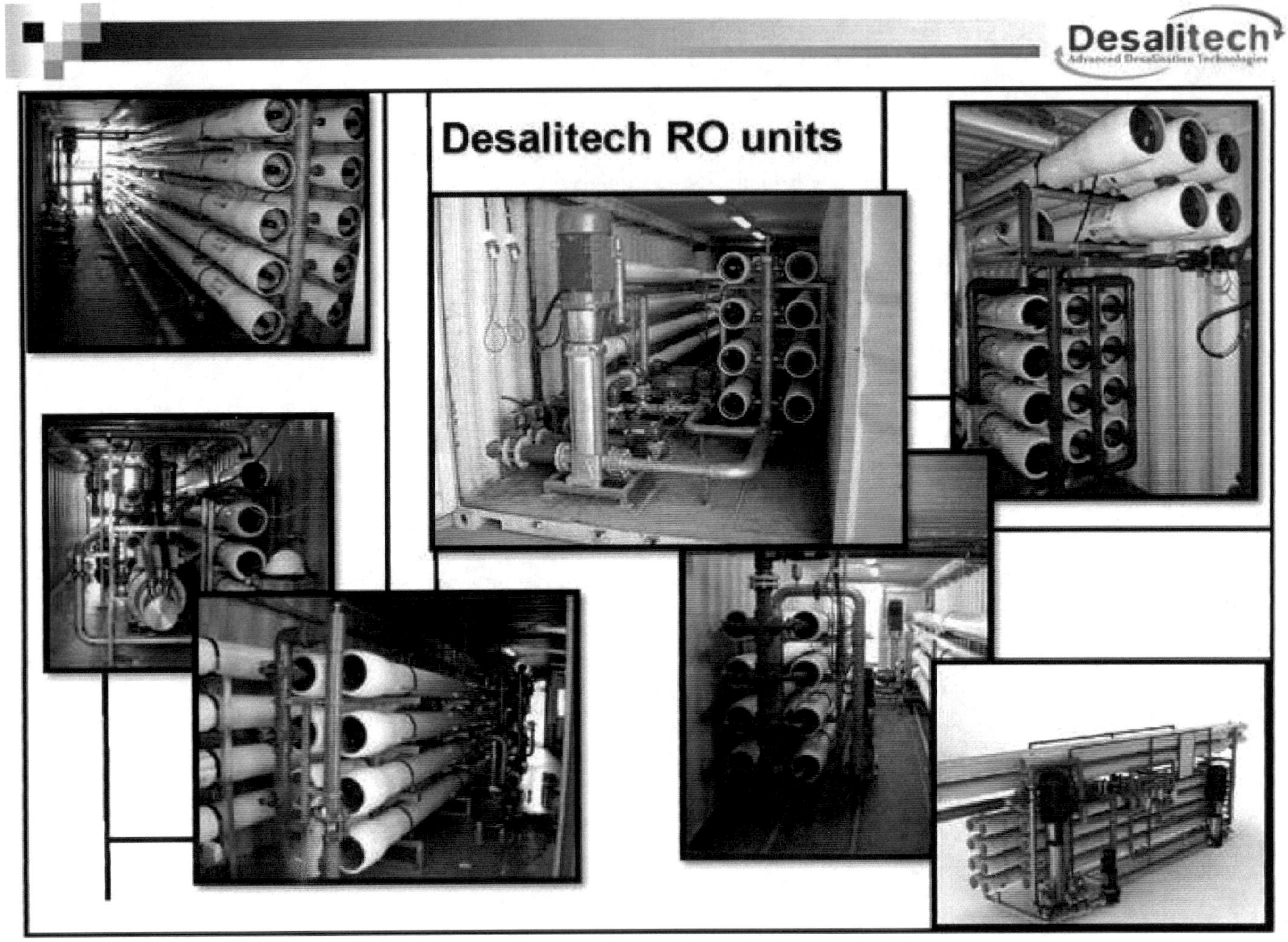
Desalitech
Advanced Desalination Technologies
Desalitech RO units

# (378d) A Detail Analysis of the Performance of Different Antiscalants in Calcium Sulfate Supersaturated Brackish Water

Anthony Tarquin and **Guillermo Delgado**, Civil Engineering, University of Texas at El Paso, El Paso, TX

**A Detail Analysis of the Performance of Different Antiscalants in $CaSO_4$ Supersaturated Brackish Water**

**Anthony Tarquin, PhD.; Guillermo Delgado, PhD Candidate**

**University of Texas at El Paso**

Reverse osmosis (RO) desalination is an economic method to remove dissolved solids from brackish water. However, the presence of sparingly soluble salts in the water (e.g. $CaSO_4$, $CaCO_3$) reduces the recovery of these systems. If the solubility limit of these salts is reached, precipitation is likely to occur causing fouling and scaling of the membranes. It is at this point where the use of antiscalants greatly improves the recovery of desalination systems.

This research presents the evaluation of different antiscalants in brackish water supersaturated with calcium sulfate. A total of eight different antiscalants from five different manufacturers were tested. Each antiscalant was tested at different calcium sulfate concentrations and different antiscalants dosages. Brackish water from Alamogordo, New Mexico was used for this project. The brackish water was saturated using 1 Molar solutions of sodium sulfate and calcium chloride. The efficiency of the antiscalants was measured by the extent that the antiscalants increased the induction time of calcium sulfate precipitation in the water. The turbidity of the water was monitored in time intervals to identify the time of calcium sulfate precipitation. A saturated solution without antiscalant served as a control. The results showed that the type of antiscalant is an important factor that significantly affects the induction time of calcium sulfate precipitation. Figure one shows the results from one test done using different types of antiscalants.

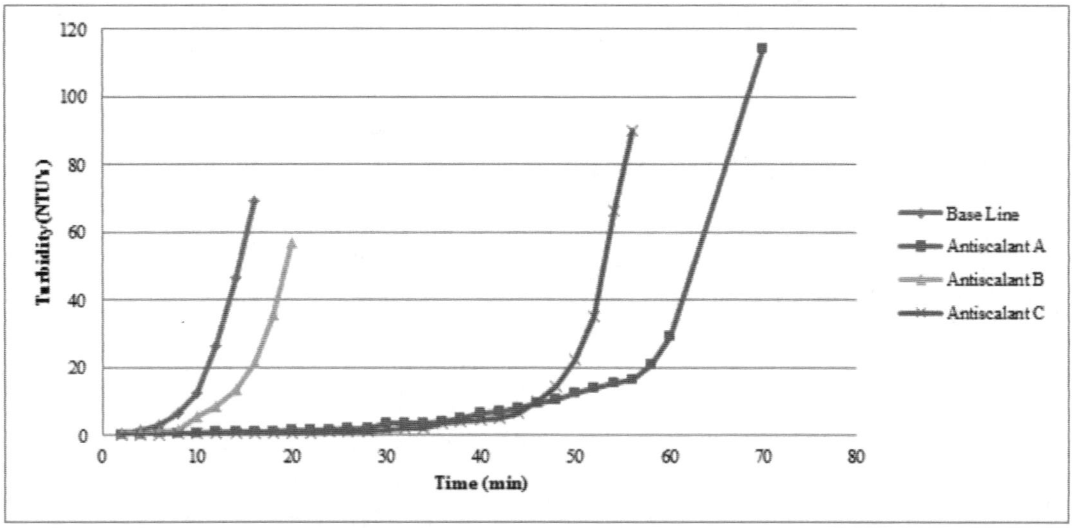


Figure 1: Antiscalant Performance at 0.03 M $CaSO_4$ Concentration and 0.8 ppm of Antiscalant

Additionally, the results showed that the dosage of the antiscalant is a factor for increasing induction time. However, the results showed that antiscalants are not efficient at high supersaturation concentrations (0.1 M) of calcium sulfate, no matter the dose. In the same way, the results showed that antiscalants only increase the induction time of calcium sulfate precipitation, but they do not increase the solubility of calcium sulfate. Figure 2 shows the results of a test done using different concentrations of the same antiscalant.

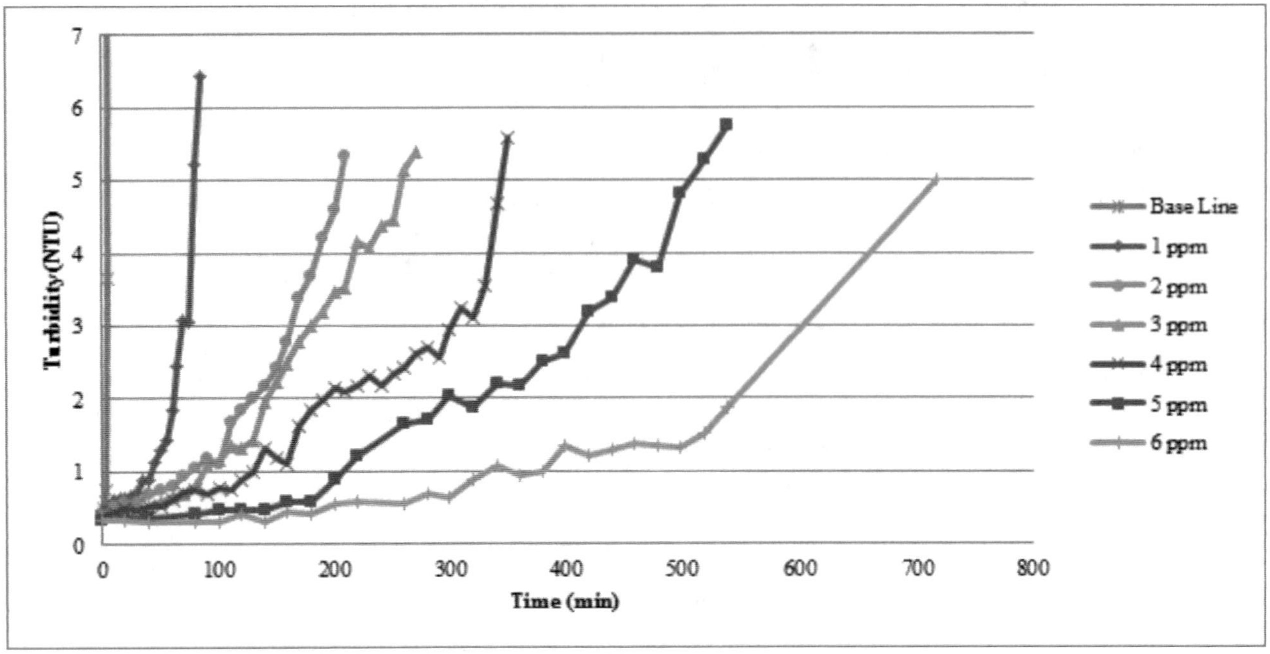


Figure 2: Different Concentrations of the Same Antiscalant at 0.05 M $CaSO_4$ Concentration.

The selection of the best antiscalant is a very important step during the design of a membrane desalination system, especially batch processes such as the CERRO process. This research showed that the selection process should include factors such as type and dosage of antiscalant. Recovery rates can be increased and operation costs can be reduced by optimizing system design.

# (378e) Purity of NaCl in the Electrodialysis Metathesis (EDM) Process for Water Desalination

**Lucy Mar Camacho**[1], Tom Davis[2] and Malynda A. Cappelle[1], (1)Center for Inland Desalination Systems, University of Texas at El Paso, El Paso, TX, (2)Center for Inland Desalination Systems (CIDS), University of Texas at El Paso, El Paso, TX

Electrodialysis metathesis (EDM) is a component in the zero discharge desalination (ZDD) process that maximizes water production and salt recovery from desalination. In the EDM process the concentrated effluent from reverse osmosis (RO) or nanofiltration (NF) is treated to recover minerals, such as gypsum and magnesium chloride, as well as additional pure water. The minerals are separated and collected in two of the four compartments of the EDM stack that are separated by ion exchange membranes. One compartment will contain a concentrated mixture of sodium salts and another compartment will contain a concentrated mixture of chloride salts. The third compartment will contained the minerals-rich feed water to be treated. The fourth compartment is filled with sodium chloride to provide the ions for the metathesis reaction. The concentrated mixtures are generated when a voltage is applied to the repeated four-compartment units of the EDM stack. Once the concentrated streams exit the system they can combine to precipitate gypsum.

The early precipitation of calcium sulfate in one of the concentrate compartments can adversely affect the potential recovery of other useful salts. Furthermore, the EDM stack can be clogged and damaged. Commercially available sodium chloride may contain sodium sulfate and other minerals that will promote the formation of gypsum in the stack. In this study we are adding impurities to the sodium chloride compartment to determine the lowest purity allowed by the system. Preliminary experiments have shown that sodium chloride with concentrations of 1,000 mg/L of sodium sulfate doesn't affect the performance of the salt separation process. This results suggest that the cost of EDM can be further reduced by reducing the need of adding high quality chemicals to the process.

# (385a) Assessing the Potential for Large-Scale Utilization of Fossil-Fuel Derived CO2

Jared Ciferno, U.S. Department of Energy, National Energy Technology Laboratory, Pittsburgh, PA and **John J. Marano**, JM Energy Consulting, Inc., Gibsonia, PA

Barring the imposition of a carbon tax (or equivalent cap-and-trade program) and corresponding international agreements, it is highly unlikely that carbon capture and storage/sequestration (CCS) will be implemented in any meaningful way in the near future. This is due to inherent inefficiency and higher costs associated with this option for mitigating climate change. Unfortunately, the science of climate change suggests that the world community must begin to act sooner rather than later. From an Industrial Ecology viewpoint, a preferable approach to carbon storage in perpetuity would be to re-use the carbon associated with the fossil-fuel derived $CO_2$ already being omitted. However, options for directly utilizing $CO_2$ are quite limited, with enhanced oil recovery (EOR) being the only physical use of $CO_2$ of any appreciable size. The concept of converting $CO_2$ to useful products has been explored in the past, but has fallen from favor in many policy circles due to the lack of a clear vision of how re-utilization of fossil carbon could be implemented. Given current in-action, it is worth re-visiting this concept.

Energetically, it is clear that the large-scale conversion of $CO_2$ to useful products requires a source of external energy that is not fossil-fuel derived; otherwise, the conversion process becomes itself a source of $CO_2$ emissions. Sources for the required energy could be direct solar or nuclear energy, or electricity generated from these primary energy resources. Obviously, a source of $CO_2$ is also required; however, all sources are of equal value. Atmospheric and mobile emissions are of little value due to the problems associated with collecting and concentrating them. Whereas, $CO_2$ from stationary point sources, and in particular large point sources such as coal and natural gas-fired power plants, are the "low hanging fruit" for implementing carbon re-use. Finally, useful products from $CO_2$ must be identified. It is suggested here that instead of pursuing a piecemeal approach consisting of an assortment of finished chemicals and material goods, a holistic approach be taken. Currently, the only existing commodity market that is big enough to absorb the large scale re-utilization of $CO_2$ lies in the petroleum sector. Thus, the basis for any carbon re-utilization platform should be the production of a petroleum substitute, a "synthetic" oil that may be refined and converted into a multitude of finished products.

This paper first compares and contrasts carbon storage with carbon utilization, and then moves on to quantify the scale of any carbon re-utilization effort; as well as, $CO_2$ life-cycle limitations of re-use. After all, in this approach much of the synthetic oil, like petroleum, will end up being (re-) combusted as an end-use transportation fuel. Thus, carbon re-use can only be considered as an interim strategy for cost effectively transitioning away from fossil fuels. We also briefly touch on the implications of large scale carbon re-use for CCS R&D, since the requirements for $CO_2$ capture, transport and short-term storage will be quite different than what is anticipated for long-term $CO_2$ sequestration. Finally, we briefly discuss the possible economic implications of this strategy. A follow-on paper at this conference identifies and analyzes what the present and best future options for producing a $CO_2$-derived synthetic oil might be.

# (385b) Some Perspectives On the Use of Solar and Other Renewable Energy to Facilitate Carbon Management

**Robert Wegeng**, Pacific Northwest National Lab, Richland, WA

The recent and ongoing development of economical solar and wind energy systems presents a number of new options for carbon management. The purpose of this presentation will be to discuss how engineered solar thermochemical processes might be used to reduce carbon emissions from fossil-fuel powerplants and gas-to-liquids plants and how concentrated solar energy might be used to capture carbon from industrial emissions or the atmosphere for sequestration or use. Perspectives will include economics and Second Law efficiencies. Particular attention will be applied to technologies and strategies that might be applied in the U.S. Pacific Northwest.

# (385c) Carbon Capture and Storage Directly From Flue Gas: The Effects of Chemical and Biological Catalysts On Magnesium Carbonate Formation

**Edward J. Swanson**, Chemical Engineering, Columbia University, New York, NY, Patrick V. Brady, Sandia National Laboratories, Albuquerque, NM and Ah-Hyung Alissa Park, Earth and Environmental Engineering & Chemical Engineering, Columbia University, New York, NY

The direct adsorption of $CO_2$ from flue gas and precipitation of mineral carbonates presents a transformative alternative to carbon capture and storage (CCS) schemes involving conventional post-combustion capture of $CO_2$ and its geologic storage. $CO_2$ capture processes generally involve energy intensive solvent regeneration and CO2 compression steps. Absorbing the carbon into an aqueous system and precipitating it as a thermodynamically stable solid carbonate in a single technology avoids both of these energy intensive steps in CO2 capture, while providing inherent storage of CO2. However, in order for this novel CCS technology to be viable, the kinetics of both CO2 absorption and mineral dissolution need to be sufficiently fast, particularly at the low partial pressure of CO2 commonly found in flue gas. In this study, both chemical (i.e., catechol and oxalate) and biological (i.e., carbonic anhydrase) catalysts were used to enhance the rates of CO2 hydration and serpentine dissolution. While these catalysts showed significant improvement over natural carbon mineralization kinetics, it was also important to investigate the effect of this catalytic system on the final formation of magnesium carbonate in terms of its chemistry and morphology. Experiments were performed in a lab-scale three-phase fluidized bed reactor, and optimum reaction conditions for carbonate precipitation were investigated.

# (385d) Environmentally Benign Processes for Absorption and Chemical Fixation of CO2

**An-Hua Liu**, Ran Ma and Liang-Nian He Sr., State Key Laboratory of Elemento-Organic Chemistry, Nankai University, Tianjin, China

Carbon dioxide as the main green house gas is widely blamed to pay most of the responsibility for global warming. Hence, controlling $CO_2$ emission and further reducing its accumulation emerges to be an urgent and challenging research topic. $CO_2$ capture and storage/sequestration (CCS) to control its emission as well as chemical utilization of the captured $CO_2$ (CCU) has been considered as the rational technique to reduce or mitigate the so-called green-house effect. However, being the sustainable carbon source, $CO_2$ as a safe, cheap and abundant $C_1$ building block in organic synthesis also comes into light [1-3]. So chemical fixation of $CO_2$ into industrially useful materials and high value-added products turns to be one of the most attractive methods in recent years.

Amine-functionalized $CO_2$ absorbents have been intensively investigated in numerous studies due to the easy preparation, high $CO_2$ affinity and robust reversibility, whereas the intrinsic nature of 2:1 absorption stoichiometry could restrain their further development for pursuing of superior capture efficiency. In present work, such inherent disadvantage was circumvented by employing modified amino acid salts in polyethylene glycol, affording approximative 1:1 ratio $CO_2$ absorption. The used absorbent can be readily reverted and recycled without loss of its capture capacity at least 5 turns. The structure of captured absorbent species was identified by means of NMR and in situ IR spectroscopy, as well as computational calculations.

Meanwhile, using commercial available and recyclable basic ion-exchange resins, one kind of polystyryl supported tertiary amines, an efficient fixation of $CO_2$ by aziridines under mild conditions were achieved, leading to formation of 5-aryl-2-oxazolidinone with excellent regio-selectivities. Notably, neither solvent nor any additive was required, and the catalyst can be recovered by simply filtration and directly reused at least five times without significant loss of catalytic activity and selectivity [4]. The present protocol was successfully applied to cycloaddition reactions of epoxides/propargyl amines with $CO_2/CS_2$. In general, this solvent-free process thus represents environmentally friendly catalytic conversion of $CO_2$ into value-added chemicals and has great potential to be applied in various continuous flow reactors in industry.

**References**

[1] Aresta, M.; Dibenedetto, A. *Dalton Trans*. 2007, 28, 2975.

[2] Yang, Z.-Z.; He, L.-N.; Zhao, Y.-N. *Energy Environ. Sci*., 2011, 4, 3971-3975.

[3] He, L.-N.; Wang, J.-Q.; Wang, J.-L. Carbon dioxide chemistry: examples and challenges in chemical utilization of carbon dioxide, *Pure Appl. Chem*. 2009, 81(11), 2069-2080.

[4] Liu, A.-H.; He, L.-N.; Peng, S.-Y.; et al. *Sci. China Chem.*, 2010, 53(7), 1578-1585.

# (385e) Process Ecoefficiency of Methanol and Dmc Production From CO2: A Comparative Analysis

Camila Ribeiro[1], José L. Medeiros[2], **Ofélia Araújo**[2], Paulo Coutinho[3] and Rita M.B. Alves[4], (1)Escola de Química (H2CIN), UFRJ - Universidade Federal do Rio de Janeiro, Rio de Janeiro, Brazil, (2)Escola de Química (H2CIN), UFRJ - Universidade Federal do Rio de Janeiro, Rio de Janeiro, Brazil, (3)management of innovation, Braskem, São Paulo, Brazil, (4)Braskem, São Paulo, Brazil

**Process Ecoefficiency of Methanol and DMC Production from $CO_2$: A comparative analysis**

Camila F. Ribeiro[1], Ofélia de Q. F. Araújo[1], José Luiz de Medeiros[1], Rita[2], Paulo Coutinho[2]

[1]Departament of Chemical Engineering, Escola de Química da Universidade Federal do Rio de Janeiro. Sala E-201, Bloco E, Centro de Tecnologia, Ilha do Fundão, CEP: 21949-900 Rio de Janeiro, Brazil.

E-mail adress: camila@eq.ufrj.br

Environmental performance of products and processes has gained importance as the industry is trying to minimize its environmental impacts. Of special concern to society, are climate effects associated to the rise of $CO_2$ atmospheric concentration. To mitigate emissions effects, processes are searched that use $CO_2$ as raw material. It is worth noting that an emission mitigating process should necessarily be a commodity as large production scale are required. With this orientation, methanol and DMC are promising alternative products: the first is a very attractive additive to diesel, and it blends with gasoline to yield a high octane fuel with low Reid vapor pressure. Methanol, on the other hand, is considered as a promising energy alternative besides displaying a wide range of industrial applications. However, in order to claim that a given technology is clean, its ecoefficiency should be addressed with quantitative sustainability indicators. In this sense, the present work selects a set of environmental metrics along with safety and economical (CAPEX and OPEX) indicators. A comparative analysis is presented for two potentially green processes: (i) methanol synthesis; and (ii) DMC synthesis. Furthermore, the later synthesis is approached in two alternatives, a direct and an indirect reaction path. In the methodology used, process simulation and optimization are employed to compare process performances at maximum yields.

# (385f) High Efficiency Direct Electrochemical Conversion of CO2 to Liquid Fuel

**Jingjie Wu**[1], Frank Risalvato[2], Fusheng Ke[2], Perry Pellechia[3] and Xiao-Dong Zhou[2], (1)Chemical Engineering, University of South Carolina, Columbia, SC, (2)Department of Chemical Engineering, University of South Carolina, Columbia, SC, (3)Department of Chemistry and Biochemistry, University of South Carolina, Columbia, SC

In this paper, we will discuss a design of an electrochemical cell for $CO_2$ reduction implementing a proton exchange membrane (PEM) by adding a liquid buffer layer between the membrane and the cathode electrode. This configuration enables production of liquid fuels from $CO_2$ on Sn gas diffusion electrode at high efficiencies. Specifically, the formate was produced from the electrochemical reduction of $CO_2$ with an overpotential lower than 0.2 V. At such a low overpotential, the Faradaic efficiency and energetic efficiency of formate production with a buffer layer of 0.1 M $KHCO_3$ reaches 78.5% and 66.4%, respectively. Moreover, a maximum Faradaic efficiency of ~95% can be achieved. The production rate of formate increases with increasing cell potential to approximately 10 μmol $min^{-1}$ $cm^{-2}$ at a cell potential of -3.0 V. In contrast, hydrogen evolution reaction dominates at the cathode side in the PEM electrochemical cell without a buffer layer.

The enhancement of $CO_2$ reduction in the PEM electrochemical cell with a buffer layer is attributed to the optimization of $H^+$ and $CO_2$ concentration at the reactions sites in the cathode. The effects of $H^+$ concentration on the Faradaic efficiency were confirmed by the results of $CO_2$ reduction with a liquid buffer layer with different pH values. As a result, the Faradaic efficiency towards formation of formic acid can be further improved by optimizing the buffer layer thickness and subsequently tuning the $H^+$ concentration at the cathode surface. The optimized thickness of the buffer layer was found ~ 3 mm. The buffer layer thickness dependency of $H^+$ concentration at the cathode electrode will be reported.

# (385g) Electrochemical Conversion of Carbon Dioxide to Useful Products

**Arun Agarwal**[1], Shan Guan[2], Davion Hill[2], Edward Rode[2] and Narasi Sridhar[2], (1)Research & Innovation, Det Norske Veritas (DNV), Dublin, OH, (2)Det Norske Veritas (DNV)

Utilization of captured $CO_2$ provides an advantageous alternative especially in locations where $CO_2$ sequestration in geologic media is not feasible. Making high value chemical products from $CO_2$, although not sufficient to convert a significant portion of the GHG emissions, provides an opportunity of extensive production of green chemicals when combined with renewable sources to feed the energy requirements. At DNV, we are developing the technology to economically convert $CO_2$ to formate salts/ formic acid using an electrochemical process. Tin is the preferred electrocatalyst due to high selectivity of $CO_2$ reduction to formate ion and high overpotential for cathodic hydrogen evolution. Gaseous $CO_2$ is continuously purged into the catholyte (2M KCl) through a porous tin electrode on which it is cathodically reduced. This cathode chamber is separated from the anodic chamber using a selective ion exchange membrane for allowing positive ion transport but not allowing the product (formate) ions to get oxidized on the anode. Oxygen evolution reaction occurs on a mixed metal oxide anode in either acidic or alkaline anolyte solutions to complete the electrochemical reaction. Electrode development in terms of high surface area of catalyst to match existing commercial electrochemical processes and reduce capital costs related to reactors will be presented. A process was developed to electroplate tin on high surface area carbon fiber substrate. Electroplating solution and substrate properties were evaluated for achieving maximum current densities at optimal product selectivity. Another focus area is understanding and optimizing the process chemistry. Selection of anolyte and catholytes, along with the applied potential regime, affects product selectivity, electrode degradation, expenditure due to chemicals consumption and the ability to reach high enough product concentrations before feasibly employing separation processes. In this presentation, a technical and economic analysis of the aforementioned factors affecting the commercial feasibility will be presented and evaluated as a guide for performance criteria of the proposed technology.

# (399a) Process Intensification by Membrane-Assisted Reactive Separation Processes: A Design Tool Based On Process Analysis and Optimization

**Johannes Holtbruegge**[1], Deenesh K. Babi[2], Philip Lutze[3], Raifqul Gani[4] and Andrzej Górak[1], (1)TU Dortmund University, Department of Biochemical and Chemical Engineering, Laboratory of Fluid Separations, D-44227 Dortmund, Germany, (2)Technical University of Denmark, CAPEC-Department of Chemical and Biochemical Engineering, DK-2800 Lyngby, Denmark, (3)Department of Biochemical and Chemical Engineering, Laboratory of Fluid Separations, TU Dortmund University, D-44227 Dortmund, Germany, (4)Denmark Technical University, Department of Chemical and Biochemical Engineering, DK-2800 Lyngby, Denmark

**Process Intensification by Membrane-Assisted Reactive Separation Processes: A Design Tool based on Process Analysis and Optimization**

Johannes Holtbruegge[a)], Deenesh K. Babi[b)], Philip Lutze[a)], Rafiqul Gani[b)], Andrzej Górak[a)]

a) TU Dortmund University, Laboratory of Fluid Separations

Emil-Figge-Strasse 70, D-44227 Dortmund, Germany

Phone: +40 (0) 231 755-2342, E-Mail: johannes.holtbruegge@bci.tu-dortmund.de

b) Technical University of Denmark, Computer Aided Process Engineering Center

Soltofts Plads 229, DK-2800 Kongens Lyngby, Denmark

Due to decreasing fossil fuel reserves, sustainability is an important factor in the chemical industry that has to be considered when new processes are developed. One opportunity to reach higher efficiencies in the production of bulk chemicals is the use of intensified processes. Two promising examples from the field of process intensification are (1) reactive distillation (RD) combining reaction and distillation within one apparatus and (2) hybrid separation processes, for instance the external integration of distillation with membrane separation. In both cases, synergetic effects lead to an increased sustainability in means of economics and ecologics. The use of a multifunctional reactor such as reactive distillation within a hybrid separation process promises even higher benefits. Despite all advantages, the potential of those and other intensified processes is hardly exploited in industry. The main reasons are a lack of process know-how and missing process synthesis tools.

Here, a combined methodology for process intensification is proposed consisting of four integrated blocks which are starting with the general design problem as input followed by "system identification & analysis", a "synthesis/design" and finally a "model-based validation/verification" to identify a near optimal intensified process[1)]. Whenever and wherever necessary, knowledge from experiments are generated and transferred by the block "experimental validation/verification".

In the "system identification & analysis", the process under investigation is analysed to identify limitations and/or identify critical phenomena which need to be intensified for the targeted performance improvement. Based on these results, criteria are used to identify which workflow(s) is/are followed for "synthesis/design" of the intensified process. The synthesis/design workflows are a knowledge based approach (KBS), a unit-operation based workflow (UBS) and/or phenomena based approach (PBS). While the complexity and the time spent to obtain a solution arises from KBS to PBS, also the novelty and potentially the achieved process improvement increases. The results of the "synthesis/design" which are a small number of potentially promising (intensified) process options are verified in the "model-based validation/verification" to identify the best process option.

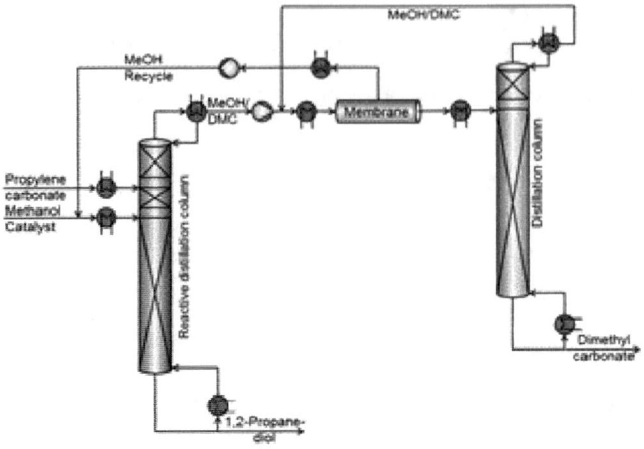


Figure 1: Simplified flowsheet of the developed membrane-assisted reactive separation process

In this contribution, the whole combined methodology will be presented and highlighted through the application to a case study which is the chemical equilibrium-limited transesterification of propylene carbonate with two molecules of methanol to produce the target product dimethyl carbonate and the byproduct 1,2-propanediol. The objective is to minimize the production costs per ton of dimethyl carbonate. Two synthesis/design workflows will be highlighted which are KBS and UBS. It will be shown that the use of the UBS workflow lead to the development of a membrane-assisted reactive separation processes that consists of an RD column, a vapor permeation membrane and a conventional distillation column (see Figure 1). Finally, in the "model-based validation/verification" an evolutionary algorithm based on the "modified differential evolution" approach is used to solve the reduced optimization problem of the remaining process options from the UBS and KBS workflow by determining apparatus dimensions and operating conditions for the minimal objective function. The results for the membrane-assisted reactive separation process compared to the base-case process obtained from KBS for the production of dimethyl carbonate comprises high savings in capital and energy use leading to lower production costs.

References:

[1)]: Lutze, P. et al.: Phenomena-based Process Synthesis and Design to achieve Process Intensification; Proc. 21$^{st}$ European Symposium on Computer Aided Process Engineering – ESCAPE 21, 2011, Greece

*Acknowledgements:*

*The financial support of the German Federal Ministry of Education and Research for the project "Energy Efficiency Management and Benchmarking in the Process Industry" is gratefully acknowledged.*

# NEW VAPOR COMPRESSION ALTERNATIVES IN THERMALLY COUPLED DISTILLATION

José A. Caballero*; Miguel A. Navarro; Juan .A. Reyes-Labarta
*Department of Chemical Engineering. University of Alicante. Ap. Correos 99. 03080 Alicante. Spain

## Introduction

The costs of chemical process are often dominated by the separation costs. Between the different separation techniques, distillation is the most important and commonly used in all chemical and petrochemical industries. Distillation handles more than 90% of separations and and this trend seems unlikely to change in the near future. The capital investment for these distillation systems was estimated in $8x10^9$ US$ [1]. Soave & Feliu [2] calculated that distillation accounts about the 3% of the total US energy consumption which is equivalent to $2.87x10^{18}$ J or 54 million tons of crude oil.

A renew interest in Thermally Coupled Distillation (TCD) appeared in, say the last 15-20 years, due the important potential savings in energy: typical values around 10% to 50 has been reported [3-5] when compared with conventional distillation sequences. Although it has been proved that fully thermally coupled systems are the arrangements that requires the minimum energy in a sequence of columns [6] it is possible to identify situations in which some column sections are operating far away the optimal conditions. For the sake of simplicity, but without losing generality, we will focus on the special case of a three component Petlyuk configuration or its thermodynamically equivalent Divided Wall Column (DWC). It will be evident that the extension to other thermally coupled configurations with more complex arrangements is straightforward.

The simulation of a DWC or its thermodynamically equivalent Petlyuk configuration can be carried out by decomposing it in the three separation tasks that form the sequence (See Figure 1) - each one of these separation tasks can be considered a distillation column- If we optimize each one of these columns independently, when we put together columns 2 and 3 (referred to Figure 1) it is evident that we have to modify the liquid or vapor flows in column 2 or column 3 to satisfy the mass balance - See Figure 1b-. (Usually it is better to perform a simultaneous optimization of all the system –included column 1- to get the optimal performance of the configuration). But the adjustment of flows produces that some sections work in suboptimal conditions (at least when compared to the individual separations tasks). If the dominant column is column 3 we have to increase flows in column 2 and then condenser duty increases. If the dominant column is column 2 the flow adjustment in column 3 produces an increase in the reboiler duty.

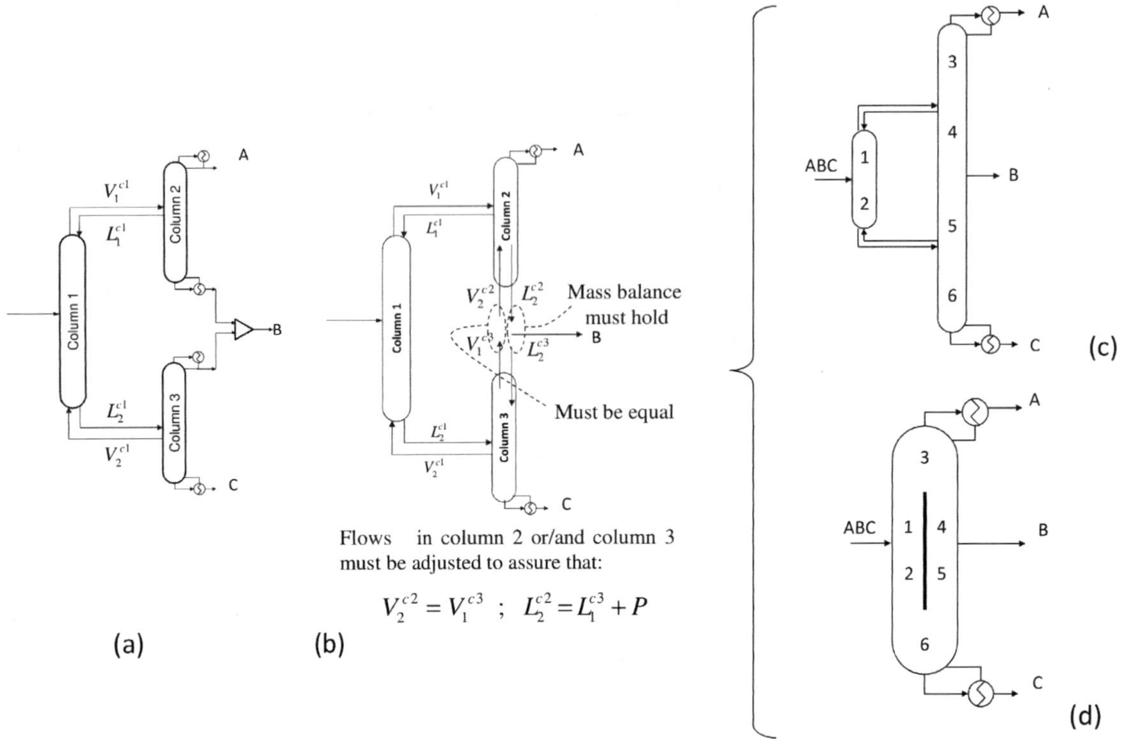


***Figure 1.*** *Generation by decomposition in its basic separation tasks of a Petlyuk configuration (c) and the equivalent Divided Wall Column (d).*

Instead of adjusting the flows, it is possible to introduce a heat exchanger (reboiler or condenser), but the extra heat added/removed is approximately equal to the extra heat if the flows are adjusted, and it is necessary to include an extra heat exchanger. Alternatively, it is possible to withdraw the extra vapor/liquid flow and use this stream as utility in the rest of the plant, what not always is feasible. An interesting alternative, presented in this paper, consists of implementing a vapor compression cycle using this extra vapor/liquid stream as shown in Figure 2.

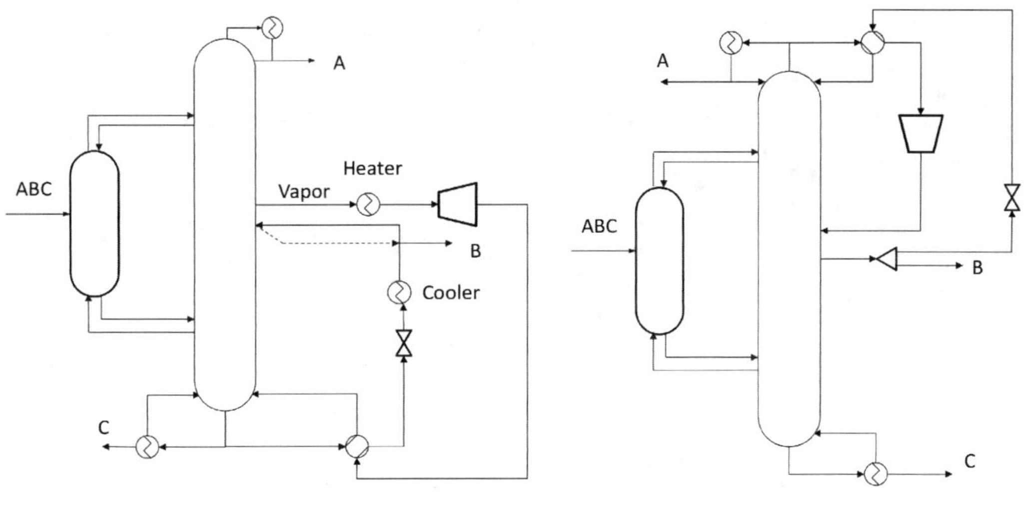


***Figure 2.*** *Vapor compression and integration with the column reboiler (left) or condenser (right).*

Some of the characteristics of these new arrangements are the following:

1. In Petlyuk/DWCs is usually not economically attractive to implement a vapor compression cycle between condenser and reboiler due to the large difference of temperatures (there is at least one component with an intermediate boiling point) and therefore large compression ratios. But with this arrangement the difference of temperatures is smaller and then the alternative could be economically attractive.

2. Both the heat duties in reboiler and condenser are reduced: the first one is due to the heat integration in the vapor compression; the other due to the reduction of internal vapor and liquid flows in the corresponding section.

3. There is a tradeoff between the savings in energy consumption in reboilers and condensers and, the investment and operation of the new equipment, mainly the compressor.

## Examples

We present here two examples. Corresponding to each one of the options presented in Figure 2. The first one consists in the separation of a mixture of Benzene, Phenol and o-Cresol (Data and main results are showed in Tables 1 and 2). All simulations were carried out in Aspen-Hysys using SRK equation of state and default values. Cost correlations are from Turton et al [7], and updated to 2012 cost. In this example it is necessary increase the boilup ratio (column 3 in Figure 1) to avoid a very large number of stages in the column, which produces large energy consumptions in the reboiler (this is an intrinsic problem of this system independently of the column sequence).

It is interesting remark that the savings in energy in the reboiler are greater than 80%. There is also a similar reduction in the energy consumption in the condenser. The investment is amortized during the first year of operation. After the amortization the savings in utilities cost is around 2.9 MM€/year (~ 76% in reduction in cost of utilities).

*Table 1. Data and main results for the example 1 without compression cycle.*

| **DWC/Petlyuk** | **Feed** | **Product A** | **Product B** | **Product C** |
|---|---|---|---|---|
| Flow (kmol/h) | 200 | 60.00 | 60.04 | 79.96 |
| Composition (mol frac.) | | | | |
| Benzene | 0.3 | 1.000 | 0.000 | 0.000 |
| Phenol | 0.3 | 0.000 | 0.995 | 0.001 |
| o-Cresol | 0.4 | 0.000 | 0.005 | 0.999 |
| Pressure (atm) | 1 | 1 | 1 | 1 |
| Condenser Duty (kW) | 13096 | | | |
| Reboiler Duty (kW) | 13607 | | | |
| Utilities annual cost (k€/year) | 3846.153 | | | |

*Table 2. . Data and main results for the example 1 with compression cycle.*

| **DWC/Petlyuk + Compression Cycle** | **Feed** | **Product A** | **Product B** | **Product C** |
|---|---|---|---|---|
| Flow (kmol/h) | 200 | 60.03 | 60.00 | 79.97 |
| Composition (mol frac.) | | | | |
| Benzene | 0.3 | 1.000 | 0.0005 | 0.000 |
| Phenol | 0.3 | 0.000 | 0.9925 | 0.0003 |
| o-Cresol | 0.4 | 0.000 | 0.007 | 0.9997 |
| Pressure (atm) | 1 | 1 | 1 | 1 |
| Condenser Duty (kW) | 1741 | | | |
| Reboiler Duty (kW) | 2686 | | | |

Table 2 (cont.). Data and main results for the example 1 with compression cycle.

| | | | |
|---|---|---|---|
| Extra Reboiler | | Compressor | |
| Heat exchanged (kW) | 10921 | Power (kW) | 836.2 |
| Cost (k€) | 540.40 | Pout (atm) | 2.0 |
| | | Investment Cost (k€) | 658.88 |
| | | Electricity Cost (k€/year) | 305.85 |
| Heater | | Cooler | |
| Heat exchanged (kW) | 215.9 | Heat exchanged (kW) | 1382 |
| Investment Cost (k€) | 15.998 | Investment Cost (k€) | 78.672 |
| Steam Cost (k€/year) | 48.296 | Water Cost (k€/year) | 14.066 |
| Total Investment Cost (k€)* | | 2149.437 | |
| Utilities Annual Cost (k€/year) | | 908.609 | |

*Referred to new equipment.

The second example corresponds to a mixture of ethylene, ethane and propane. Here the main cost is due to the necessity of refrigeration in the condenser, and therefore the implementation of a reverse cycle (expansion previous compression) is of interest (second case in Figure 2). Tables 3 and 4 shows main data and results

In this example the energy savings in condenser is around 39%. The investment is amortized in less than two years. After this time the benefit is around 2.8 MM€/year.

*Table 3. Data and main results for the example 2 without compression cycle.*

| **DWC/Petlyuk** | **Feed** | **Product A** | **Product B** | **Product C** |
|---|---|---|---|---|
| Flow (kmol/h) | 2000 | 600 | 600 | 800 |
| Composition (mol frac.) | | | | |
| Ethylene | 0.3 | 0.9996 | 0.0065 | 0.0000 |
| Ethane | 0.3 | 0.0004 | 0.9920 | 0.0002 |
| Propane | 0.4 | 0.000 | 0.0016 | 0.9998 |
| Pressure (atm) | 20 | 20 | 20 | 20 |
| Condenser Duty (kW) | 13902 | | | |
| Reboiler Duty (kW) | 16475 | | | |
| Utilities annual cost (k€/year) | 8939.917 | | | |

Table 4. . Data and main results for the example 2 with compression cycle.

| **DWC/Petlyuk + Compression Cycle** | **Feed** | **Product A** | **Product B** | **Product C** |
|---|---|---|---|---|

| | | | | |
|---|---|---|---|---|
| Flow (kmol/h) | 2000 | 600 | 600 | 800 |
| Composition (mol frac.) | | | | |
| Benzene | 0.3 | 0.9997 | 0.0013 | 0.0000 |
| Phenol | 0.3 | 0.0003 | 0.9939 | 0.0004 |
| o-Cresol | 0.4 | 0.0000 | 0.0048 | 0.9996 |
| Pressure (atm) | 20 | 20 | 20 | 20 |
| Condenser Duty (kW) | 9410 | | | |
| Reboiler Duty (kW) | 10415 | | | |
| Extra Condenser | | Compressor | | |
| Heat exchanged (kW) | 5483 | Power (kW) | | 1533 |
| Cost (k€) | 1906.393 | Pin (atm) | | 6.67 |
| | | Investment Cost (k€) | | 1207.516 |
| | | Electricity Cost (k€/year) | | 560.710 |
| Total Investment Cost (k€)* | | 3113.909 | | |
| Utilities Annual Cost (k€/year) | | 6078.407 | | |

## Acknowledgements

The authors wish to acknowledge support from the Spanish Ministry of Science and Innovation under project (CTQ2009-14420-C02).

### References

1. Humphrey, J., *Separation processes: playing a critical role.* Chemical Engineering Progress, 1995. **91**(10): p. 43-54.
2. Soave, G. and J.A. Feliu, *Saving energy in distillation towers by feed splitting.* Applied Thermal Engineering, 2002. **22**(8): p. 889.
3. Caballero, J.A. and I.E. Grossmann, *Structural considerations and modeling in the synthesis of heat-integrated-thermally coupled distillation sequences.* Industrial & Engineering Chemistry Research, 2006. **45**(25): p. 8454-8474.
4. Fidkowski, Z.T. and R. Agrawal, *Multicomponent thermally coupled systems of distillation columns at minimum reflux.* AIChE Journal, 2001. **47**(12): p. 2713-2724.
5. Rudd, H., *Thermal Coupling For Energy Efficiency.* Chemical Engineer-London, 1992(525): p. S14-S15.
6. Halvorsen, I.J. and S. Skogestad, *Minimum energy consumption in multicomponent distillation. 3. More than three products and generalized Petlyuk arrangements.* Industrial & Engineering Chemistry Research, 2003. **42**(3): p. 616-629.
7. Turton, R., et al., *Analysis Synthesis and Design of Chemical Processes*2003, New York: McGraw-Hill.

# Development of new designs of oscillatory mesoscale baffled reactor for continuous biodiesel production

*Anh N. Phan[*] and Adam P. Harvey*
*Newcastle University, Newcastle Upon Tyne, United Kingdom*
*Corresponding: anh.phan@ncl.ac.uk*

## Abstract

Three novel designs of mesoscale oscillatory baffled reactors (integral, central or axial and helical baffled reactors) were constructed and characterised at low net flow rates varying from 0.3ml/min to 8ml/min. High degree of plug flow, expressed via the number of tanks-in-series, could be easily achieved in these reactors when imposing an oscillatory flow upon a net flow. The number of tanks-in-series could be up to 40 at a ratio of oscillatory flow to net flow of above 50 for the helical baffled reactor.

These designs were used to investigate two-phase liquid reaction in continuous mode such as biodiesel production. High methyl ester content was produced at a consistent quality and there were rapid responses to variations in input conditions. More than 97% methyl ester content can be produced within 5mins, only 10% of the time required for biodiesel production in conventional stirred tank reactors. Start-up time to reach steady state is not an issue for continuous applications in these reactors as a steady state condition was easily achieved, within 1.5-2 times residence time. A clear step-change between various steady state molar ratios of methanol to oil and dynamic screening was also observed in these mesoscale oscillatory baffled reactors. Rapid screening such as this has the potential to significantly reduce process development, operating costs and environmental impact.

Keywords: continuous biodiesel production, steady state, dynamic screening, plug flow.

## 1. Introduction

Mesoscale oscillatory baffled reactors (mesoscale OBRs), have received considerable attention due to their small volume and ability to operate at low flow rates, therefore reducing reagent requirements and waste. Fluid mechanics inside a mesoscale OBR are governed by geometrical parameters (baffle spacing, L, and baffle open area, S) and dynamic dimensionless groups (net flow Reynolds number, $Re_n$, if operated continuously, oscillatory Reynolds number, $Re_o$, and Strouhal number, St) as described below. The oscillatory Reynolds number, $Re_o$, describes the intensity of mixing inside the OBRs. It is similar to net flow $Re_n$ but the net flow velocity is replaced by a maximum velocity of the sinusoidal oscillation, $2\pi f x_o$. Strouhal number measures the propagation of eddies generated each baffle cavities. In order to correlate the interdependence between oscillatory flow and net flow, a velocity ratio ($\varphi$) is introduced, which is defined as a ratio of oscillatory velocity and superficial velocity

Net flow Reynolds number: $$Re_n = \frac{\rho d u}{\mu} \quad (1)$$

Oscillatory Reynolds number: $$Re_o = \frac{2\pi f x_o \rho d}{\mu} \quad (2)$$

Strouhal number: $$St = \frac{d}{4\pi x_o} \quad (3)$$

Velocity ratio: $$\varphi = Re_o/Re_n = 2\pi f x_o/u \quad (4)$$

Where: f (Hz) is the frequency of oscillation; $x_o$ (m) is the centre-to-peak amplitude of oscillation; u (m/s) is the superficial velocity of the liquid with its density ρ ($kgm^{-3}s^{-1}$) and viscosity μ (Pa.s) through a tube diameter d (m).

Biodiesel has been considered as an alternative for diesel because of its biodegradable and cleaner fuel. It is commercially produced by transesterification of vegetable oil/fat and methanol using base catalysts. Transesterification is a reversible reaction, which requires a super-stoichiometric ratio of methanol to oil (above 3:1) to drive the reaction towards the product side and improve the productivity. It involves various transitions stage, of which the first stage is mass transfer limited due to low solubility of methanol and oil. The reaction becomes slow in the final stage as the equilibrium state is approached. If the mixing is insufficient, the difference in densities between glycerol and methyl esters causes stratification, leading to the incomplete reaction because most of catalyst resides in glycerol phase. There was no reaction between melted beef and methanol in the presence of sodium hydroxide when no mixing was applied [1]. The mass transfer limitations at the beginning of the reaction becomes less important if the impeller speed (using a helix stirrer) was around 600 rpm for a 250ml three-neck batch reactor [2].

In this study, three designs of mesoscale OBRs were constructed and characterised at low net flow rates, corresponding to $Re_n$<40 over a wide range of oscillation conditions in order to identify an operating window for individual designs of reactors in which high degree of plug flow can be achieved. Continuous two-phase liquid reaction such as biodiesel production was also studied in these mesoscale reactors to demonstrate rapid production biodiesel and to investigate the effects of oscillation conditions on the induction time to reach steady state.

## 2. Experimental Method

### 1.1. Characterisation of mesoscale OBRs

Experimental set-up for residence time distribution is shown in Fig.1a. A set of three syringe pumps (Eurodyne Ltd.) were used to provide a net flow (deionised water), inject tracer (KCl) and oscillate the fluid.

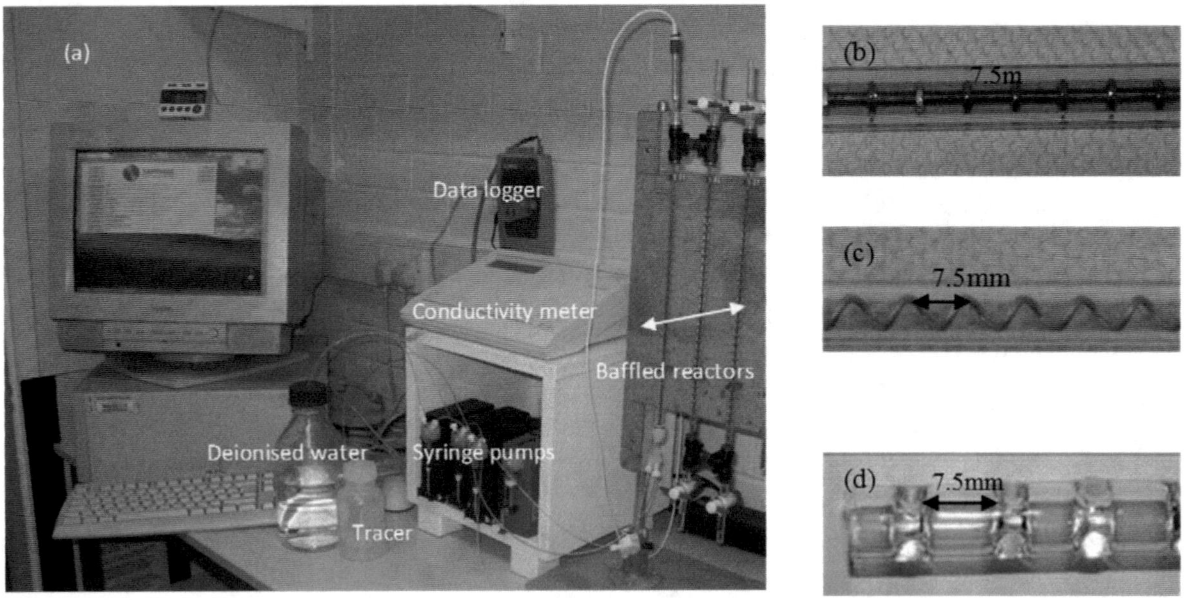


Fig.1: (a) Experimental set-up for characterisation (b) central (axial circular) baffles, (c) helical baffles and (d) integral baffles [3].

Deionised water was dispensed continuously at low rates from 0.3ml/min to 8.0ml/min, corresponding to net flow Reynolds number of ~1-34. A known amount of the 0.1M KCl tracer

(Sigma-Aldrich) was rapidly injected into the system at the bottom of the column within 2seconds to produce a pulsed injection. At the outlet, the corresponding conductivity versus time of the tracer was measured using a 4mm diameter and 103mm long E61M014 conductivity probe connected to a CDM210 conductivity meter (Hach-Lange Ltd). The data was recorded and logged on a computer via a DaqPro 5300 data logger.

Three baffles are engineered at the baffle spacing is 1.5 times the tube diameter and open cross sectional area of 25-40%. Central baffles are hexagonal solid discs placed along a studding rod (Fig.1b). Helical baffled design was a 1.1mm diameter wire embedded continuous along the tube wall with a helical pitch of 7.5mm (Fig.1c). Integral baffles are smooth periodic baffles with a thickness of around 3mm, diameter of the orifice of 2.5mm (Fig.1d).

Flow characteristics in the reactors were quantified by using dimensionless statistical moment methods such as distribution curve (E-curve) and variance [4] as described below. The tanks-in-series model was used to characterise the designs because it is the simplest and most robust model with the parameter of the model N (number of tanks-in-series) derived from an RTD curve, particularly it represents the flow patterns inside mesoscale OBRs [3, 5, 6]. A RTD curve approaches a Gaussian form (plug flow behaviour) when $N \geq 10$.

Distribution curve:

$$E(\theta) = \tau E(t) = \tau \frac{C_i}{\sum_i C_i \Delta t_i} \tag{5}$$

Normalised variance:

$$\sigma(\theta)^2 = \frac{\sigma(t)^2}{\tau^2} = \frac{\sum_i (t-\tau)^2 E(t) \Delta t}{\tau^2} \tag{6}$$

Where $\tau = \frac{\sum_i t_i C_i \Delta t_i}{\sum_i C_i \Delta t_i}$, mean residence time

Where, $C_i$ is the existing tracer concentration at time $t_i$, and $\Delta t_i$ is the interval between two measurements.

The age exit distribution curve (E-curve) for N tanks-in-series under the pulsed injection is determined as follows:

$$E(\theta) = \tau E(t) = \frac{N(N\theta)^{N-1}}{(N-1)!} e^{-N\theta} \tag{7}$$

The number of tanks (N) can be estimated as:

$$N = \frac{1}{\sigma(\theta)^2} \tag{8}$$

An initial estimated value of N was defined from experimental data by using equations 6 and 8. A RTD curve obtained from equation 7 with the initial estimated value of N was compared to that obtained from experiment data (equation 5) in terms of shape of the curve, spread of distribution and height of the distribution. The value of N was changed until the best correspondence was achieved.

### 1.2. Continuous biodiesel production

Continuous transesterification of oil with methanol was carried out using a set-up as shown in Fig.2 [7, 8]. The reactor is a series of 5mm inner diameter and 340mm length tubes connecting together by Swagelok connectors. Three syringe pumps were used for dispensing pre-heated rapeseed oil (from a 500ml reservoir inserted in a water bath), methanol and a mixture of methanol and potassium hydroxide whereas the fourth syringe pump was used for oscillating the fluid. The measured viscosity and density of oil at 60°C were 25.5mPas and 890kg/m$^3$, respectively. The water bath was set at around 60°C. Total flow rate (rapeseed, methanol and the solution of methanol and KOH) was fixed at 1.74ml/min-3.48ml/min ($Re_n<3$), depending on the residence time (reaction time) required.

Approximately 0.5-1ml samples were taken at various intervals (i.e. every minute for dynamic screening test, or every five minutes for steady state test) by a 10ml syringe. The sample was transferred into a 2ml vial and quenched immediately using 0.1N HCl. The sample was then allowed to settle (for several hours) before analysis. There were two separate phases: methyl ester-rich phase (top layer) and a denser glycerol-rich phase. About 50 μl was withdrawn from the top-layer (methyl ester rich phase) of the vial and mixed with an internal standard for fatty acid methyl esters (FAME) analysis (based on British standard BS14103). A 5890 Hewlett Packard Series II with flame ionization detector was used to quantify the FAME. A 30 m length, 0.25 mm diameter and 0.25 μm thickness CP wax capillary (BPX70) column was used, with helium gas as carrier.

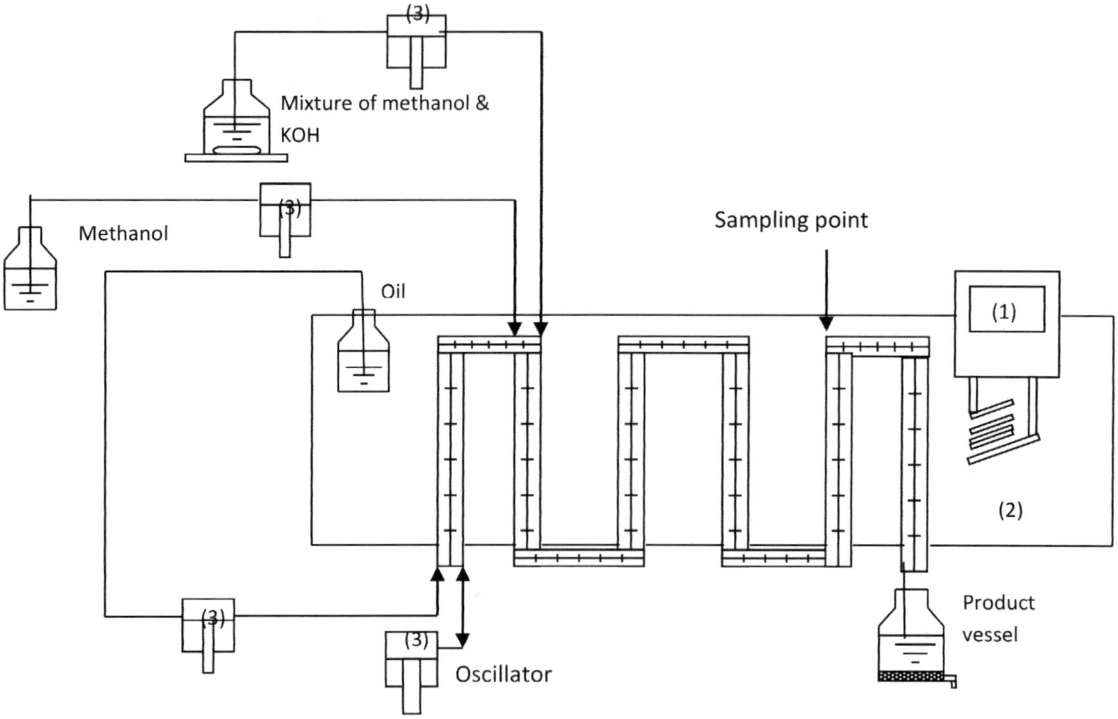


Fig.2: Schematic experimental set-up for continuous biodiesel production; (1): temperature controller; (2) water bath and (3) syringe pumps [7, 8]

## 3. Results and Discussion

### 3.1. Characterisation of mesoscale baffled reactors

Under a near pulse injection, residence time distribution (RTD) profiles obtained from the mesoscale OBRs were well-defined when a velocity ratio of $Re_o$ to $Re_n$ in a range of 4-12 but become skewed with a further increase in a velocity ratio. As shown in Fig.3a, the number of tanks-in-series, N, was calculated above 15 for a length of 340mm baffled reactor at $\psi$=4-12. However, oscillation conditions had a little influence on the mixing inside the mesoscale OBRs when $Re_n$ was sufficient high, e.g. $Re_n$=34, as the number of tanks in series remained almost constant at around 10-15 as shown in Fig.3b over a range of tested $Re_o$ from 100-400.

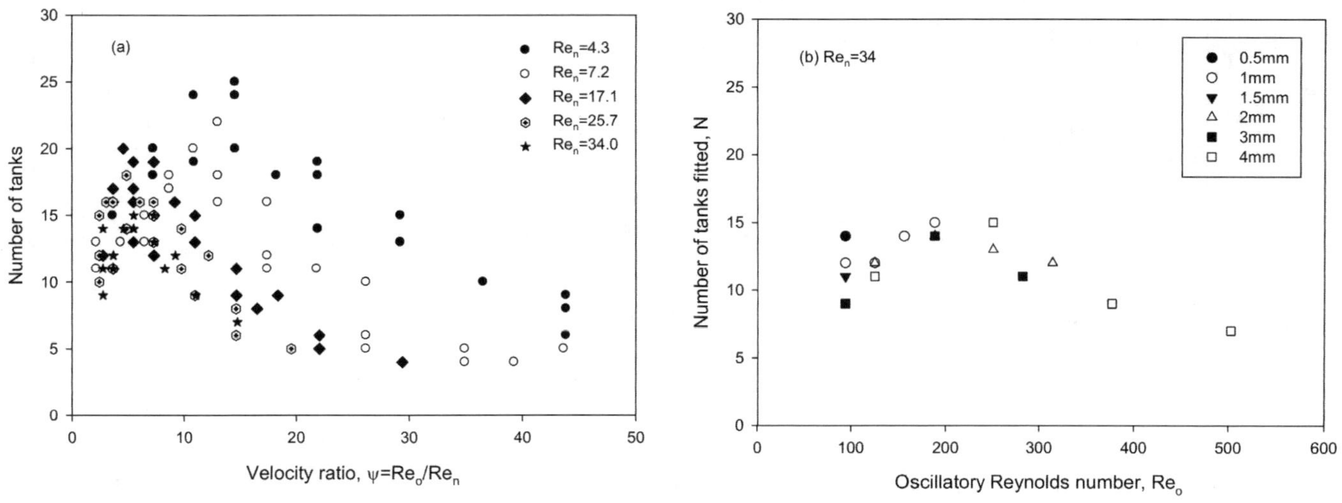


Fig.3: Number of tanks-in-series obtained in central baffled design [3]

For the integral baffled reactor, the fluid mixing was little affected by oscillation conditions at low net flow rates, $Re_n<4.3$ since the RTD curves were similar to that of a stirred tank reactor (Fig.4a). At $Re_n=7.2$, the fluid mixing improved significantly at oscillation amplitudes of 0.5-1mm (St=0.4-0.8). Although $Re_o$ remained the same, the RTD profile broadened when oscillation amplitude was 2mm (St=0.2) as presented in Fig.4(b).

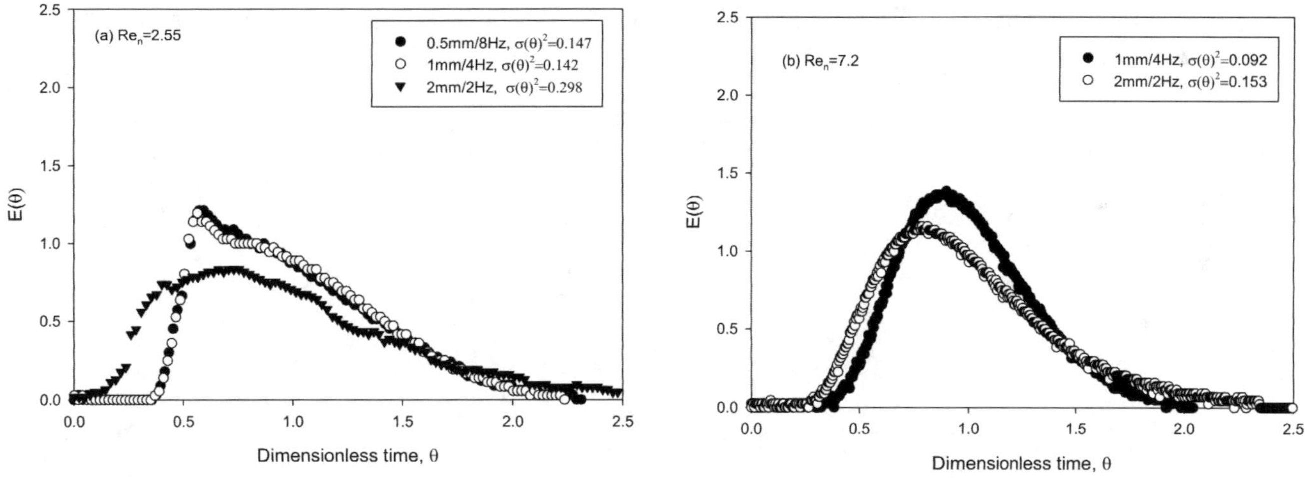


Fig.4: Residence time distribution (RTD) obtained from the integral baffled designs at oscillatory Reynolds number, $Re_o$= 126 and (a) $Re_n=2.55$; (b) $Re_n=7.2$

For helical baffled design, a high degree of plug flow achieves at a wide operating window. Above 30 tanks-in-series were obtained at various velocity ratios depending on net flows at an oscillation amplitude of 2mm or St=0.2 (Fig.5a). Approximately 40 number of tanks-in-series could be obtained in a range of velocity ratio of 100-200 when oscillation amplitudes was 3mm (St=0.13) for a 340mm length of the reactor (Fig.5b). The difference in mixing behaviour between the helical baffled design and the integral and central baffled designs is due to the differences in open section area and baffle shape. For the central and integral baffled designs, the uniform mixing is due to the periodic dissipation and formation of vortices inside each baffle cavity when the oscillatory flow superimposes upon the net flow. In contrast, in the helical baffled design, the fluid patterns inside each turn are more complex, they are combination between swirling flow and vortices.

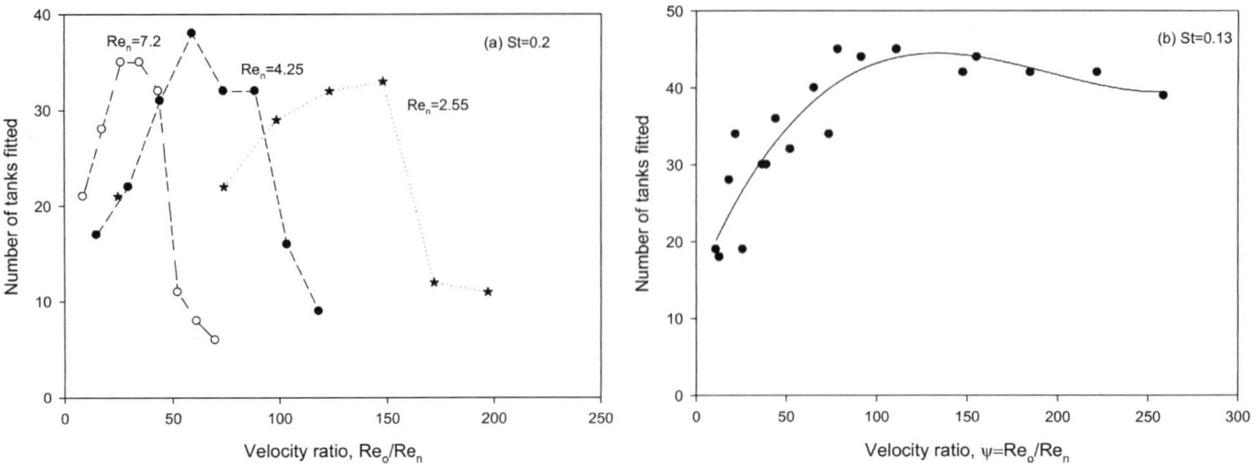


Fig.5: Number of tanks-in-series obtained in helical baffled designs at (a) St=0.2 and (b) St=0.13 [6]

### 3.2. Continuous biodiesel production

#### 3.2.1. Effect of mixing conditions on the induction time to establish a steady state and the conversion of oil into their corresponding methyesters

Continuous two phase liquid-liquid biodiesel production was carried out in the integral and sharp-edged helical baffled reactors. The effect of mixing conditions on the conversion of oil into FAME and the induction time (a period that the conversion reaches its steady state) was investigated. The variance in the steady state and induction time must be minimised to reduce un-desired products. As shown in Fig.6, increasing mixing conditions from $Re_o$ of 36 to 160 decreased the induction time and the variance.

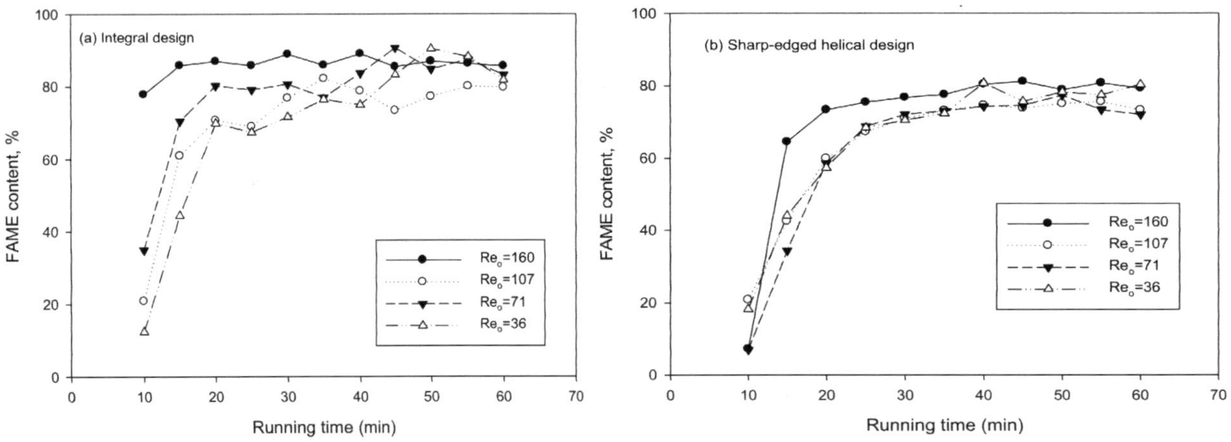


Fig.6: Effect of oscillation conditions on FAME content in continuous biodiesel production at a molar ratio of methanol to oil of 6:1, reaction temperature of 60°C, reaction time (residence time) of 10mins ($Re_n$=1.74) and 1.5%wt KOH obtained from (a) integral baffled design and (b) Sharp-edged helical design [8]

For the integral design (Fig.6 (a)), the conversion reached its steady state at 15mins, corresponding to 1.5 times reaction time (residence time) at $Re_o$=160, with the FAME content of 86.74±1.28%; whilst it took 25mins (2.5 times residence time) at $Re_o$=107, with FAME content of 78.30±2.63%. At lower $Re_o$, the steady state was not achieved after 60mins running time. Consequently, the FAME content fluctuated from 80-90% at $Re_o$=71 and 70-90% at $Re_o$=36.

As shown in Fig.6(b), the slightly longer induction time and lower FAME content (approximately 6-8%) were observed for the sharp-edged helical design than for the integral baffled design at the same $Re_o$. the FAME content reached its steady state of 78.75±1.91% at 25mins or 2.5 times residence time at $Re_o$=160. At Reo=107, the steady state was obtained at 30mins (3 times residence time) with FAME content of 72.91±1.35%.

Inconsistency in the FAME content during the running time at low $Re_o$<107 was due to the insufficient mixing inside the reactor. The glycerol phase separated with methyl ester phase due to the stratification as shown in Fig.7, causing incomplete reaction because most of catalyst remains in the glycerol phase.

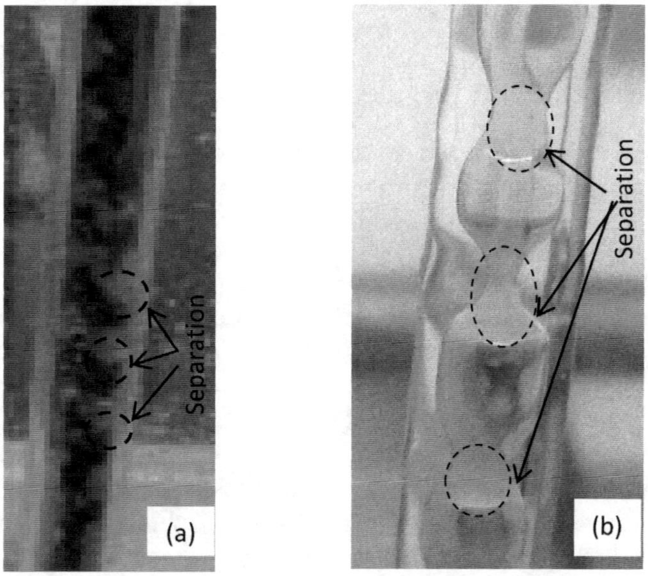


Fig.7: Separation occurred in (a) sharp-edged helical baffled design and (b) integral baffled design at $Re_o$=36, $Re_n$=1.74, corresponding to reaction time of 10mins, reaction temperature of 60°C, 1.5%wt KOH and a molar ratio of methanol to oil of 6:1

Generally, using the mesoscale OBRs in continuous process could reduce significantly amount of waste generated due to the quickly obtaining the steady state in mesoscale OBRs (1.5-2 times of residence times) compared to conventional continuous stirred tank reactors, which was at least 5 times residence time [9].

### 3.2.2. Intensified biodiesel production

As shown in Fig.8, continuous transesterification reached its equilibrium, with FAME content >97% at only 5 mins reaction time (residence time) for 1.5%wt KOH and 10mins reaction time for 1.5%$NaOCH_3$. Referring to alkaline catalysed transesterification in conventional stirred tank reactors, which takes at least 60mins to achieve the European biodiesel standard EN14212 (FAME content ≥96.5%), the mesoscale OBRs can produce biodiesel within less than 10% of the required time. The higher FAME content produced in a short time was due to significant mass transfer enhancement when using mesoscale OBRs.

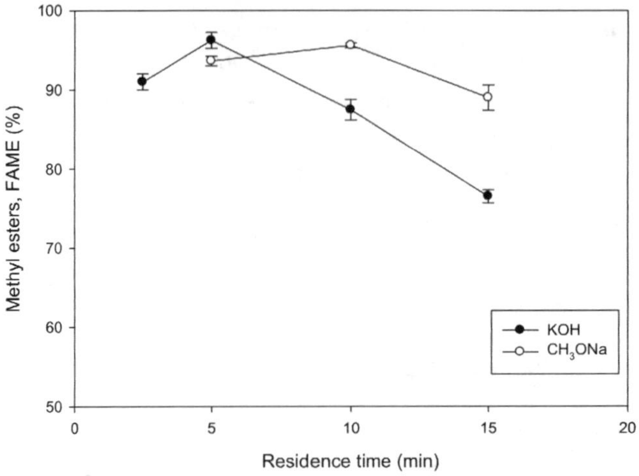


Fig.8: FAME content versus residence time (reaction time) obtained from the integral baffled reactor at operating conditions of 6:1 molar ratio of methanol to oil, reaction temperature of 60°C and an oscillation condition $Re_o$ of 160 [8]

3.2.3. Dynamic screening using the mesoscale OBRs

A set of steady states of molar ratios of methanol to oil was imposed into the reactor at different durations, e.g. 10mins for 4:1 and 5:1 and 5mins for 8:1 and 12:1. Directly after this sequence, the reactor was operated in a dynamic screening mode in which each molar ratio lasted for only one minute. The reactor was sampled at regularly intervals, e.g. every 5 mins for steady state and every minute for dynamic screening section. As shown in Fig.9, a series of stable states were established corresponding to the change of molar ratios (designed as "A"). Clear step changes at the 10th minute, the 20th minute and 25th minute were observed which are in good agreement with the time of the steady state 4:1, 5:1 and 8:1 coming through the reactor. It can be also seen that the FAME content in dynamic screening mode (designed as "B") was in good agreement with that in the step change steady states for individual molar ratios. From experimental results, it can be concluded that the mesoscale OBRs produced a near enough plug flow behaviour to allow for high throughput dynamic screening for two-phase liquid-liquid reactions in continuous mode.

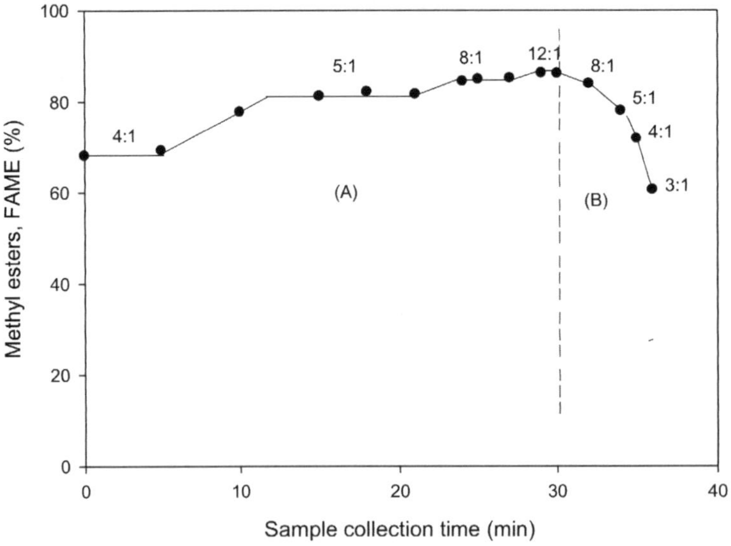


Fig.9: FAME content obtained in a sharp-edged helical baffled reactors: (A) steady states screening and (B) dynamic screening [7]

## 4. Conclusion

Three designs of mesoscale OBRs can provide high degree of plug flow at low net flow rates, $Re_n<50$ over a wide range of oscillation conditions. For the central and integral baffled designs, the number of tanks-in-series was above 15 for a 340mm length of a reactor when velocity ratios were in a range of 4-12 and net flow $Re_n>4.3$. At lower $Re_n$, the oscillatory conditions had little effect on the mixing inside the integral designs as the number of tanks-in-series remained at 5-10. For the helical baffled design, a higher number of tanks in series were observed at a wider operating window, e.g. the number of tanks-in-series was 40 at a range of velocity ratios of 50-150 at St=0.13.

These designs were used in screening processes involving immiscible liquids such as biodiesel as they allow plug flow to be achieved at the lower oscillatory Reynolds numbers. More than 96.5% methyl ester could be produced in a very short reaction time, e.g. 5mins for 1.5%wt KOH catalyst and 10%wt for $NaOCH_3$ catalyst, which reduced nearly 90% of the time required for biodiesel production in conventional stirred tank reactors. These designs produced high and consistent FAME content in a short induction time, e.g. 1.5-2 times residence time compared to 5-6 times in conventional stirred tanks. This is an enormous advantage when operating in continuous mode as it reduces significantly un-desired products.

A niche application of mesoscale OBRs is to convert multi-phase long reactions (residence time from minutes to hours) from batch to continuous mode. Choosing a design of mesoscale OBRs is dependent of applications. The integral baffled design is favourable for shear-sensitive applications because of smooth periodic constrictions. Helical baffled design is more suitable for solid-liquid reactions because they have larger opened-cross sectional areas and provide a wider operating window compared to the central and integral baffled designs.

## References


1. Ma, F., L.D. Clements, and M.A. Hanna, *The effect of mixing on transesterification of beef tallow.* Bioresource Technology, 1999. **69**: p. 289-293.
2. Vicente, G., et al., *Kinetics of sunflower oil methanolysis.* Ind.Eng.Chem.Res, 2005. **44**: p. 5447-5454.
3. Phan, A.N. and A.P. Harvey, *Development and evaluation of novel designs of continuous mesoscale oscillatory baffled reactors.* Chemical Engineering Journal, 2010. **159**: p. 212.
4. Levenspiel, O., *Chemical reaction engineering*1999: John Wiley & Sons.
5. Phan, A.N. and A.P. Harvey, *Characterisiation of fluid mixing in novel designs of mesoscale oscillatory baffled reactors operating at low net flow rates (0.3-0.6ml/min).* Chemical Engineering and Processing: Process Intensification, 2011. **50**: p. 254-263.
6. Phan, A.N. and A.P. Harvey, *Characterisation of mesoscale oscillatory helical baffled reactor-Experimental approach.* Chemical Engineering Journal, 2012. **180**: p. 229.
7. Phan, A.N., A.P. Harvey, and M. Rawcliffe, *Continuous screening of base-catalysed biodiesel production using new designs of mesoscale oscillatory baffled reactors.* Fuel Processing Technology, 2011. **92**: p. 1560-1567.
8. Phan, A.N., A.P. Harvey, and V. Eze, *Rapid production of biodiesel in mesoscale oscillatory baffled reactors.* Chemical Engineering and Technology, 2012. **35**: p. 1214-1220.
9. Darnoko, D. and M. Cheryan, *Continuous production of palm methyl esters.* Journal of the American Oil Chemists' Society, 2000. **77**(12): p. 1269-1272.

# Nano-$TiO_2$ sol-gel synthesis in the spinning disc reactor

*Somaieh Mohammadi* and Kamelia V.K. Boodhoo*

School of Chemical Engineering & Advanced Materials,

Newcastle University, Merz Court,

Newcastle Upon Tyne NE1 7RU, UK

Corresponding author: s.mohmmadi@ncl.ac.uk

Tel: +44 191 222 7621

Fax: + 44 191 222 5292

Abstract

The purpose of this experimental study is to determine the feasibility of producing nanoparticles of titanium dioxide ($TiO_2$) in a continuous spinning disc reactor (SDR) as an example of a process intensification technology. The primary characteristics of the SDR relevant to this study are the thin film flow of the liquid on its surface, the very vigorous mixing within the film and the extremely short and controllable residence times of the liquid. A liquid-liquid sol gel technique was chosen to investigate experimental parameters of rotational speed, flowrate and reagents ratio. The nanopowders of $TiO_2$ synthesized by sol gel method in the SDR are compared with the nanopowders produced by stirred tank reactor. It has been found that the higher the flowrate and rotational speed in the SDR, the smaller the particle size and narrower particle the size distribution (PSD). In contrast, under optimized conditions, the nanopowders produced in a stirred tank reactor had a wider PSD.

Keywords

$TiO_2$, nanoparticles, spinning disc reactor, particle size distribution

# 1. Introduction

Nowadays there is an increasing number of industries that rely on nano titanium dioxide ($TiO_2$) which is employed in a variety of products such as white pigment, white food colouring, pharmaceutical and cosmetic products, skin care products and photocatalysts for use under ultraviolet light and surface coatings amongst others. Although the processes for producing titanium dioxide are well-established and developed, new processing methods which are more economical and have the potential to offer better product characteristics are sought after to drive the $TiO_2$ industry forward in order for it to continue to be profitable. In recent years, spinning disc reactors (SDRs) have been developed as a process intensification technology, where rapid mass and heat transfer rates can be obtained from the thin film of liquid produced as a result of the high centrifugal acceleration on the rotating disc. In developing these characteristics, the SDR is considered a tool for process intensification due to its compactness, flexibility (as an inherently safe and continuous reactor technology) and capability to deliver better product quality. SDR appears to be an excellent candidate for industrial applications due to its very short and controllable residence time compared to conventional reactors. SDR is also a highly intensified reactor when dealing with rapid exothermic reactions and viscous liquids such as bulk polymerization [1], solution [2], condensation [3] and also cationic polymerization [4] of styrene. The SDR thin films offer higher mixing intensity and high heat and mass transfer rates [5] which result in tighter molecular distribution and higher quality of polymer product [2, 3]. SDR is an attractive and viable technology for drug manufacture and crystallization [6]. Crystallization processes in the pharmaceutical industries are intended to achieve crystals with controlled size, shape, purity, and polymorphic form [7]. An imperfect crystallization is caused by less intensive mixing, as fast processes, like nucleation in crystallization, are typically highly dependent on mixing. Boodhoo et al [8] showed the potential of the SDR as an alternative device for processes where a high degree of mixing is desirable. A plug flow crystalliser provides impressively productive crystallization methodology and consistent product can only be achieved in plug flows. Mohammadi et al provided a valuable insight into the operating conditions of the SDR which generate flow profiles approaching plug flow behavior [9]. Spinning disc reactors were previously investigated in the precipitation of barium sulphate crystals [10, 11], calcium carbonate [12] and silver nanoparticles [13, 14]. It was found that high rapid mixing coupled with high levels of supersaturation lead to very small crystals with a tighter size distribution being produced. This was considered to produce better quality products than conventional process techniques actually used in the pharmaceutical industry [6]. The main factor controlling the crystal size was the very high rates of mixing experienced on the spinning disc, which lead to the rapid depletion of supersaturation, and much higher nucleation rates. Cafiero et al [10] demonstrated that the energy input in the spinning disc process was much lower than the use of a T-mixer arrangement, suggesting that operating costs would also be reduced along with attaining better control of crystal size. With the understanding that the rapid mixing and high supersaturation leads to very small crystals being produced along with very short residence times and tighter CSD, it was hypothesised that a similar circumstance could exist for $TiO_2$ precipitation routes. The morphology of titanium dioxide is more complicated than that of barium sulphate, capable of forming three different crystal forms, and therefore emphasis on making the right size, shape and form is important in the final product. Understanding the morphology from this rapid precipitation method is a factor in the present study.

This work aims to develop and demonstrate a novel processing method on the basis of the spinning disc reactor (SDR) for the production of $TiO_2$ nanoparticles in order to enhance the process efficiency and the product quality. The spinning reactor seems more desirable due to its considerably lower specific energy consumption. Additionally this technique is utilized to give high flow rate and very short residence time in continuous mode of operation [10].

2. Materials and Methods

In the present work, the production process of the $TiO_2$ nanoparticles consisted of the reaction between distilled water and titanium tetra isopropoxide (TTIP) at 50°C under nitrogen and the subsequent precipitation of the produced $TiO_2$. The adopted apparatus is equipped with a disc of 100 mm of diameter rotating at speeds ranging between 400 and 2400 rpm. The two reagent solutions were fed separately at the centre of the disc as indicated in Figure. 1. The total feed flow rate of reagent solutions was equal to 2-8 mL/s. A pH value of the distilled water of 1.5 was achieved by the addition of nitric acid to the distilled water reservoir. The ratio between the two reagents were from 4-20 of water/TTIP. For sake of comparison the same procedure was applied by using a 250 mL capacity baffled stirred vessel.

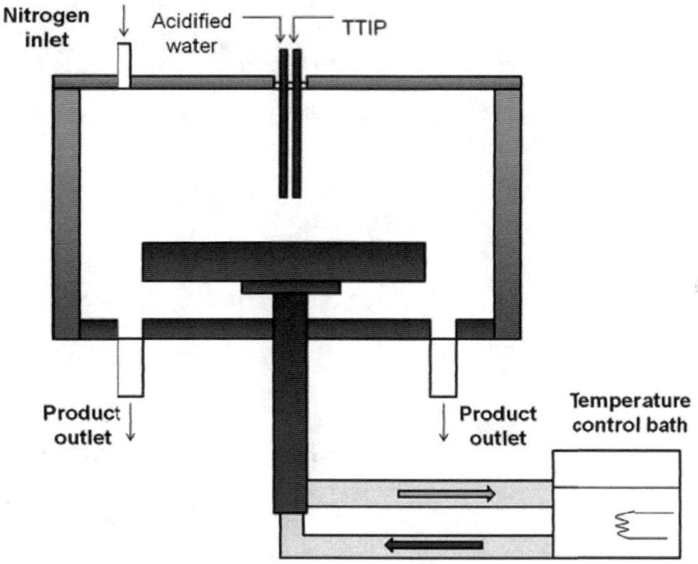


Figure1. Schematic set up of $TiO_2$ experiments by spinning disc reactor

3. Highlights

Exploratory works and primary studies show that the important advantage of the SDR technology is the increased number of particles with controlled size, shape and size distribution that reach quality requirements. It has been seen that at the same ratio of Water/ TTIP (=4), applying higher rotational speed provides more uniform and smaller nanoparticles (Figure 2).

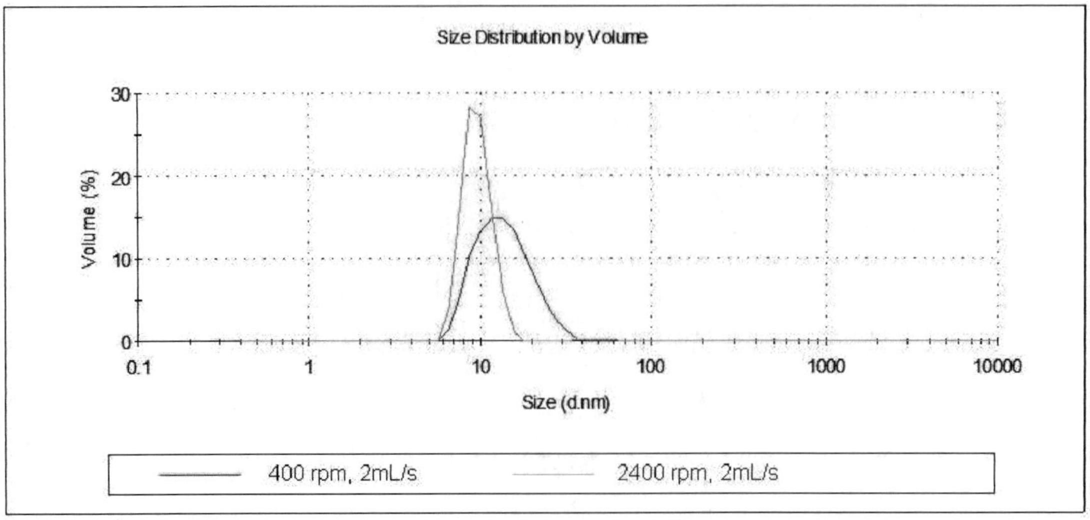


Figure 2. Particle size distribution of TiO2 produced by SDR (water/ TTIP =4)

Similarly an increase in flowerate from 2 mL/s to 8 mL/s results in smaller particles whereas the mean size of particles decreased from 38.5 nm to 19.5 nm.

References


1. Moghbeli, M.R., S. Mohammadi, and S.M. Alavi, *Bulk free-radical polymerization of styrene on a spinning disc reactor.* Journal of Applied Polymer Science, 2009. **113**(2): p. 709-715.
2. Boodhoo, K.V.K. and R.J. Jachuck, *Process intensification: spinning disk reactor for styrene polymerisation.* Applied Thermal Engineering, 2000. **20**(12): p. 1127-1146.
3. Boodhoo, K.V.K., W.A.E. Dunk, and R.J. Jachuck, *Continuous photopolymerization in a novel thin film spinning disc reactor.* Photoinitiated Polymerization, 2003. **847**: p. 437-450.
4. Boodhoo, K.V.K., et al., *Classical cationic polymerization of styrene in a spinning disc reactor using silica-supported BF3 catalyst.* Journal of Applied Polymer Science, 2006. **101**(1): p. 8-19.
5. Aoune, A. and C. Ramshaw, *Process intensification: heat and mass transfer characteristics of liquid films on rotating discs.* International Journal of Heat and Mass Transfer, 1999. **42**(14): p. 2543-2556.
6. Paul Oxley, C.B., Francois Ricard, Norman Lewis, and Colin Ramshaw, *Evaluation of Spinning Disk Reactor Technology for the Manufacture of Pharmaceuticals.* Ind. Eng. Chem, 2000. **39(7)**: p. 2175–2182.
7. Alvarez, A.J. and A.S. Myerson, *Continuous Plug Flow Crystallization of Pharmaceutical Compounds.* Crystal Growth & Design, 2010. **10**(5): p. 2219-2228.
8. Boodhoo, K.V.K. and S.R. Al-Hengari, *Micromixing Characteristics in a Small-Scale Spinning Disk Reactor.* Chemical Engineering & Technology, 2012. **35**(7): p. 1229-1237.
9. Mohammadi, S. and K.V.K. Boodhoo, *Online conductivity measurement of residence time distribution of thin film flow in the spinning disc reactor.* Chemical Engineering Journal, 2012( In press).
10. Cafiero, L.M., et al., *Process intensification: Precipitation of barium sulfate using a spinning disk reactor.* Industrial & Engineering Chemistry Research, 2002. **41**(21): p. 5240-5246.
11. Dehkordi, A.M. and A. Vafaeimanesh, *Synthesis of Barium Sulfate Nanoparticles Using a Spinning Disk Reactor: Effects of Supersaturation, Disk Rotation Speed, Free Ion Ratio, and Disk Diameter.* Industrial & Engineering Chemistry Research, 2009. **48**(16): p. 7574-7580.
12. Burns, J.R. and J.J. Jachuck, *Monitoring of CaCO3 production on a spinning disc reactor using conductivity measurements.* Aiche Journal, 2005. **51**(5): p. 1497-1507.
13. Tai, C.Y., Y.-H. Wang, and H.-S. Liu, *A green process for preparing silver nanoparticles using spinning disk reactor.* Aiche Journal, 2008. **54**(2): p. 445-452.
14. Tai, C.Y., et al., *Preparation of Silver Nanoparticles Using a Spinning Disk Reactor in a Continuous Mode.* Industrial & Engineering Chemistry Research, 2009. **48**(22): p. 10104-10109.

# Shared research for process intensification – an example project

*Leon Geers, Axel Lexmond, Murk van Rooijen, TNO, Delft, The Netherlands*

In the last decade, process technology took a giant leap forwards with the arrival of new manufacturing technology and design methodologies to create micro-structured reactors. Chemical industry already reacts by changing their thinking about process design and the use of continuous flow reactors. However, they are often not well equipped (regarding facilities, time, and/or money) to test all kinds of equipment to assess what works and what doesn't for their chemical process. Secondly, development or optimization of equipment is not their core business. Similarly, equipment manufacturers are well equipped to develop the most intricate process equipment conceivable, but they lack detailed knowledge of the specifics of industrial chemical processes.

The Netherlands Organization for Applied Scientific Research TNO is currently running a program called CoRIAC in cooperation with the Institute for Sustainable Process Technology (ISPT). In total, 11 partners from chemical industry (DSM, Janssen Pharmaceutica, P&G), equipment manufacturers (ESK, Chemtrix, Mettler Toledo, Bronkhorst HighTech), a pilot plant builder (Zeton), and academia (Eindhoven University of Technology) actively work together to develop and benchmark different reactor designs and analytical tools for a number of specific industrial chemical processes on a small scale (1-10 kg/hr). A small selection of these processes will be demonstrated on a pilot plant scale (10-100 kg/hr).

CoRIAC focuses on processes that are notoriously difficult to execute in continuous flow equipment due to the presence of multiple phases, high exothermicity, or dynamic viscosity changes. The approach of this shared research program will be explained on the basis of one of the industrial cases that was addressed in the project. The experience that was gained during the project will be presented, together with the benefits for all partners involved. It will be shown that the form of shared research is an effective way to speed up the implementation of novel process equipment concepts in new or existing industrial processes.

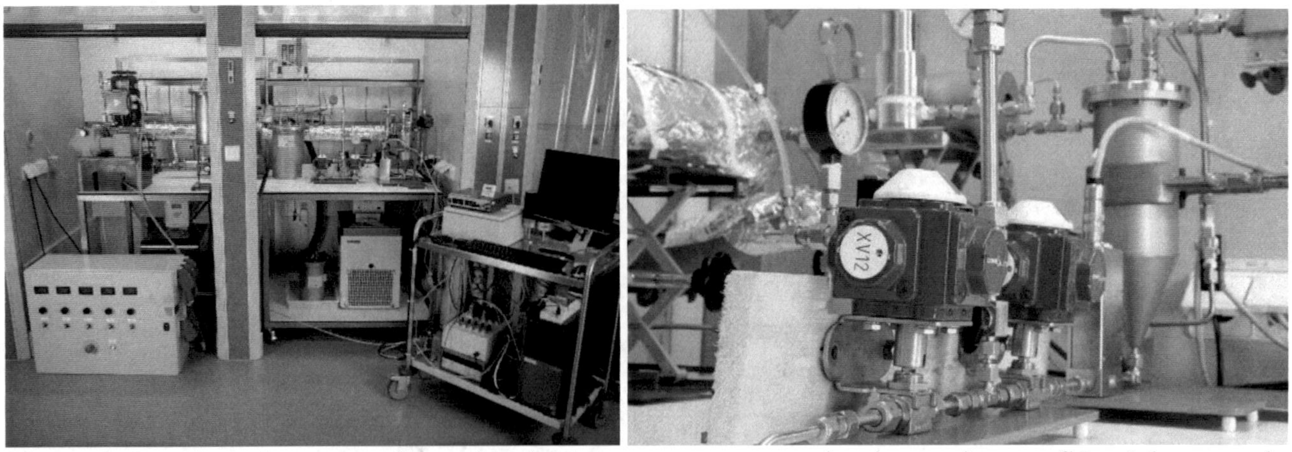

Figure 1: Bench scale test configuration for heterogeneous reaction in continuous flow (photographs by H.J. Siemer).

# (399f) Effect of Sulfur Poisoning On A Cobalt-Based Fischer-Tropsche Catalyst in A Fixed BED Compact HEAT Exchange Reactor

Hani Gadalla[1], Steven J. Vallee[1], Zhijun Jia[1], Burtron H. Davis[2] and Dennis E. Sparks[2], (1)Chart Energy & Chemicals, La Crosse, WI, (2)Center for Applied Energy Research, University of Kentucky, Lexington, KY

Fischer-Tropsch (FT) is a highly exothermic reaction that is very important in most XTL technologies. Integrating compact heat exchange technology with chemical reactors, Chart's Compact Heat Exchange Reactor (CHER) design has been successfully demonstrated to remove the high heat of reaction generated by a Fischer-Tropsch catalyst. The integrated heat exchange of CHER provides process intensification to the FT process.

In this study, the performance of a commercial Fischer-Tropsch cobalt catalyst in a Compact Heat Exchange Reactor is tested. Hydrogen sulfide is injected with the syngas in concentrations in the order of 100ppm by volume. The deactivation of the catalyst as well as the change in methane selectivity is calculated and compared to the baseline operating conditions. The flow rate of the syngas is varied to alter the residence time. The rate of deactivation is calculated at the different flow rates. We attempt to measure the amount of sulfur on the catalyst along the axial direction of the CHER, to understand the magnitude at which sulfur is absorbed onto the catalyst. Also, the rate of deactivation is compared to that of a CSTR using the same commercial FT catalyst. CHER has shown a higher tolerance to sulfur poisoning over the CSTR, further increasing the viability of process intensification in an FT process.

# (409a) Thermophiles for Biofuels Production

Lew P. Christopher, Center for Bioprocessing Research & Development, South Dakota School of Mines & Technology, Rapid City, SD

Environmentally friendly alternatives for sustainable biofuels production are presented through utilization of renewable lignocellulosic biomass such as switchgrass and biomass waste including agri-waste (corn stover, prairie grasses, etc.). Switchgrass is viewed as one of the most promising energy crops for the U.S. conditions and its production is predicted to significantly increase within the next ten years. Production of biofuels from biomass would reduce dependence on fossil fuels, enhance energy security, contribute to the effort of preventing global warming, and alleviate the acute and growing pollution problem in broad perspectives. To date, enzymatic hydrolysis of cellulose presents one of the major technological challenges to the economics of biofuels production due to the relatively high enzyme costs, slow reaction rates, low specific activity, end-product inhibition, time- and temperature-dependant loss of activity, etc. From this perspective, the search for and the discovery of novel thermostable enzymes with unique properties and enhanced capabilities for cellulose degradation could lead to significant improvements in the bioethanol process. The advantages of employing thermostable enzymes in lignocellulose bioprocessing to bioethanol at elevated temperatures are multi-fold: 1) Increased reaction/conversion rates due to improved mass transfer rates and improved substrate accessibility allowing the use of lower enzyme dosages and resulting in higher yields of fermentable sugars; 2) Improved substrate solubility and reduced viscosity of feedstock allowing the use of higher solids loadings; 3) Decreased risk of contamination resulting in increased robustness of process; 4) Decreased cost of energy for cooling when thermal pretreatment of feedstock is desirable; 5) Improved yields due to reduced ethanol intolerance of fermenting strains; 6) Improved integration, heat recovery and flexibility in process configuration. The use of higher solids loadings and temperatures during enzymatic hydrolysis prove beneficial for the significant reduction of hydrolysis times and enzyme loadings with important implications for reduced capital and operating costs of ethanol production. Thermophiles isolated from unique environments such as Yellowstone National Park (WY, USA) and Homestake Mine (SD, USA) and their thermostable enzymes have been employed for production of bioethanol, biohydrogen and biodiesel. Results on biofuels production utilizing new thermophilic isolates will be presented and potential benefits discussed.

# (409b) Using a Recombined Microorganism Consortium for One-Step Production of Ethanol and Hydrogen Directly From Cellulose

RAN DU, Institute of New Energy Technology, Tsinghua University, Beijing, China and Shizhong Li, Institute of New Energy Technology, Tsinghua.University, Beijing, China

Consolidated bioprocess (CBP) is referred to use a super strain to realize one-step cellulosic ethanol production. It has been proven to be impossible in a short term by many literatures. The utility of multifunctional reconstructive consortium referred by our research team is a promising solution to overcome this bottleneck. Consortium H which can convert cellulose into ethanol and hydrogen under anaerobic fermentation was isolated from farm soils in Shandong Province, China. Construction information of consortium H was analyzed by Denaturing Gradient Gel Electrophoresis (DGGE) and Terminal Restriction Fragment Length Polymorphism (T-RFLP). Based on these data, strains with outstanding capability of conversion of cellulosic ethanol were isolated from consortium H. A new consortium reconstructed by the selected strains and the original consortium H was found to significantly increase ethanol conversion rate and yield of hydrogen production. Its ethanol conversion rate was increased from 61% to 81%, while the highest hydrogen production rate of this consortium was increased from 200mL/day L culture to around 240mL/day L culture. T-RFLP was used to monitor the construction changes of this reconstructive consortium, and the results showed obvious changes between the new consortium and consortium H during fermentation process. Further investigation relating to mechanism will be performed on consortium construction and metabolism flux based on metagenomics and metatranscriptomics sequencing. Our group named this method which to reconstruct a novel consortium consulted by bioinformatics as constituted biology, and a series of experiments on how to realize precise regulation on consortium based on this method are being held.

# low cost and low energy consumption process to convert soluble sugar and homocelluloses in sweet sorghum stalks into ethanol

Jihong Li, Bing Han, Shizhong Li*

Institute of New Energy Technology, Tsinghua University

Sweet sorghum is a promising feedstock for ethanol production due to its high fermentable sugar and homocellulose content in the stalks. The soluble sugar in sweet sorghum stalks can be converted into ethanol by a solid-state fermentation technology which uses the crushed stalks as culture substrates directly. Ethanol produced during the solid-state fermentation can be separated by distillation of these fermenting stalks. And the residue solid is called SSB which is rich with homocelluloses (70%). In present research, a process which combined ethanol separation stage after solid-state fermentation and alkaline pretreatment stage of SSB was designed for ethanol production with sweet sorghum. The process was optimized according to the yield of sugar-ethanol and amounts of monosaccharide released from SSB. In our novel process, energy consumption of mechanically squeeze and pretreatment of SSB was avoided. So it is efficient to reduce the cost of ethanol production with sugar crops and implement bio-conversion of all carbohydrates in sweet sorghum stalks. It is considered a promising process to produce ethanol at commercial scale.

Keywords：

bio-ethanol；alkaline pretreatment，solid-state fermentation，sweet sorghum

# (409d) Xylose Fermentation and Respiratory-Deficiency in Saccharomyces Cerevisiae

Adel Ghaderi, Benjamin L. Wang, Hang Zhou and Gregory N. Stephanopoulos, Chemical Engineering, Massachusetts Institute of Technology, Cambridge, MA

For several practical reasons, the yeast *Saccharomyces cerevisiae* is the preferred workhorse for industrial bioethanol production, however this organism lacks native pathways to utilize xylose – one of the more abundant sugars in lignocellulosic biomass. Heterologous expression of xylose-assimilation pathways confers the ability to metabolize this substrate upon *S. cerevisiae*, though several studies report such recombinant yeasts exhibit a respiratory-response to this carbon-source. It has been thought that respiratory-deficient strains cannot grow on xylose in spite, quite remarkably, of their ability to ferment it, however a recent study demonstrated that extensive adaptive evolution can confer growth phenotypes on these strains. As such, balancing characteristics of respiratory and fermentative metabolism may prove crucial for efficient xylose fermentation.

We hope to elucidate the xylose metabolism of recombinant *S. cerevisiae* in the context of respiration and fermentation. Our approach involves the isolation and characterization of mutants from genomic and respiration-rescuing libraries expressed in petite backgrounds (respiratory-deficient yeasts). The most desired mutants include strains which exhibit high fermentation rates while limiting the loss of carbon to respiration and production of biomass. Such strains are difficult to isolate from diverse libraries by virtue of their growth characteristics. To address such challenges, we have developed a method employing microfluidics capable of measuring the production and consumption of bulk-phase metabolites on a single-cell basis. We have previously demonstrated the application of this approach in order to identify genomic changes that result from adaptive evolution in xylose-consuming *S. cerevisiae* which confer superior xylose assimilation. By extending the use of this platform to the study of genomic and respiratory libraries expressed in petite backgrounds, we hope to identify targets for inverse metabolic engineering that will allow for the rational design of strain which exhibit superior xylose-fermentation characteristics.

# (409e) Evaluation of a Pilot-Scale Facility for Biocrude Production by Activated Sludge Cultures Via Lignocelullose Sugar Fermentation and Downstream Recovery Processes

Andro Mondala[1], Rafael Hernandez[1], Todd French[1], William Holmes[1], Magan Green[1], Linda McFarland[1], Larry Pearson[2], Jesse Simmons[1], Robert Glenn[1] and Brady Biggers[1], (1)Dave C. Swalm School of Chemical Engineering, Mississippi State University, Mississippi State, MS, (2)Institute for Clean Energy Technology, Mississippi State University, Starkville, MS

This paper will present results on fermentation experiments and downstream product recovery processes in a pilot-scale facility designed for the production of biocrude by activated sludge cultures and oleaginous microorganisms via lignocellulose sugar fermentation. The pilot plant consists of three mixing/holding tanks for sugar, media, and clean-in-place (CIP) solutions; twin 500-L bioreactors with full pH, dissolved oxygen, aeration, agitation, and temperature control designed for batch fermentation and can be operated under semicontinuous feeding mode; and the downstream processing equipment consisting of a 500 L/h membrane separator used for increasing biomass concentration in the fermentation broth, a spray dryer, a solvent extraction vessel, and a solvent recovery/biocrude concentration unit. Samples will be obtained from the bioreactors for laboratory analysis of biomass and lipid yields, and residual sugar and nutrient concentrations. Samples from the membrane separator and spray dryer will be analyzed for moisture content and moisture removal efficiencies will be calculated. The dried biomass that will be obtained from the spray dryer will be subjected to solvent extraction to separate the biocrude from the biomass, and the solvent-biocrude mixture will be concentrated using the vacuum-operated solvent recovery unit. Overall and component mass balances will be conducted for the whole pilot facility and for each piece of equipment. The experimental data will then be compared with kinetic model predictions to evaluate the performance of the pilot-scale operations compared with the bench-scale trials.

# (409f) Optimization of Extracellular Polymeric Substance (EPS) Production by Activated Sludge Via Fermentation of Glucose

**Hien Nguyen**, Dave C. Swalm School of Chemical Engineering, Mississippi State University, Mississippi State, MS

***Optimization of Extracellular Polymeric Substance (EPS) Production by Activated Sludge via Fermentation of Glucose***

***Authors with affiliation: Hien Nguyen, Patrisha Pham, Rafael Hernandez, Todd French, Andro Mondala***

**Abstract**

The use of activated sludge micro flora to convert sugars, derived from the hydrolysis of lignocellulose biomass, into lipids for biofuel production is of current research interest in our group. Lipid accumulation in this system has been enhanced by manipulating the carbon-nitrogen (C:N) ratio and initial sugar loading on activated sludge bioreactors with glucose as the main carbon source. Activated sludge is a flocculated suspension composed of microorganisms that can also produce extracellular polymeric substances (EPS) *in situ* during fermentation. Microbial EPS, which are secreted largely by bacteria in harsh environments, are a matrix rich in polymers including polysaccharides, glycoproteins, proteins, nucleic acids and phospholipids. The nature and composition of EPS makes it potentially useful as a biofloculant and/or a matrix for other bioproducts. Studying EPS production along with lipid formation could result in a process to produce feedstocks for biofuel production and multiple specialty products.

In this study, samples were obtained from aerobic batch fermentation inoculated with 20% (v/v) activated sludge. The total incubation time for the batch fermentation experiment was fourteen days. EPS from the collected samples were extracted using a Dowex cation exchange resin in phosphate buffered saline solution and precipitated with 95% (v/v) cold ethanol. Fourier transform-infrared spectroscopy of freeze-dried EPS revealed characteristic IR peaks for polysaccharides and proteins. Total polysaccharide and protein concentrations were estimated in each EPS sample using phenol-sulfuric and modified Lowry methods, respectively. An increase in lipid % (g/g cell dry weight) and a decrease in EPS % (g/g cell dry weight) were observed during the incubation period in all treatments. Increasing the carbon loading while maintaining the same C:N ratio increased lipid % and increased specific EPS %. However, for a low C:N ratio, lipid concentration remained constant while specific EPS concentration increased. These results will be useful for manipulation of EPS and lipid production and consequently enable development of valuable strategies to enhance the technical and economic feasibility of the activated sludge to biofuels and bioproducts process.

# (409g) Enzymatic Hydrolysis of Cellulosic Biomass in Nonaqueous Solvents

**Vance Jaeger** and Jim Pfaendtner, Chemical Engineering, University of Washington, Seattle, WA

Cellulases and xylanases are two classes of glycoside hydrolases that act upon biomass in order to produce simple sugars for later fermentation or catalytic upgrading into biofuels. Ionic liquids are organic salts that by definition have melting points below 100°C. Ionic liquids are of interest to those concerned with cellulosic biomass because they help to unravel the complex mixture of cellulose, hemicellulose, and lignin that constitute the substantial portion of most biomass. After pretreatment of the substrate with ionic liquids, cellulase and xylanase gain additional interfacial over which they can more easily and more quickly hydrolyze long polysaccharide chains. Yet, this type of pretreatment requires an additional separation step in which the ionic liquid is separated from the mixture before enzymatic hydrolysis can occur. If one could design compatible ionic liquid-enzyme systems, this separation step may be incorporated into existing downstream processes. Certain glycoside hydrolases have been known to still yield appreciable amounts of sugar as product in low concentrations of ionic liquids in water. One study discovered two hyperthermophilic cellulases that retain nearly complete activity in low concentrations of [Emim][OAc] in water even at highly elevated temperatures [1]. We have discovered that the family 11 xylanase from Trichoderma longibrachiatum also retains enzymatic activity with reduced yield at concentrations of up to 20 wt% [Emim][$EtSO_4$] or [Emim][OAc] in water. We have employed molecular dynamics (MD) simulation in order to uncover the mechanisms by which these enzymes are affected by the addition of non-aqueous cosolvents.

Our simulations and the relevant experimental work of other groups do not show that enzyme denaturation is the cause of the lost activity. On the contrary, our simulations indicate that the highly ionic systems display much more fidelity to crystallographic structure when compared to water, as measured by root mean square displacement (RMSD) on time scales of up to 500ns. Our simulations suggest that ionic liquids, even in low concentrations, dampen the enzymes' dynamic motions that may have implications in proper substrate binding and enzyme activity. In systems that remain more active in ionic liquids, these motions tend to be less affected by the presence of the solvent. In more aqueous solvents, key features of the enzymes fluctuate more rapidly and with a higher magnitude, as measured by root mean square fluctuations (RMSF). For the systems in which ionic liquids reduce the yield of the enzymes, the mixed solvents appear to disrupt a network of coordinated motions and change the pattern of intrinsic fluctuations. Along with disruption of dynamic motions, we found that cations tend to become kinetically trapped near the catalytic residues of the enzymes. Therefore, inhibition might also play a role in lowering catalytic yield. For these systems of glycoside hydrolases, our simulations suggest that dampened dynamics and inhibition are the two most likely causes of hindered enzymatic function in these ionic liquid systems, while denaturation may not play a role at all.

[1] S. Datta et al., "Ionic liquid tolerant hyperthermophilic cellulases for biomass pretreatment and hydrolysis", Green Chemistry, **2010**, 12, 338-345.

# (466a) Application of Quality by Design Principles for An Existing Product for the Control of the Dimer Impurity in the Manufacturing Process

**Gautam Samanta**[1], Praveen K Sharma[2], S Srividhya[2], Kamlesh Pai Fondekar[2], K R Janardan Sarma[2], Saurabh Gupta[2], Grb Ravish[2] and Y Raghuram[3], (1)Dr. Reddy's Laboratories Ltd., Hyderabad, India, (2)Dr. Reddy's Lab, Hyderabad, India, (3)DR. Reddy's Lab, Hyderabad, India

Application of Quality by Design principles for an existing product for the Control of the Dimer Impurity in the Manufacturing Process

Gautam Samanta, Praveen K Sharma, S Srividya, Kamlesh Pai Fondekar, K R Janardan Sarma, Saurabh Gupta, GRB Ravish, and Y Raghuram

Dr. Reddy's Laboratories

Integrated Product Development Organization

Bachupally, Hyderabad

India

The philosophy of quality by design is all about the patients' need. The designing, developing, and manufacturing of the product must meet those needs. Most of the innovator and few generic companies have adopted the concept of QbD in their product development. Dr. Reddy's Laboratories has started implementing the QbD principles for all new APIs development. The QbD implementation for new products is well documented but not for existing products. The key drivers for the implementation of QbD for the existing products are to reduce variability in the product quality, improve yield, reduce cycle time, solve manufacturing issues, reduce cost of quality, and introduction of real time release testing for manufacturing process.

The objective of this work is to provide a perspective of QbD application for existing API manufacturing where a stepwise reverse QbD approach has been applied. The unique features of this work comprise methodology and implementation strategy in defining design space and thereafter executing the control strategy for consistent production of the API.

Irbesartan is an angiotensin II receptor antagonist used mainly for the treatment of hypertension. Dr. Reddy's Laboratories developed the generic version and filed DMF in March 2009. In stage 1, IRB-1 is formed by reaction between 2 n butyl-4-spiro cyclopentane-2-imidazoline-5-one and 4-bromo methyl-2'-cyanobiphenyl at -5 to -10 °C in the alkaline medium (Figure 1). The reaction time is 2-3 h. In stage 2, IRB-1 is converted to tetrazole by the reaction with $NaN_3$ at elevated temperature and the final product Irbesartan (IRB-2) is obtained after crystallization.

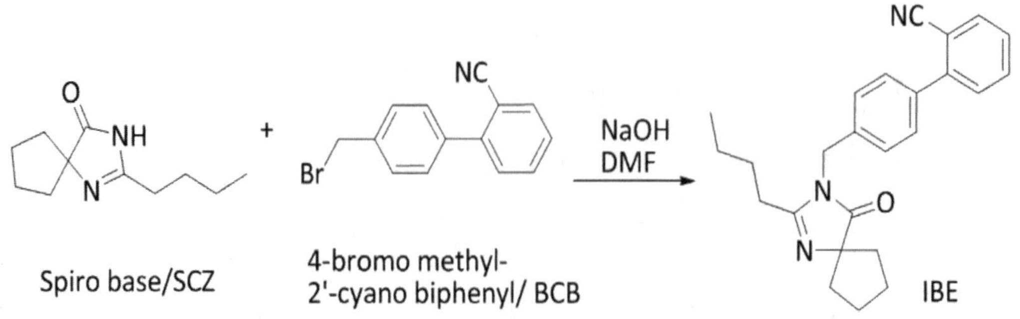


**Figure 1**: Stage 1 for the synthesis of Irbesartan and the structure of the Dimer impurity (CQA)

In stage 1, there is a significant variation (upto 5%) in the formation of a dimer impurity, which is difficult to remove in the subsequent stages. If the dimer content is more than 1%, significant yield loss occurs for its removal. So the cost of quality is high. The dimer impurity is the critical quality attributes (CQA) in this stage. The QbD principles have been applied to address the issue.

The following stepwise QbD approaches has been adopted:

1. Historical data and analysis of data to identify the problem

There is a significant variation in the formation of the dimer in the process. So some variables are playing critical role. To understand the variability in the dimer quantity experiments were conducted at pre-determined time under the same experimental conditions.Figure 2 shows the experimental results, which indicate that the formation of the dimer depends on the reaction time. Even within the reaction time 2-3 h, the dimer quantity is varying significantly and the variation is more after 3 h which indicates that reaction time is very critical. The best fit with the data is the third-order polynomial equation ($R^2$= 0.994).

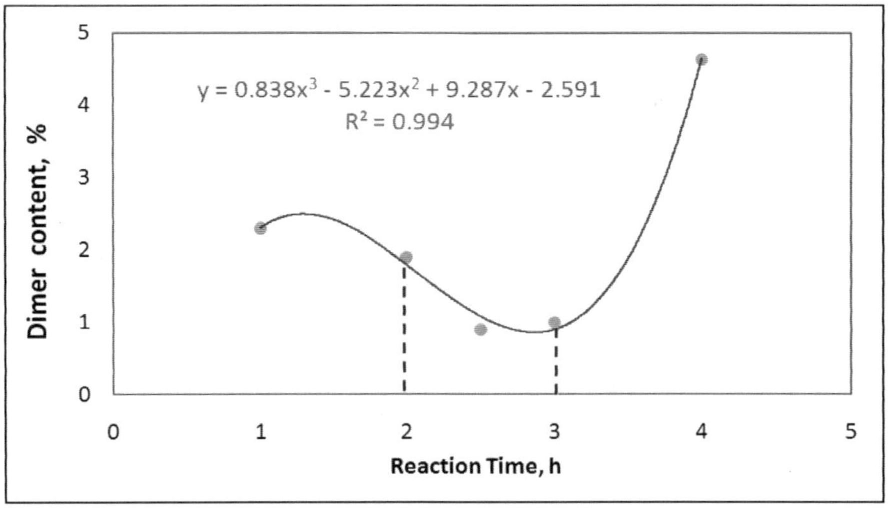


**Figure 2**: Best fitting of the experimental data

2. Quality risk assessment for material attributes and process parameters

To identify the critical variables the cause-effect diagram (Figure 3) has been used. All material attributes and process parameters have been listed in this diagram. Based on the chemistry knowledge as well as practical experience it has been concluded that material attributes are not responsible for the dimer. So only the process parameters have been considered for risk analysis.

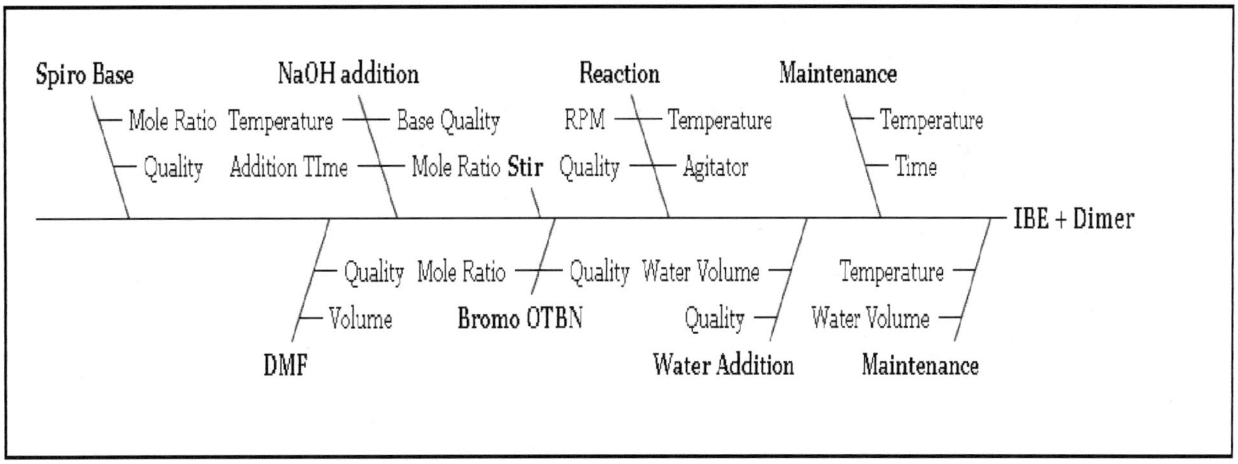


**Figure 3**: Cause-effect or fishbone diagram for MA and PP to understand the effect on CQAs

In the case of PPs, the first phase is to identify those potential CPPs that may affect the CQAs of the API (*i.e.,* the severity). The second phase of the assessment is to estimate the probability of a parameter affecting the API (*i.e.,* the risk). The parameters with a high probability of influencing the API with enough severity to produce unacceptable API are critical parameters. The overall risk is the combination of severity and probability and risk assessment is the impact of the process on the API. Table 1 shows the risk assessment of the process parameters.

**Table 1**: Risk assessment for the process parameters

| Trigger event | Severity | Probability | Overall Risk | Risk assessment |
|---|---|---|---|---|
| DMF volume | Low | Low | Low | Low risk to Dimer |
| Water volume | Low | Low | Low | Low risk to Dimer |
| $H_2O$ addition temp. | Low | Medium | Medium Low | May exceed range; no impact on Dimer |
| Reaction temp. | Low | Medium | Medium Low | Parameter is critical |
| RPM | High | Medium | Medium High | May impact on Dimer; Parameter is critical |
| NaOH | High | Medium | Medium High | May impact on Dimer; Parameter is critical |
| Reaction time | High | High | High | High impact on Dimer; Parameter is critical |

Based on the preliminary risk analysis of the process parameters, RPM, NaOH quantity, and reaction time have been identified as critical process parameters. Though the reaction temp is critical, it has not been considered for DoE study because the reaction is not completing above -5 °C and the reaction mass is freezing below -18 °C. Due to a narrow window of operation it was decided to fix it at -7.5 ± 2.5 °C. To establish the relation between the critical process parameters and CQAs the design of experiments (DoE) has been applied.

3. DoE, understanding of process, and design space

Initially, two-level factorial design was selected and twelve experiments ($2^3$ + 4 center points) were conducted in the lab considering three variables and their respective ranges as mentioned in Table 2. The normal plot for the dimer

(Figure 4) reveals that all the three variables are making significant positive impact on the dimer formation (NaOH 36%, RPM 21%, and Reaction time 13%). It is interesting to note that the three-variable interaction (ABC) has negative impact (15%) on the dimer formation. Due to significant curvature a central composite design was selected and remaining experiments were conducted.

**Table 2**: Critical process parameters and their ranges

| Variables | Low | High | Rational for Selection |
|---|---|---|---|
| NaOH | 1.2 eq. | 1.7 eq | Kinetic Effect: Higher amount of base could increase the rate of formation of dimer. |
| Reaction Time | 150 min | 200 min | Kinetic Effect: Longer reaction time could increase the formation of dimer |
| RPM | 300 | 600 | In heterogeneous system, RPM will influence the heat/mass transfer and distribution of solid in liquid medium. |

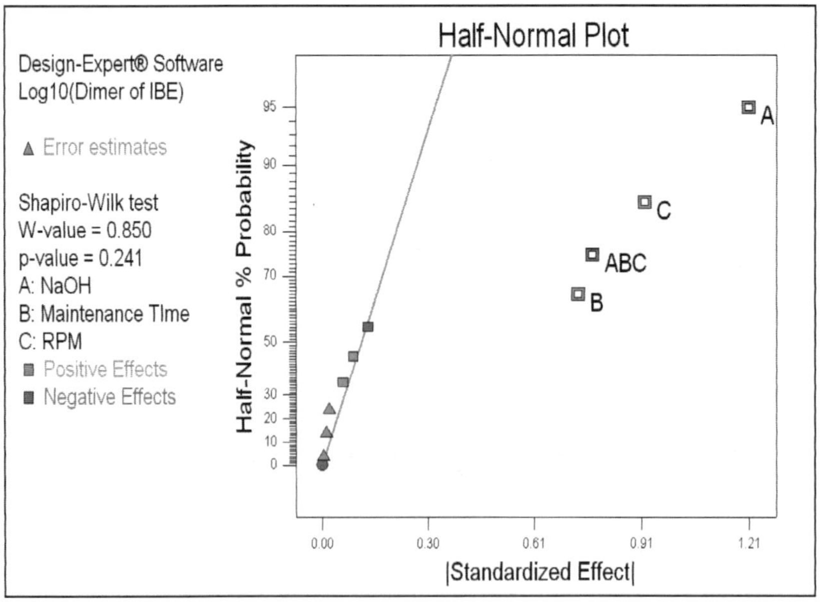


**Figure 4**: Half-normal plot for two-level factorial design

The data from CCD are fitted to a cubic model ($R^2$ = 0.9999, Adj $R^2$ = 0.9997, Pred $R^2$ = 0.9496, and PRESS = 2.76). Analysis of variance testing (Table 3) confirms the significance of three-factor interaction (ABC), as well as of other terms $A^2B$, $AB^2$, $A^2C$, $A^2$, $B^2$, $C^2$ and two-factor interactions AB, AC, and BC. The three-dimensional surface plot, illustrated in Figure 5 is generated according to the cubical model and demonstrates the dependence of dimer on the amount of NaOH and RPM at optimal level of reaction time (2.5 h). To see the influence of three-factor interaction

(ABC), the model term (ABC) has been excluded and the analysis of variance indicates that in the absence of that term almost all the terms become insignificant (Table 4) and the lack of fit becomes significant. This indicates that the three-factor interaction term is playing very critical role in the formation of dimer.

Figure 6 is an overlay plot of yield, dimer, and other QAs, setting the foundation for continued improvement of the process. The yellow region is design space where the desired QA levels will be obtained. Laboratory validation experiments were performed under the experimental conditions (NaOH 1.3 eq, Reaction time 200 ± 10 min, RPM 300 ± 20) chosen from the design space and the dimer (0.29 ± 0.06) and other quality attributes were within the specifications.

**Table 3**: ANOVA table and associated statistics for response surface cubic model for dimer

**ANOVA for Response Surface Reduced Cubic Model**

**Analysis of variance table [Partial sum of squares - Type III]**

| Source | Sum of Squares | df | Mean Square | F Value | p-value Prob > F | |
|---|---|---|---|---|---|---|
| Model | 54.68 | 13 | 4.21 | 5578.23 | < 0.0001 | significant |
| *A-NaOH* | *1.33* | *1* | *1.33* | *1761.66* | *< 0.0001* | |
| *B-Maintenance* | *0.95* | *1* | *0.95* | *1262.71* | *< 0.0001* | |
| *C-RPM* | *0.31* | *1* | *0.31* | *413.81* | *< 0.0001* | |
| *AB* | *0.065* | *1* | *0.065* | *85.93* | *< 0.0001* | |
| *AC* | *4.74* | *1* | *4.74* | *6289.96* | *< 0.0001* | |
| *BC* | *0.22* | *1* | *0.22* | *288.82* | *< 0.0001* | |
| $A^2$ | *1.82* | *1* | *1.82* | *2408.79* | *< 0.0001* | |
| $B^2$ | *1.30* | *1* | *1.30* | *1724.81* | *< 0.0001* | |
| $C^2$ | *0.38* | *1* | *0.38* | *506.63* | *< 0.0001* | |
| *ABC* | *3.67* | *1* | *3.67* | *4869.51* | *< 0.0001* | |
| $A^2B$ | *0.11* | *1* | *0.11* | *151.83* | *< 0.0001* | |
| $A^2C$ | *1.26* | *1* | *1.26* | *1671.22* | *< 0.0001* | |
| $AB^2$ | *0.55* | *1* | *0.55* | *732.34* | *< 0.0001* | |
| Residual | 4.525E-003 | 6 | 7.541E-004 | | | |
| *Lack of Fit* | *2.241E-003* | *1* | *2.241E-003* | *4.91* | *0.0776* | *not significant* |
| *Pure Error* | *2.283E-003* | *5* | *4.567E-004* | | | |
| Cor Total | 54.69 | 19 | | | | |

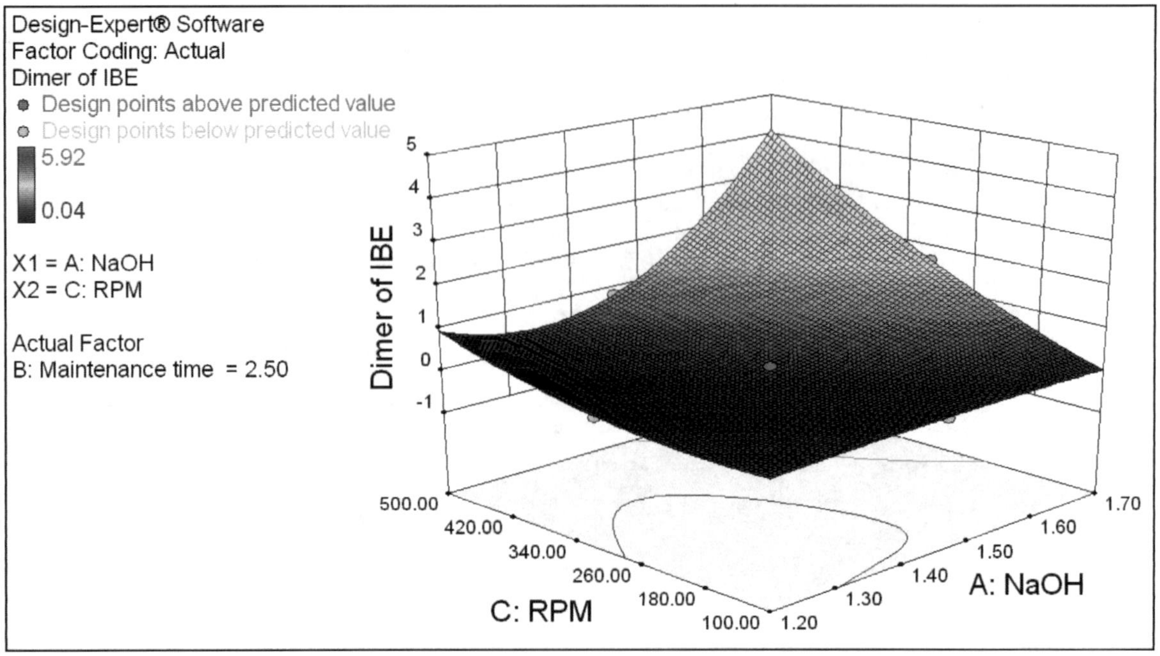


**Figure 5**: Influence of NaOH and RPM on the formation of dimer impurity at reaction time 2.5 h

**Table 4**: ANOVA table and associated statistics for response surface cubic model for dimer excluding ABC interaction term

**ANOVA for Response Surface Reduced Cubic Model**

**Analysis of variance table [Partial sum of squares - Type III]**

| Source | Sum of Squares | df | Mean Square | F Value | p-value Prob > F | |
|---|---|---|---|---|---|---|
| Model | 51.01 | 12 | 4.25 | 8.09 | 0.0052 | significant |
| *A-NaOH* | *1.33* | *1* | *1.33* | *2.53* | *0.1558* | |
| *B-Maintenance* | *0.95* | *1* | *0.95* | *1.81* | *0.2201* | |
| *C-RPM* | *0.31* | *1* | *0.31* | *0.59* | *0.4660* | |
| *AB* | *0.065* | *1* | *0.065* | *0.12* | *0.7357* | |
| *AC* | *4.74* | *1* | *4.74* | *9.03* | *0.0198* | |
| *BC* | *0.22* | *1* | *0.22* | *0.41* | *0.5401* | |
| $A^2$ | *1.82* | *1* | *1.82* | *3.46* | *0.1053* | |
| $B^2$ | *1.30* | *1* | *1.30* | *2.48* | *0.1596* | |
| $C^2$ | *0.38* | *1* | *0.38* | *0.73* | *0.4219* | |
| $A^2B$ | *0.11* | *1* | *0.11* | *0.22* | *0.6548* | |
| $A^2C$ | *1.26* | *1* | *1.26* | *2.40* | *0.1653* | |
| $AB^2$ | *0.55* | *1* | *0.55* | *1.05* | *0.3393* | |
| Residual | 3.68 | 7 | 0.53 | | | |
| *Lack of Fit* | *3.67* | *2* | *1.84* | *4022.95* | *< 0.0001* | *significant* |
| *Pure Error* | *2.283E-003* | *5* | *4.567E-004* | | | |
| Cor Total | 54.69 | 19 | | | | |

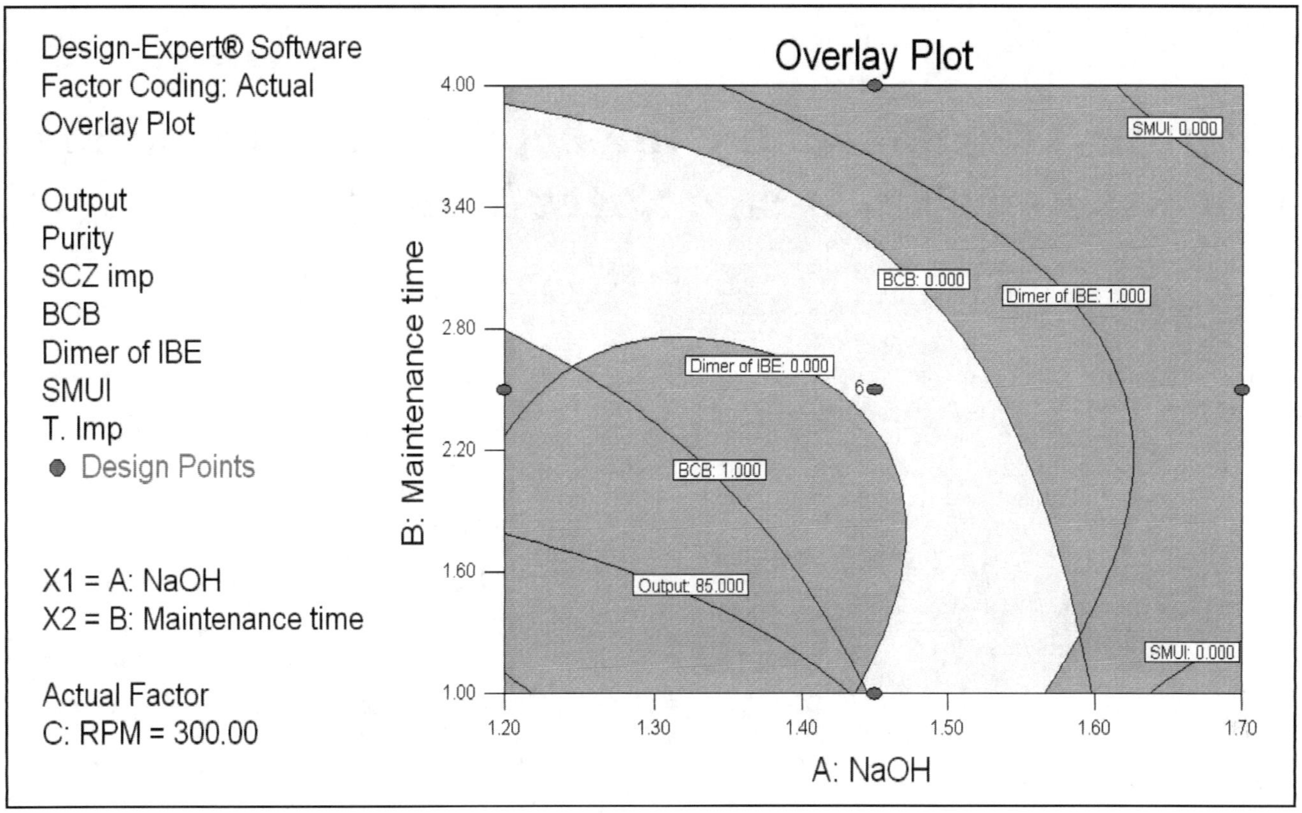


**Figure 6**: Overlay plot of Yield, starting materials, purity, dimer impurity, and single maximum unknown impurity for stage 1, which provides a process operating window

4. Control strategy

Does the output from the laboratory-scale multivariate design get reflected during scale-up? It is important to do the appropriate scale-up for the scale dependent variables before manufacturing. Based on the process understanding and criticality a control strategy has been developed to consistently ensure product quality. To control the formation of the dimer in the plant the scale dependent variable, RPM was scaled-up based on tip-speed and control on other variables was achieved through automation. In-process analysis was not available to monitor the formation of the dimer at that time. The consecutive six batches data in the plant indicates that the dimer is under control within the specification (<1%).

The QbD principles provide a structured approach for gaining process knowledge and developing robust manufacturing process. The application of the QbD principles for the existing products is very limited. This has been applied successfully to reduce the variability of the dimer impurity in the process and reduces the cost of quality. On the basis of the process understanding, it has been demonstrated that the control of the dimer impurity can be achieved by controlling the critical process variables.

# (466b) Investigation of the Impact of Mode of Operation (Batch vs Flow) On Design Space Map

**Shankar Vaidyaraman**, Sze Wing Wong and Shekhar K. Viswanath, Chemical Product R&D, Eli Lilly and Company, Indianapolis, IN

This talk will discuss how the mode of operation (batch vs flow) impacts the design space map for impurities thus impacting the robustness of the process. A specific case study involving a reaction model will be used for this investigation that builds upon previous work on failure mode analysis in continuous processes (Wong *etal* 2011). There are three impurities of interest formed in this reaction. A reaction kinetic model developed from batch experiments is used to design for both batch and flow processes. The key questions that this investigation addresses is:

- How much does the mode of operation impact the size of the design space?
- Is the link between the uncertainties due to routine variability and model uncertainty and the output impurity levels same or different depending on the mode of operations?
- Can certain generalizations be drawn based on the results of this case study?

[1] Sze-Wing Wong, Prashant Kokitkar, K. Derek Berglund, Martin D Johnson, and Shekhar K. Viswanath, Failure Mode Analysis of Continuous Process with Process Models, AIChE Annual Meeting, MN 2011

# (466c) A QbD Approach to the Process Development of Tofacitinib

**Bob Dugger**, Chemical Research and Development, Pfizer Inc, Groton, CT

Tofacitinib citrate (CP-690,550-10) is currently being evaluated for the treatment of rheumatoid arthritis, psoriasis and other diseases. An NDA has recently been submitted to the FDA for the treatment of rheumatoid arthritis. The Pfizer Tofacitinib API Team has worked to improve the efficiency of the process and ensure that the quality of the drug substance will meet the standards of the regulatory agencies. Quality by Design is important not only from a regulatory perspective, but also from a process development perspective since it enforces a much more thorough degree of process understanding. DOE is often the focus of QbD but there are often other techniques (experimental and modeling) that can be used in the development of a process. This talk will provide examples of how we use these various techniques to ensure we have a robust process that can be run at a wide range of scales and consistently meet our quality standards.

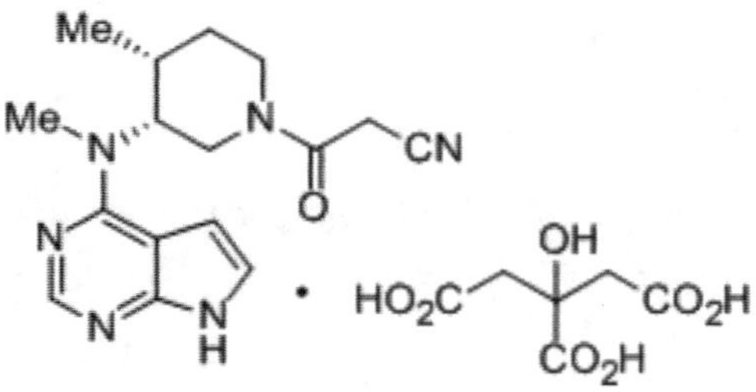


Tofacitinib citrate

# (466d) Design of a Pharmaceutical Manufacturing Process Using QbD Principles: The Challenge of Technical Transfer to Manufacturing and Lifecycle Development

**Conor Collins** and Steve Ball, GlaxoSmithKline, Cork, Ireland

The route of manufacture for product X is a convergent, multi-stage chemical synthesis that has been developed using QbD principles.

A description of the approach to chemical process development utilising QbD methodologies, to determine and control quality critical process parameters is provided. A workflow from process assessment, through process understanding to development of a control strategy, providing a platform for technical transfer to commercial manufacturing is described.

The process involved in the development of a commercial manufacturing strategy to utilise the thorough process understanding, in order to ensure the provision of a robust commercial manufacture is also illustrated.

Rising to the challenge of providing a robust chemical synthesis, adequately described in the registered manufacturing description in combination with supporting compliance during the product's commercial lifecycle is essential.

# (466e) Strategy and Process for Effective Scale-up On the Removal of Pd During the Synthesis of A Drug Substance

**Alexander Chu-Kung**, Global Pharmaceutical Research and Development, Abbott Laboratories, North Chicago, IN and Hsien-Hsin Tung, Global Pharmaceutical Research & Development, Abbott Laboratories, North Chicago, IL

Metal catalyst, for example Pd or its derivatives, are commonly used during the synthesis of small molecule drug substance. Due to the toxic nature of these catalysts, their levels are highly controlled at the final drug substance stage. Generally, rejection of metal catalyst is not easy and requires multiple steps of purifications. However, from time to time the approach of metal removal could be highly empirical with a minimum understanding on the fundamentals. Given than, it would be very desirable to address the issue through a more fundamental approach to ensure the efficiency and consistency of catalyst removal upon scale-up.

In our presentation, we share our experience during the development of drug substance for the removal of Pd catalyst. A dual approach, incorporating both the solubilization method and adsorption method are employed and fully integrated into the process. For solubilization, the critical control points are identified to be mixing and cake washing. For adsorption, the most cost-effective adsorbents are shown to be carbon and filtrol. Pd rejection consistency and efficiency are fully demonstrated at the pilot plant scale after implementing the proper procedures with the findings.

# (466f) Implementation of Reaction Modeling and Process Analytical Technology in Design Space Development

**Daniel Hallow**, Michael Fenster, Antonio Ramirez, Robert Wethman and Nathan Domagalski, Bristol-Myers Squibb Company, New Brunswick, NJ

A key aspect of Quality by Design (QbD) in pharmaceutical manufacturing is the development of a design space for multivariate processes. In this case study, we explored the application of mechanistic modeling and process analytical technology (PAT) to define the design space of a reaction that generates the pharmaceutical intermediate of a drug substance in development.

The reaction involves a nucleophilic aromatic substitution that affords the desired material along with two critical impurities. The main parameters impacting the formation of the impurities are the excess base and the temperature of the reaction. A reaction model based on mechanistic studies was built using data captured by HPLC and then used to conduct simulations that generated the response surfaces corresponding to the conversion of the starting materials and the formation of the critical impurities. In addition, the reaction progress was analyzed with Raman spectroscopy and a chemometric model was built to allow real-time monitoring of the starting material conversion. Combining the knowledge of the reaction response across multiple parameters with Raman spectroscopy for real-time monitoring of the reaction progress allows for relaxation of processing constraints and resulting in the expansion of the design space.

In summary, this work shows the synergistic benefits of combining mechanistic modeling and PAT in a QbD approach to enhance the design space and robustness of a process.

# Application of Quality by Design in API Process Development

*Umetrics – An MKS company*
*Erik Johansson, Senior Application Specialist*
*Nirav Shah, Senor Application Specialist*

Design of Experiments (DOE) is the most effective method to achieve product and process efficiency and optimization. Design of Experiment (DOE) studies can help develop *process development knowledge* by revealing relationships, including multifactorial interactions, between the variable inputs (e.g., component characteristics or processing parameters) and the resulting outputs (e.g., in-process material, intermediates, or the final product).

QbD should by definition involve DoE to identify critical interactions and noise. There are many steps in establishing a Design Space for QbD. This presentation will focus on the actual calculation and determination of a Design Space with many factors. The ICH Q8 document gives an illustrative example of how to handle this for two factors and broadly describes how additional factors should be treated. In an example based presentation we will demonstrate how the design space can be calculated for 5 factors and also illustrate the fundamental difference between a design space approach and a process set-point with a maximum process window. The process set-point with 5 factors gives a high and low value for each factor corresponding with ICH Q8. The Monte Carlo derived design space approach is more in accordance with ICH Q11 where the actual design space is given by an equation. The advantage with this approach is that it gives the possibility for control- meaning that the best value for one factor is changed on the basis of the set point of another factor. In the graph below we can see that when the Substrate Concentration is 1.7, the best values for NaOH and CaCl2 are 1.6 and 1.4. Similarly at Substrate Concentration 1eq the corresponding values are 1.55 and 1.2.

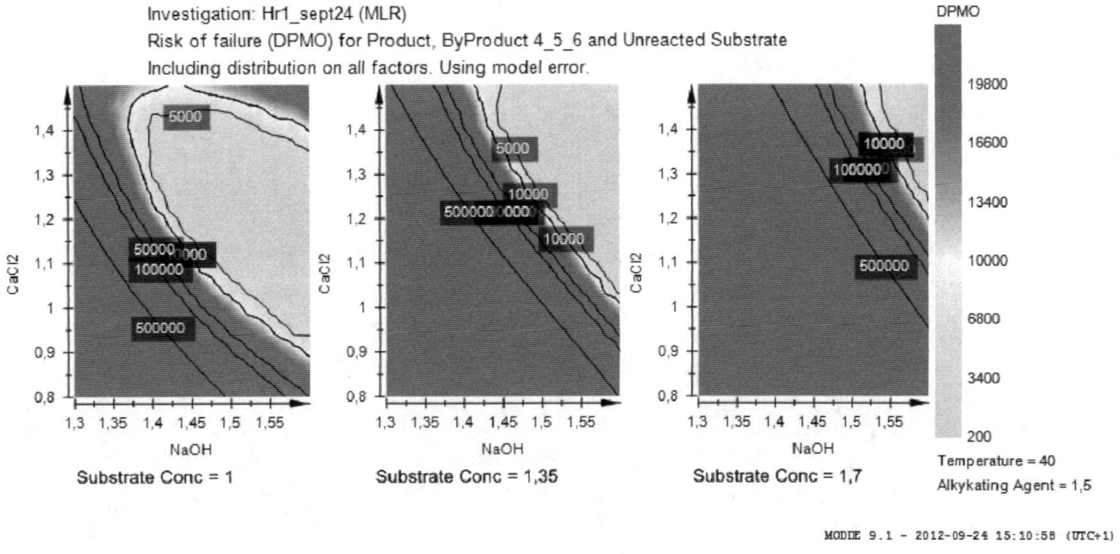

# (474a) Utilizing Thermochemical Process Streams to Upgrade Bio-Char Into Value Added Activated Carbon Through Physical Activation

**Joseph Polin**, South Dakota State University, Brookings, SD and Zhengrong Gu, Agricultural and Biosystems Engineering, South Dakota State University, Brookings, SD

Fast pyrolysis involves the thermochemical conversion of biomass at 400C – 600C in the absence of oxygen to produce three main products: liquid bio-oil, solid bio-char, and synthesis gas (syngas). Current biofuel research focuses on upgrading the bio-oil further into renewable transportation fuels, while the co-products bio-char and syngas are considered less valuable and generally undesired. Bio-char possesses many potential low value end-uses such as a soil amendment to improve soil quality, as a carbon sequestration method to reduce environmental greenhouse gases, and as a solid fuel for producing additional process heat. Syngas comprising a large percentage of carbon monoxide could also be burned to supply process heat. Using a heat recovery combustion unit to provide the heat required for the endothermic fast pyrolysis reaction results in flue gas combustion products, carbon dioxide and water vapor. This project aims to upgrade thermochemical bio-char into higher value activated carbon using syngas and flue gas process streams as activating agents.

Activated carbon materials have many high value commercial uses including adsorbents for gas or liquid purification processes, catalyst supports for catalytic chemical processes, and advanced energy storage materials. Water treatment and purification of potable water is a significant use for activated carbon materials. The increasing demand for clean water in highly populated developing countries is expected to increase the global market demand for activated carbon materials for water purification. Traditional feed stocks for activated carbon include coal, lignite, wood, and coconut shells. Future biorefineries using herbaceous or woody biomass feed stocks will produce large quantities of bio-char that may need to be upgraded differently, depending on their composition and characteristics.

Chemical and physical activation methods both introduce an activating agent to the carbon precursor and typically perform the reaction at very high temperatures in the range of 750°C – 900°C. This study focuses on the physical activation process and chose carbon dioxide ($CO_2$), carbon monoxide (CO), and steam ($H_2O$) as the activating agents to simulate syngas and flue gas process streams. The syngas from fast pyrolysis could be used directly as the activating agent or burned to produce flue gas components and process heat, both necessary for the activation reaction. These activating agents would be very economical since they are inexpensive and readily available gas streams from the pyrolysis unit and heat recovery combustion process. These syngas and flue gas atmospheres are investigated for upgrading the solid bio-char into higher value activated carbon.

Physical activation experiments were conducted using a fixed bed reactor (FBR) and a plug flow reactor (PFR) coupled with a gas flow system for metering the activating agent. Mixtures of high purity $N_2$, $CO_2$, CO, and $H_2O$ were used as model compounds to simulate syngas and flue gas streams. This project investigated the influence of process parameters such as temperature, reaction time, flow rates, and gas composition to produce highly porous activated carbons. The response variables include burn off %, BET surface area, total pore volume, and measurements of methylene blue adsorption capacity and iodine number.

The burn off % is calculated by measuring the bio-char weight before and after activation. The surface area and total pore volume were measured using a Micromeritics TriStar 3000 automated gas adsorption analyzer. Samples were first degassed using a Micromeritics FlowPrep 060 unit at 200°C for a minimum of 5 hours to ensure any moisture was removed prior to nitrogen isotherm analyses. Specific surface areas were calculated using the Brunauer–Emmett–Teller (BET) model for the nitrogen isotherm. Methylene blue adsorption capacity was measured following the American Standard Test Method (ASTM D3860) – Determination of Adsorptive Capacity of Activated Carbon by Aqueous Phase Isotherm Technique. The iodine number was measured following ASTM D4607 – Determination of Iodine

Number of Activated Carbon.

The physical activation experiments exposed dry bio-char to gas mixtures for 20, 40, 60, and 80 minute durations and temperature's influence was studied from 800°C, 850°C, and 900°C. An intermediate temperature of 850°C generally produced better activated carbons with higher surface area. The activated carbon reached a maximum surface area at 50% burn off after 40 minutes. After this time, it is believed that pore destruction becomes more prevalent during later stages of the reaction than the pore creation typically found in the beginning stages.

The activating agent's total flow was varied from 1.3 liters per minute (Lpm) to 0.7 Lpm and 0.25 Lpm, and its influence on the activation was studied. It was found that a lower gas flow rate produced activated carbons with higher surface areas. Linares-Solano et al. discussed similar results in their paper, *Porosity Development during $CO_2$ and Steam Activation in a Fluidized Bed Reactor*. They explained that higher flow rates provided a very reactive environment for the carbon and resulted with underdeveloped porosity from external surface burning.

The amount of $CO_2$ in the gas mixture was diluted with inert $N_2$ from 95% to 60% and 30% (volume %) while the total flow was held constant. Linares-Solano et al. explains the key to well-developed porosity is linked to the $CO_2$ gasification kinetics. During chemical reaction controlled kinetics, the reaction is limited by CO intermediates occupying reactive sites and the reaction occurs uniformly on the carbon (bio-char) particles. During diffusion controlled kinetics, the reaction readily reacts on the particle surface and results in an unreacted core with poorly developed porosity. Results from experiments with varying amounts of $CO_2$ are still under review.

The influence of syngas composition with varying ratios of $CO_2$:CO is studied to determine the optimum and tolerable amount of CO in the activating agent. The syngas process stream possesses a large fraction of CO, so it may prove to be an important parameter since Linares-Solano et al. demonstrated that CO intermediates help facilitate the porosity development. Results from these experiments are currently under review.

The activated carbon materials produced have similar pore characteristics, surface area, methylene blue adsorption capacity, and iodine number that are comparable to conventional activated carbon materials and previously reported results. Another value added co-product improves the economic potential of thermochemical conversion pathways for producing renewable transportation fuels and purification materials for water treatment. These products can be produced more economically using a synergistic combination of thermochemical, physical activation, and heat recovery processes.

# Optimum Parameter Selection and Modeling of Batch Extraction of Betulinic Acid from leaves of Vitex Negundo Linn

**S. V. Taralkar[a], S. Chattopadhyay[b*], V. G. Gaikar[c], P.V.Vijayababu[d], A.K. Ray[e]**

[a]*Chemical Engineering, MAEER, Alandi (D), Pune, India.*

[b]*Polymer and Process Engineering, I.I.T.Roorkee, Saharanpur Campus, UP, India.*

[c]*Chemical Engineering, Institute of Chemical Technology, Mumbai, India.*

[d]*Chemical Engineering, Dr.B.A.T.Univ., Lonere, Maharashtra, India.*

[b]*Pulp and Paper Technology, I.I.T.Roorkee, Saharanpur Campus, UP, India.*

*Corresponding author: Tel.:+91-132-2714374[O]; Fax: +91-132-271-4311

E-mail Address: [b*]sujay1999@rediffmail.com, [a]suyogkumartaralkar@yahoo.co.in.

**Abstract**

Extraction of betulinic acid from leaves of Vitex Negundo Linn was carried out in a series of solvents (methanol, ethanol, isopropyl alcohol, and water). Methanol was found to be the best solvent. Maximum recoverable amount of betulinic acid was determined by Soxhlet extraction using methanol as solvent. Batch extraction of betulinic acid was carried out separately, to estimate the effect of operating parameters (e. g. particle size speed of agitation, solid loading and temperature) on the recovery of betulinic acid. Concentration of betulinic acid was estimated in the extracted mass by reverse phase, RP-HPLC system with symmetry C-18 column, and UV detector (λ=210nm). Extraction parameters were optimized for betulinic acid. An unsteady state diffusion model (as a flat particle geometry for leaves) was proposed to predict the experimental data. The experimental data fitted well with the model proposed and effective diffusivities of betulinic acid through cellulose matrix of leaves were estimated (1.7 to 51.5 x $10^{-13}$ $m^2$/sec) at various temperatures. Diffusion of betulinic acid increases with temperature and thus the energy of activation required for diffusion was evaluated by Arrhenius equation and found to be 125.50 kJ/mol.

**Keywords**: Vitex Negundo Linn, Betulinic acid, Batch extraction, Diffusion model, Diffusivity, Activation energy.

## 1. Introduction

Vitex Negundo Linn (Verbenaceae family) is an important medicinal plant usually found in the south-east Asian countries. The extract from the leaves and the roots has important applications in the field of medicine. Decoction of leaves is considered as tonic and vermifuge [1]. The leaf extract is also used in Ayurvedic and Unani [2] medicine. Its mosquito repellent [3], anti-arthritic [4], analgesic [5], hepatoprotective [6], anti-inflammatory, and anti-allergic [6, 7] activities are commendable. The plant has insecticidal [1, 8], antibacterial, and antifungal activities [9] also. A long series of compounds e.g. essential oils [10], vitamin-C, carotene [11], β-sitosterol, iridoid glycosides [12], flavonoids, flavone glycosides [13, 14], and triterpenoids-ursolic acid and betulinic acid [6] are also identified from the leaves and seeds of this plant.

Betulinic acid (3β-hydroxylup-20-(29)-en-28-oic acid) is a pentacyclic triterpenoids [15] with molecular formula $C_{30}H_{48}O_3$, Molecular weight 456.68, and melting point 283-285 °C. The structure of betulinic acid is as shown in Fig. 1 [16].

The greatest potential of betulinic acid to selectively kill human melanoma cells leaving healthy cells untouched is really commendable. Due to its apparent specificity for melanoma cells, betulinic acid is a promising anti-cancer ingredient. Retarding effect (by preventing the formation of syncytia) of HIV-1 infection is also reported [16] for this acid. Thus betulinic acid has huge potential as an important pharmaceutical product. Literature review discusses chemical structure, and pharmaceutical potential and chromatographic isolation from rest of the components present in leaves.

Wongkittipong et al. [17] had modeled solid-liquid extraction of andrographolide from Andrographis paniculata while, Simeonov et al. [18] had investigated kinetics and modeling of extracts from tobacco leaves and oak bark. Simeonov et al. had correlated experimental kinetics and porosity to estimate effective diffusion coefficient. The commercial method of extraction, parametric optimization, and mathematical model has not been looked into for extraction of betulinic acid from Vitex Negundo Linn. In the present report an effort has been made to look into parametric optimization of experimental parameters.

The first part of the work deals with the study of different parameters such as solvent (methanol, ethanol, water, and isopropyl alcohol), speed of agitation (500, 700 and 950rpm), size of particle (0.106-0.420, 0.420-0.600, 0.600-0.850mm), solid loading (5%, 7.5%, 10%) and temperature (28°C, 36°C, 40°C, 45°C, 50°C). In the second part an unsteady state mass diffusion model is proposed to account for the experimental data. Activation energy required for diffusion of betulinic acid was determined.

## 2. Experimental

### 2.1. Materials

The leaves of Vitex Negundo Linn were collected locally (Raigad district of Maharashtra, India). Leaves were washed with water and shade dried (at 30°C for 168hrs) and powdered. The powdered mass was then sorted in to various sizes by screening technique and 0.106-0.42, 0.42-0.60, and 0.60-0.85mm sizes were prepared. Standard betulinic acid (90%) was obtained from Fluka, USA, and used for calibration purposes. All solvents (acetonitrile, methanol, isopropyl alcohol, and ethanol) used in the experiments were of HPLC grade of S. D. Fine, India.

### 2.2. Soxhlet extraction

The total content of recoverable betulinic acid was obtained by soxhlet extraction method using methanol as solvent. 10gms of screened solid material (0.106-0.42mm) was taken in soxhlet apparatus with 200ml methanol, and extraction was carried out for various time durations (2, 4, 6, 8, 10, and 12hrs). Samples (free from leaves) collected at the end, were stored in freeze, and analyzed in reverse phase HPLC system to determine the concentration of extracted betulinic acid. Each experiment was repeated twice to ensure reproducibility.

### 2.3. Batch extraction experiment

The objective of the batch experiments is to measure the kinetics of extraction and to analyze the influences of the operating parameters. 5gm leaves were taken in a four baffled 150ml stirred borosilicate glass vessel (5cm ID, and 9.5 cm height) along with 100 ml solvent. Four bladed turbine type agitator (2cm diameter) was used to stir the mass at 700rpm and the revolution was monitored through an online speedometer. All experiments were conducted at 28°C (±1 $^0$C) unless otherwise stated. Fig. 2 shows the schematic of batch extraction setup.

Effect of solvents of varying polarity (e.g. methanol, ethanol, isopropyl alcohol, and water), speed of agitation (500, 700, 950rpm), particle sizes (0.106-0.42, 0.42-0.60, 0.60-0.85 mm), percent solid loading (5%, 7.5 %, 10%), and temperature (28°C, 36°C, 40°C, 45°C, 50°C) were investigated to select the optimum conditions for each parameter. Samples were collected at various time intervals and extraction was carried out for 120min to ensure no further improvement in yield of betulinic acid occurs. The collected samples were freezed, and analyzed in HPLC to determine the concentration of betulinic acid using following chromatographic conditions.

## 2.4. Apparatus and chromatographic conditions

Standards and all unknown samples of betulinic acid concentrations were measured using HPLC system of Waters (USA). The system consists of Waters HPLC 600 pump, a Rheodyne 7725i injector with 5μl sample loop, on-line degasser AF (Waters, USA), Waters 2487 Dual λ absorbance detector (at λ=210nm, for betulinic acid) with 0.05aufs sensitivity. Betulinic acid was analyzed with a Symmetry® C-18 (4.6×250mm, 5μm) column equipped with automatic temperature controller (±0.1°C). The isocratic mobile phase (acetonitrile: methanol = 80:20) with an elution volume of 0.5 ml/min was selected for identification of betulinic acid. The column temperature was maintained at 35°C (±0.1°C).

## 2.5. Calibration of betulinic acid using HPLC

Standard solutions of betulinic acid (0.003, 0.006, 0.009, 0.012, 0.015, 0.018, and 0.06 mg/ml) were prepared in methanol. All standard as well as samples were filtered through 0.2μm (PALL Ultipor $N_{66}$) membrane filter and injected in the HPLC system through another syringe filter 0.2μm. The area under the peak of betulinic acid was calculated using Waters Empower-II software and that corresponds to the concentration of betulinic acid present. The plot of area against concentration of betulinic acid standard injected shows a linear fit with correlation coefficient 0.9991 (Fig. 3). The same calibration chart was subsequently used for quantification of unknown samples. The chromatograms of standard betulinic acid and methanol extract are shown in Fig. 4 and Fig. 5 respectively as reference. Retention time of betulinic acid at the chosen chromatographic condition is 10.92 min.

# 3. Results and discussion

## 3.1. Selection of solvent

Solvents of varying polarity ($CH_3OH$, $C_2H_5OH$, $i$-$C_3H_7OH$, and $H_2O$) were chosen to extract betulinic acid from leaves at room temperature. As indicated in the batch extraction, dry leaves (5 gm) were taken along with each solvent (100 ml) and stirred at 28°C (±1) for 30 min. At the end of each experiment the solution was filtered and volume was adjusted at its initial value (100ml) to avoid errors due to vaporization loss. Fig. 6 shows the concentration of extracted betulinic acid after 30mins of batch extraction at 28°C. Concentration of betulinic acid shows maximum for methanol and it decreases gradually with ethanol, IPA, and water. No extraction is recorded in water.

This is probably due to solubility values of betulinic acid decreases in the above order. Smaller size of methanol helps it penetrate better inside the cellulose matrix of leaves which gradually becomes difficult for larger alcohols e.g. ethanol, and IPA. Bulky size of them prevents easy penetration; hence, gradual lowering in extraction is noticed. Solubility of betulinic acid is negligible in water at the experimental condition, thus it does not show any recovery of acid.

## 3.2. Soxhlet extraction

The soxhlet extraction of leaves of Vitex Negundo Linn with methanol as solvent was used to find out the maximum amount of recoverable betulinic acid content. After 10hrs of soxhlet extraction the betulinic acid does not increase beyond 0.265% on dry basis, in the selected variety of leaves. Therefore, this was selected as the maximum recoverable amount of betulinic acid and later used as basis for calculation of percentage extraction.

## 3.3. Effect of speed of agitation

The extraction of betulinic acid was studied with methanol as a solvent at room temperature (28 °C) at varying speeds of agitation 500, 700, and 950rpm and Fig. 7 shows that percentage extraction increases with increase in speed. The rise is considerable for 500rpm to 700rpm, but thereafter rise is negligible for 950rpm. Increased turbulence at higher speed reduces interface mass transfer resistance but attains a limiting value at and beyond 700rpm, indicating thereby pore diffusion controls, and thus there is no further increase in mass transfer. Therefore, optimum speed of agitation was selected at 700rpm for all subsequent studies. Lovasoa et al. [19] had investigated the effect of speed of agitation (100, 200, 400, 600 rpm) and reported that at lower rpm because of sedimentation of particles extraction rate protopine is slow while between 200 to 600 rpm no difference in extraction rate occurs.

### 3.4. Effect of particle size

Batch extractions with methanol as solvent, 700rpm speed, and 28°C temperature were performed using different particle sizes (0.106-042, 0.42-0.60, 0.60-0.85 mm). As particle size is reduced, the rate of extraction for betulinic acid increased (Fig. 8). For smaller particles, diffusional resistance path decreases, and surface area increases, both of these help in increased recovery of acid. Similar observations were reported by Wongkittipong et al. [17] while investigating effect of particle size (0.1-0.3, 0.45-0.6, 0.6-0.8 mm) on the extraction of andrographolide.

### 3.5. Effect of solid loading

Influence of percentage solid loading (gm/ml) on extraction of betulinic acid was investigated in methanol at 700rpm speed, and 0.106-0.42mm particle size at 28°C (±1). Fig. 9 shows that, ultimate value (at 120min) of extracted betulinic acid (%) increases with decrease in solid loading. Lovasoa et al. [19] also reported that increased solid loading reduces the percentage recovery of protopine. During initial 40min the percentage extraction is very similar for each type of loading, this because at the beginning the solvent is free of any solute thus in any type of loading, acid molecules try to come out and get dissolved in solvent to become saturated. Thus the diffusion process is quite fast initially, and then slows down and tries to attain equilibrium. The extracted equilibrium concentration of acid while compared with maximum amount of acid (0.265% dry basis) present in the dry solid taken, the yield decreases with higher loading. As time increases beyond 50min the percentage recovery becomes distinct. At higher solid loading, concentration gradient at the solid liquid interface decreases with time as the amount of solvent remains unchanged. Therefore, after 120min the percentage of betulinic acid recovered reaches minimum for the 10% loading and gradually increases for 7.5% and 5%.

### 3.6. Effect of temperature

Effect of temperature on recovery of betulinic acid was investigated at 28°C, 36°C, 40°C, 45°C, and 50°C using methanol as solvent at 700rpm and 0.106-0.42mm particle size. Fig. 10 shows that with increase in temperature percentage recovery of betulinic acid increases. Wongkittipong et al. [17] and Lovasoa et al. [19] had investigated the effect of temperature on the extraction of andrographolide and protopine respectively and also reported an increase in extraction with rise in temperature. This increase may easily be attributed to higher solubility of betulinic acid in methanol at higher temperature. The kinetic energy as well as diffusivity of solvent molecules increases with increasing in temperature, and thus solvent penetrate better in side the cellulose matrix leading thereby easy release of the acid molecules.

Vaporization loss of solvent was taken care. Existence of large concentration gradient during initial 10min, the rate of extraction (46% at 50°C) is very rapid, and it slows down afterwards. With increase in temperature from 28°C to 50°C the ultimate value of recovered betulinic acid percentage increases from 12% (28°C) to 74% (50°C). Temperature of experiment was limited at 50°C due to methanol (boiling point 63°C) being used as solvent. The thermal stability of betulinic acid (melting point, 283-285°C) was checked in methanol and it doesn't degrade in the experimental condition even after several hours of exposure. At higher temperature the kinetic energy, diffusivity of solvent increases and they penetrate better inside the cellulose matrix, and dissolves betulinic acid easily and solubility of the acid also increases with increase in temperature.

### 3.7. Mass transfer (Diffusion) model

Steady state mass transfer model [20, 21], kinetic model [22], and unsteady diffusion models [17, 18, and 23] were used to represent solid-liquid extraction of different ingredients from plants. In the present investigation unsteady state mass diffusion is considered to account dynamicity of the process of extraction.

The mathematical model for extraction of betulinic acid from leaves (flat) was developed using the following assumptions [17, 18, and 23]:

1. All particles are flat in shape (being leaves) and symmetrical (betulinic acid is uniformly distributed throughout).
2. The solvent is perfectly mixed. The transfer resistance in the liquid phase is negligible, and thus the solute concentration in the solvent depends only on time.
3. Transport of solute is purely a diffusion process, and it is quantified by diffusion coefficient (D) and is independent of time.
4. At the interface (solid–liquid), the concentrations of solute inside and outside pores are equal.

Over a flat surface, unsteady state mass balance may be expressed as:

$$\frac{\partial C}{\partial t} = D\frac{\partial^2 C(t,x)}{\partial x^2} \quad (1)$$

Initial and boundary conditions as given below:

(i) Initial conditions (t = 0)

(ii) For the continuous phase: *Cs* = 0 (concentration of solute in the solvent),

(iii) For the dispersed phase:

Boundary conditions

At the center of a particle ($x = 0$) due to symmetry condition

$$\left(\frac{\partial C(t,x)}{\partial t}\right)_{x=0} = 0 \quad (2)$$

At the interface ($x = e$), equality of flux of solute (in mass):

The overall flux of the solute (obtained by integrating over the entire solid) is:

$$F = -DA\left(\frac{\partial c(t,x)}{\partial x}\right)_{x=e} \quad (4)$$

The incoming flux in the liquid expressed by:

$$F = V_L\frac{dC_L(t)}{dt} \quad (5)$$

where, $V_L$ is the solvent volume and *A* the specific area.

### 3.8. Numerical solution

The particle thickness was discretized in space with second-order finite difference algorithm. The boundary conditions were also expressed with the same order of accuracy. Crack–Nicolson technique was applied to solve these equations. A MATLAB program was written to solve the discretized equations' set. Experimental data of each interval were tried to fit with the model and obtain the optimum diffusion coefficient. The diffusion coefficient was estimated from the experimental data using a simple dichotomy method based on a quadratic criterion.

### 3.9. Comparison between model and experiments

Mass transfer (Diffusion) model has been used to estimate the diffusion coefficient for extraction of betulinic acid. An excellent fit between experimental and numerical data is obtained as shown in Fig. 10(MeOH),Fig.11(EtOH), Fig.12 (IPA). Initial rate of extraction of betulinic acid is very rapid till 20min which is quite obvious due to large concentration gradient exists in the solvent. The diffusion coefficient values were estimated at various temperatures (28 to 50°C) as reported in the Table 1. The coefficient of diffusion for betulinic acid varies from $1.7 \times 10^{-13}$ to $51.5\times10^{-13}$ $m^2$/sec. Wongkittipong et al. [17] estimated the diffusivities of andrographolide in ethanol in the temperature range between 22 and 60°C and found to vary in the range $0.843\times10^{-13}$ to $5.21\times10^{-13}$ $m^2$/sec.

### 3.10. Determination of activation energy

Activation energy can be calculated from the diffusion coefficients at different temperatures by using Arrhenius law:

$$D = D_0 e^{-E/RT} \qquad (6)$$

Where,

D= diffusion coefficient, $D_0$ = frequency factor, E =activation energy, R =gas constant.

From Eq. 6, by plotting lnD vs 1/T (Fig. 13), activation energy is calculated from the slope and value of frequency factor calculated from the intercept. The value of activation energy for betulinic acid extraction was obtained as 125.50 kJ/mol. The activation energy 192.88 kJ/mol is reported for extraction of andrographolide [17] from Androghaphis paniculata in ethanol. The activation energy gives a direct correlation of the bond energy associated with the ingredients and the cellulose structure of the plant matrix.

## 4. Conclusions

In this study, extraction of betulinic acid from leaves of Vitex Negundo Linn was carried out in a batch extractor. Influences of operating parameters such as solvent polarity, speed of agitation, particle size, solid loading, and temperature were investigated. Methanol has higher solvation power, and it shows greater release on extraction of betulinic acid than ethanol, isopropyl alcohol and water. Speed of 700rpm, particle size (0.1604-0.420mm), solid loading (5%), and temperature (50°C) were found to be the optimum parameters. The unsteady state diffusion model was used to estimate diffusion coefficient. The estimated value of diffusion coefficient ranges between $1.70\times10^{-13}$ to $51.5\times10^{-13}$ $m^2$/sec in the temperature range 28 to 50°C. The activation energy for diffusion of betulinic acid in methanol was calculated to be 125.50kJ/mole by linearization of Arrhenius equation.


## Acknowledgement

The financial support from AICTE Project (Grant no. 8023/RTD/RPS52/2004-05) and TEQIP (World-Bank project) for this research work is gratefully acknowledged.


## References


[1] Chadha Y. R., Publication Information Directorate, CSIR: New Delhi. 10 (1976) 522- 524.

[2] Kapur V., Pillai K. K., Hassain S. Z., Balani D. K., Indian J. Pharm. Sci. 56 (1994)160.

[3] Hebbalkar D. S., Hebbalkar G. D., Sharma R. N., Indian J. Med. Res. [A] 95 (1992) 200-203.

[4] Tamhankar C. P., Saraf M. N., Indian J. Pharm. Sci. 56 (1994) 154-160.

[5] Gupta M., Mazumder U. K., Bhawal R. S., Indian J. Exptl. Biol. 37 (1999) 143-146.

[6] Chawla A. S., Sharma A. K., Ha S. S., Dhar K. L., J. Nati. Prod. 55 (1992) 163-167.
[7] Jana U., Chattopadhyay R. N., Shaw B., Indian J. Pharma. 31 (1999) 232-233.
[8] Aswal B. S., Goel A. K., Kulshrestha D. K., Indian J. Exptl. Biol. 34 (1996) 444-467.
[9] Ragasa L. Y., Morales E., Rideout J. A., Philipp J. Sci. 128 (1999) 21-29.
[10] Taneja S. C., Gupta R. K., Dhar K. L., Atal C. K., Indian Perfumer. 23 (1979) 162-163.
[11] Basu N. M., Ray G. K., De N. K., J. Indian Chem. Sci. 24 (1947) 358-360.
[12] Dutta K. P., Chowdary S., Achari B., Pakrashi C. S., Tetrahadran. 39 (1993) 3067-3072.
[13] Liu J., J. Ethnopharmacology. 49 (1995) 57-68.
[14] Prema M. S., Misra G. S., Indian J. Chem. 16 (1978) 615-616.
[15] Chandramu C., Manohar R. D., Krupadanam D. G .L., Dashavantha R. V., Phyto. Res. 17 (2003) 129-134.
[16] Pisha, Nature Medicine. 10 (1995) 1046-1051.
[17] Wongkittipong R., Prat L., Damronglerd S., Gourdon C., Sep. and Puri. Tech. 40 (2004) 147-152.
[18] Simeonov E., Tsibranska I., Minchev A., Chem. Engg. J. 73 (1999) 255-259.
[19] Lovasoa R., Havet J., Porte C., Fauduet H., Sep. And Puri. Tech. 59 (2008) 253-261.
[20] Handayani A. D., Sutrisno, Indraswati N., Ismadji S., Bio. Tech. 99 (2008) 4414-4419.
[21] Hau K., Zheng Q., Li Y., Shen J., Hu S., Comp. and Chem. Engg. 24 (2000)1343-1348.
[22] Lovasoa R., Havet J., Porte C., Fauduet H., Sep. And Puri. Tech. 54 (2007) 253-261.
[23] Varzakas T. H., Leach G. C., Israilides C. J., Arapoglou D., Enz. And Micro. Tech. 37 (2005) 29-41.

Figures:
## List of Figures

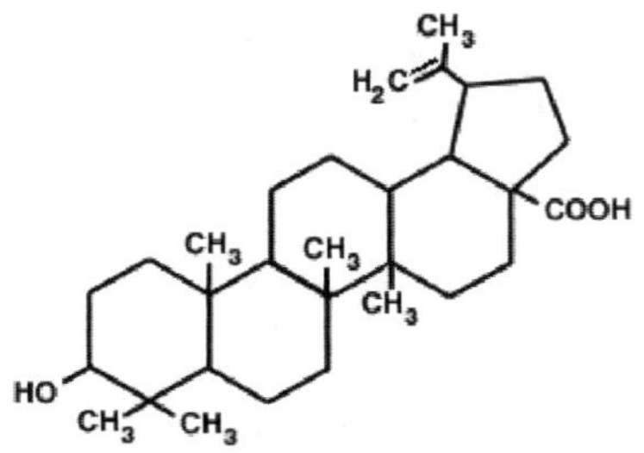


**Fig. 1. Structure of Betulinic acid**

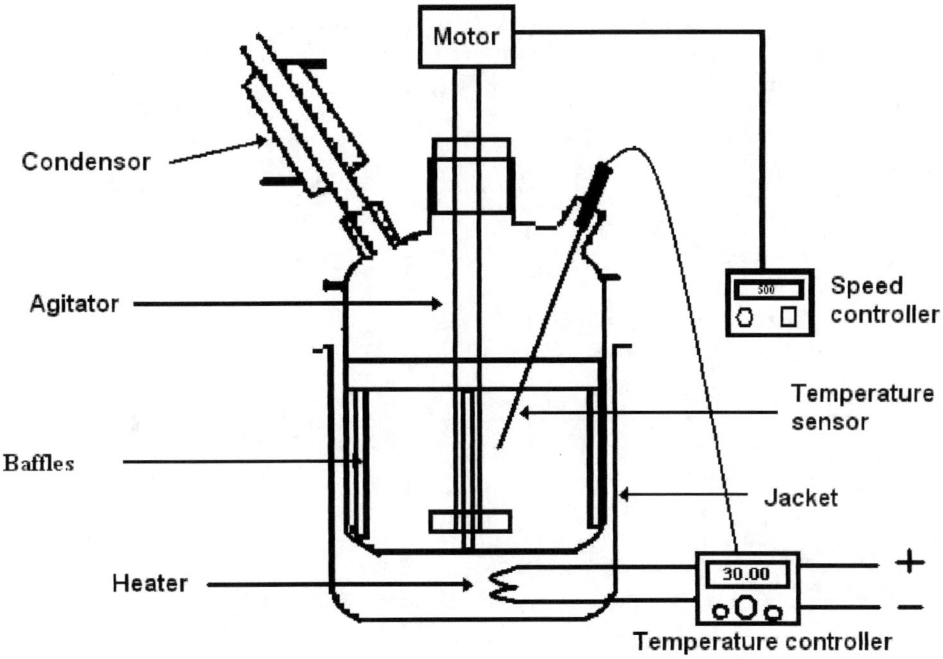


**Fig. 2. Batch extraction setup**

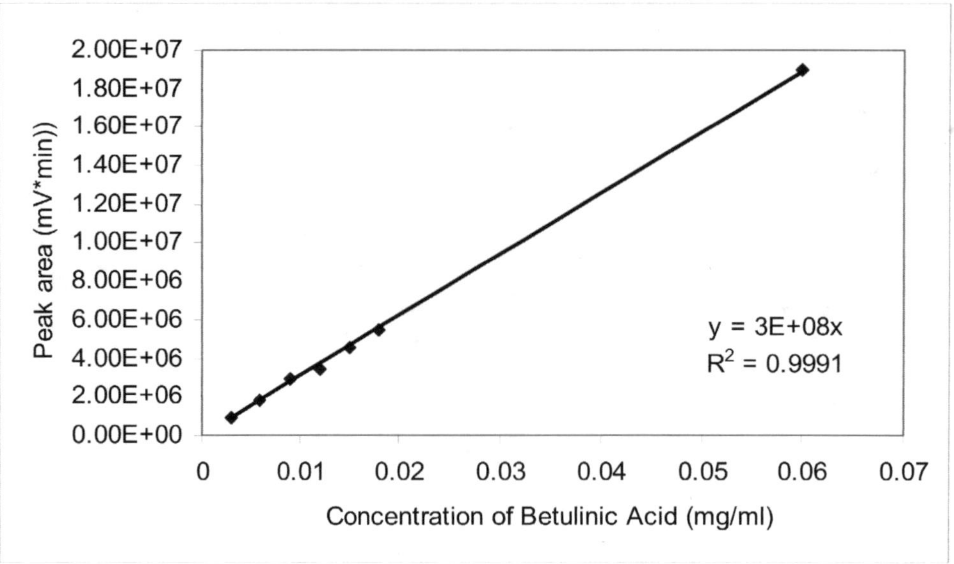


**Fig. 3. HPLC calibration of Betulinic acid [Symmetry® C-18 column (4.6×250 mm, 5 μm), acetonitrile : methanol (80:20 V/V) as mobile phase and UV detector (λ=210 nm wavelength) 0.05 aufs sensitivity and 0.5 mL/min as flowrate.**

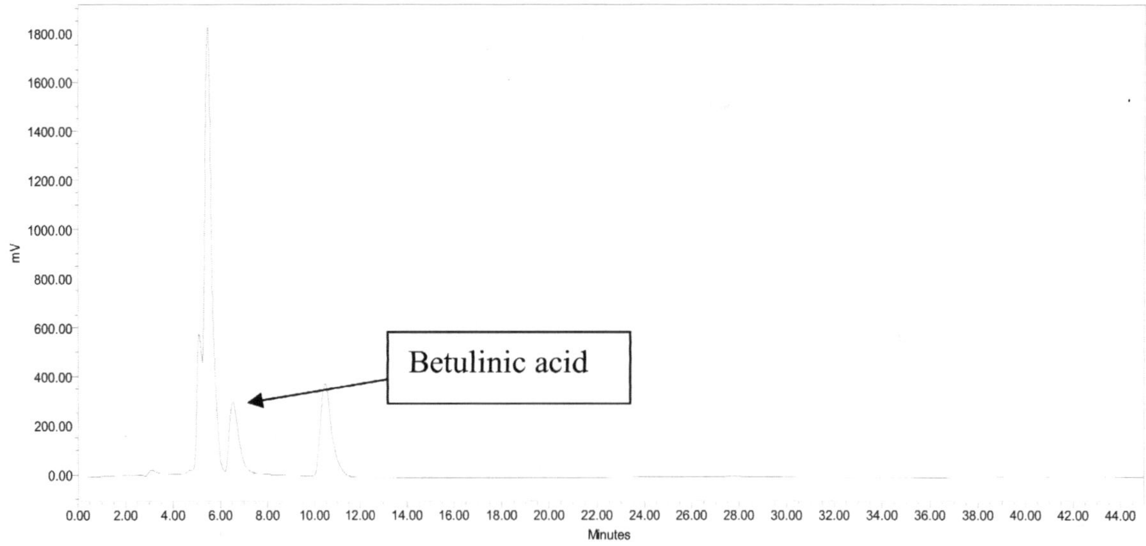


**Fig. 4. Betulinic acid from standard solution with Symmetry® C-18 column (4.6×250 mm, 5 μm), acetonitrile: methanol (80:20 V/V), 210 nm wavelength, 0.05 aufs sensitivity and flowrate of 0.5 mL/min**

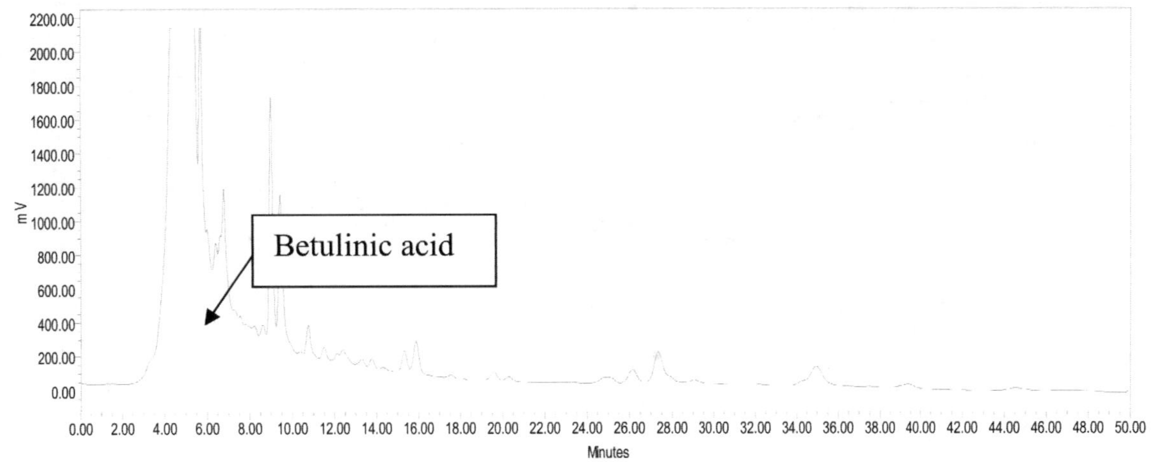


**Fig. 5. Betulinic acid obtained from crushed leaves of vitex nigunda linn xtracted mass with Symmetry® C-18 column (4.6×250 mm, 5 μm), acetonitrile:methanol (80:20 V/V), 210 nm wavelength, 0.05 aufs sensitivity and flowrate of 0.5 mL/min**

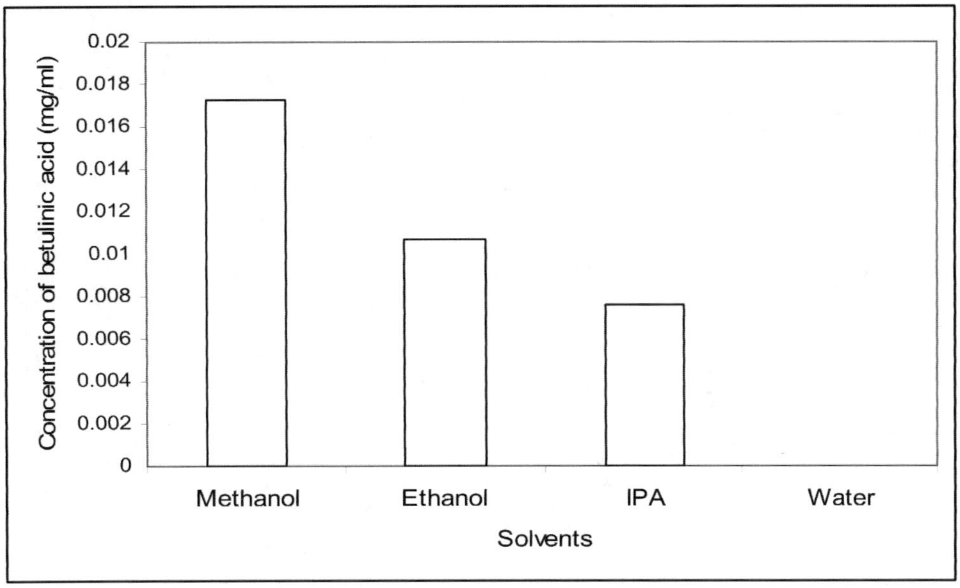


**Fig. 6. Selection of solvent for extraction of Betulinic acid with 5 % loading 0.106-0.42 mm particle size and 28°C**

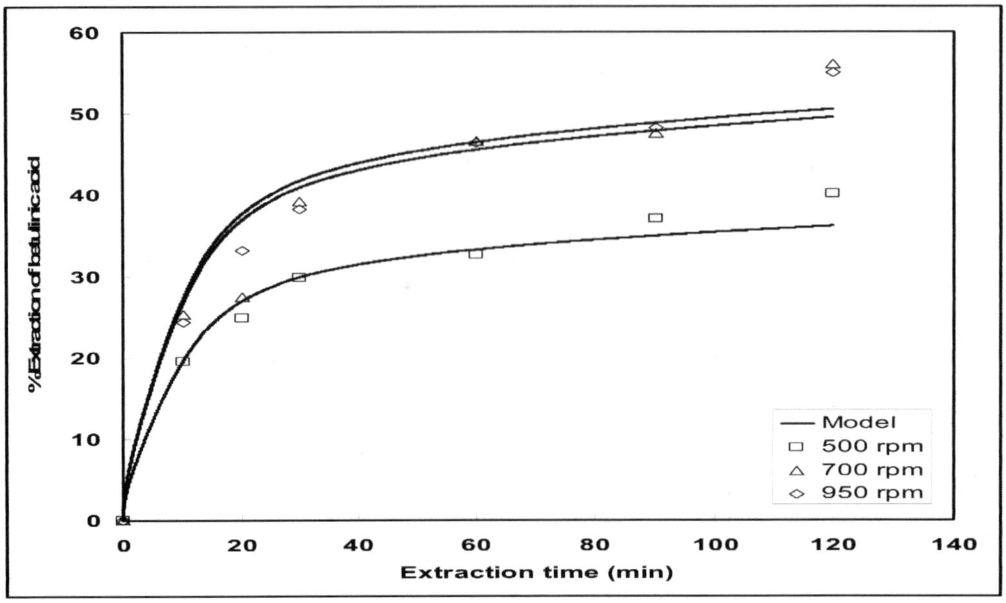


**Fig. 7: Influence of the speed of agitation on the extraction of betulinic acid at 5% solid loading, 28 °C, particle size 0.106-0.42 mm and methanol as a solvent**

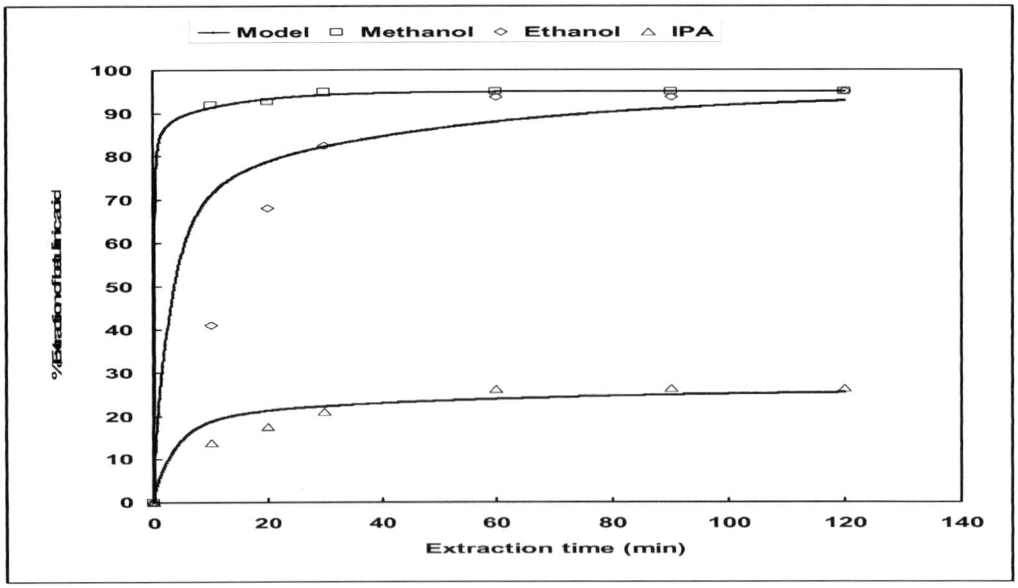


**Fig. 8: Effect of solvent at its boiling point for extraction of Betulinic acid with 5 % loading 0.106-0.42 mm particle size**

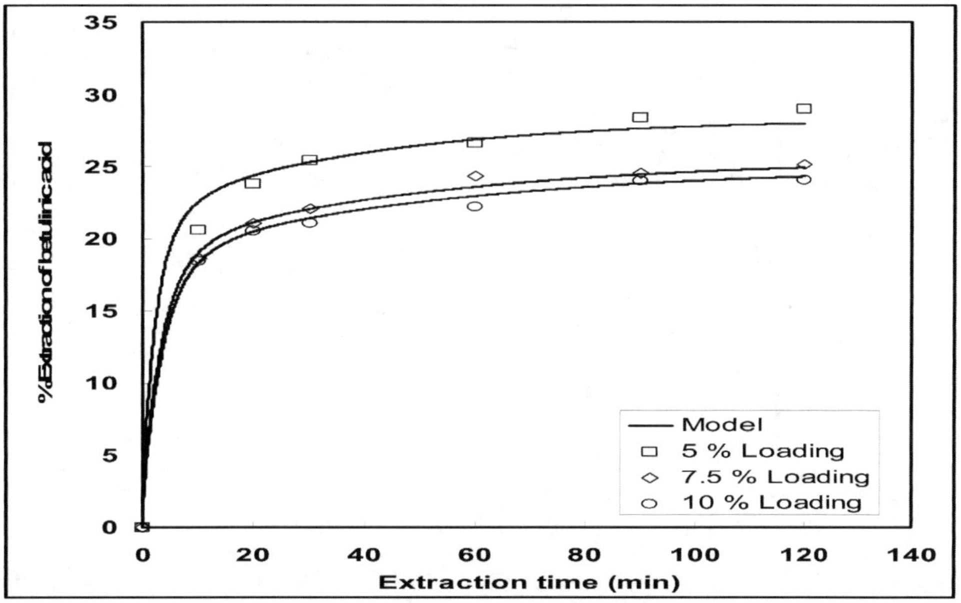


**Fig. 9: Influence of solid loading on extraction of betulinic acid with methanol as solvent at 700rpm agitation speed and 0.106-0.42mm particle size at 28°C (±1)**

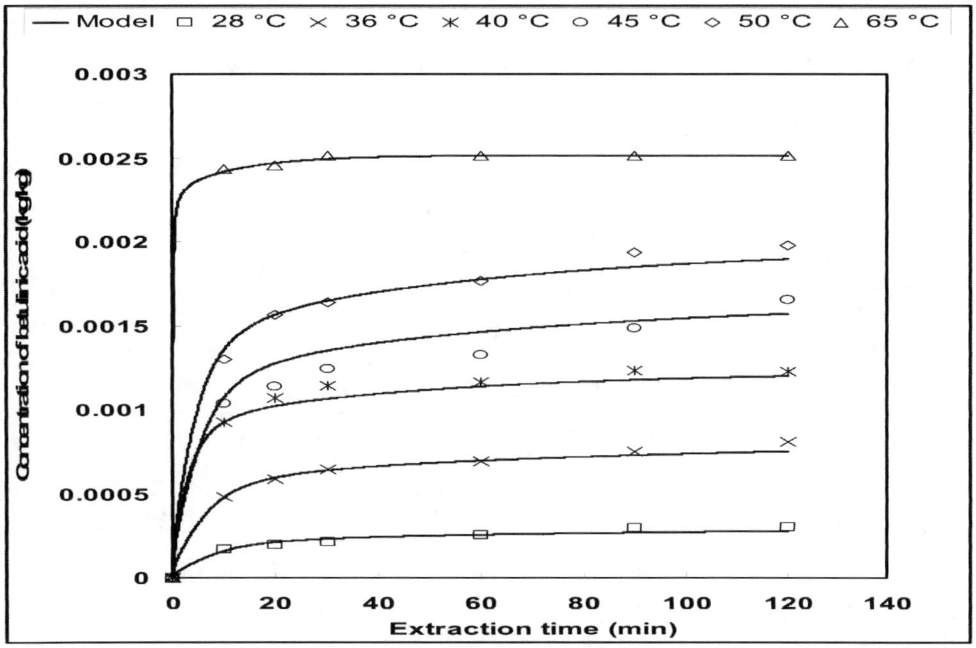


**Fig. 10: Influence of system temperature for extraction of betulinic acid with methanol as solvent, 950 rpm, 0.106-0.42 mm particle size, 5% solid loading**

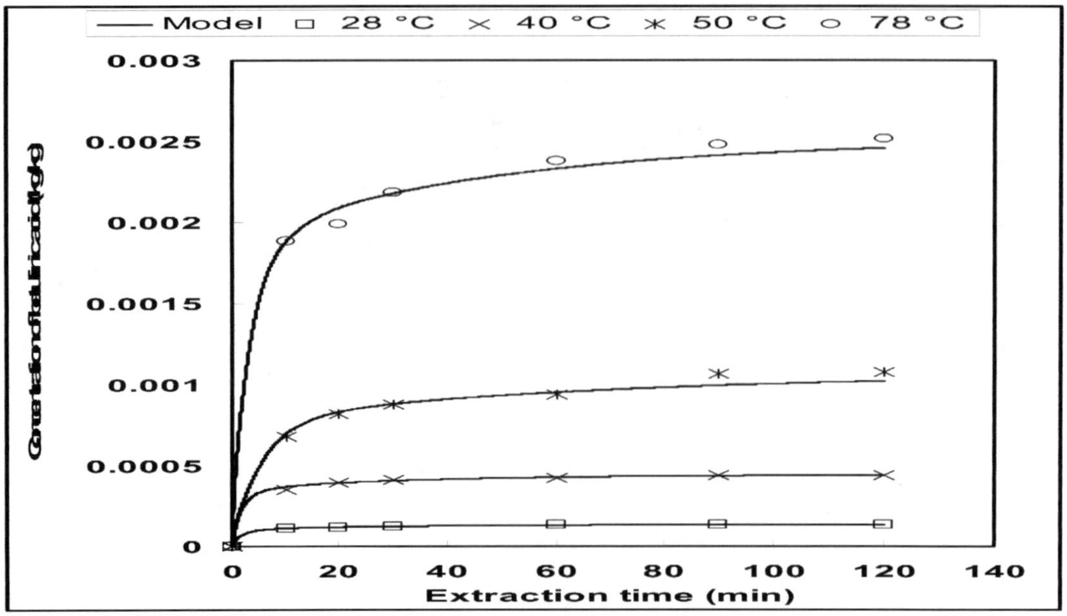


**Fig. 11: Influence of system temperature for extraction of betulinic acid with ethanol as solvent, 950 rpm, 0.106-0.42 mm particle size, 5% solid loading**

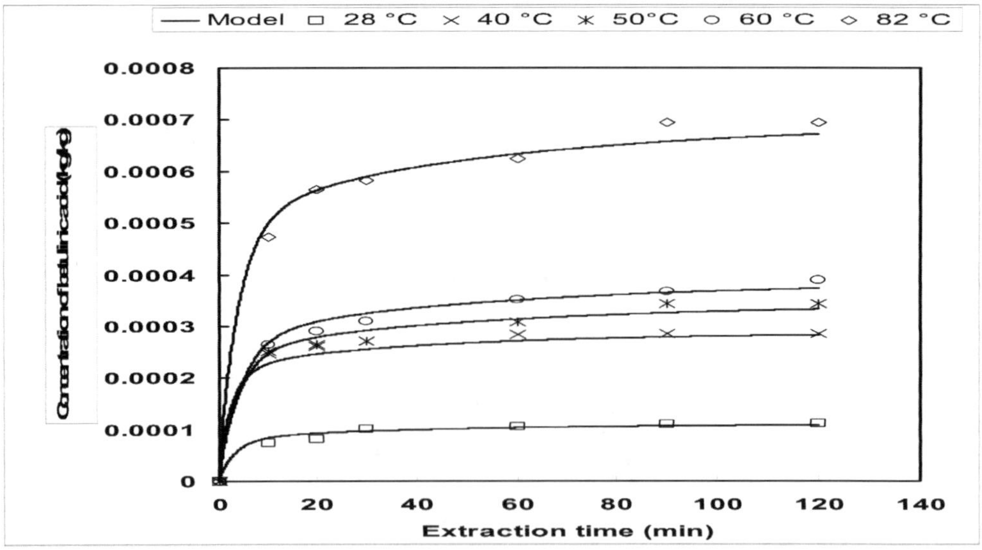


**Fig. 12: Influence of system temperature for extraction of betulinic acid with IPA as solvent, 950 rpm, 0.106-0.42 mm particle size, 5% solid loading**

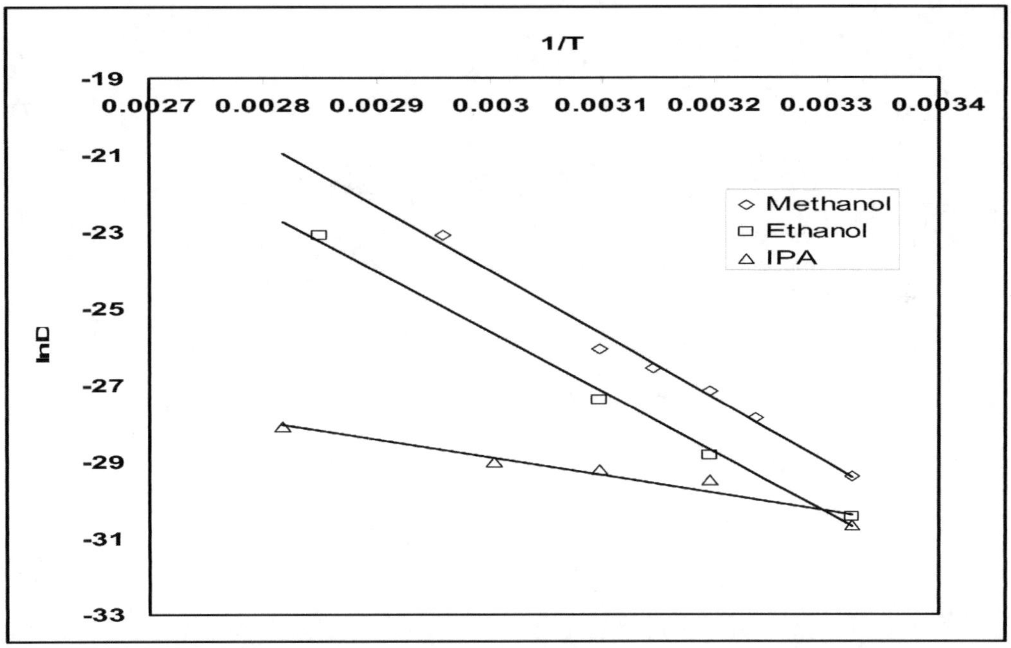


**Fig. 13: Plot of ln D vs 1/T for extraction of betulinic acid in methanol, ethanol and isopropyl alcohol with 700rpm and 0.106-042 mm particle size.**

# Efficient conversion of carbohydrates in hot water using $SO_2$ as a recoverable catalyst

Yucui Hou[a], Weina Liu[b], Shidong Tian[b], Weize Wu[b,*]

[a]*Department of Chemistry, Taiyuan Normal University, 030031, Taiyuan, China.*

[b]*State Key Laboratory of Chemical Resource Engineering, Beijing University of Chemical Technology, 100029, Beijing, China.*

## EXTENDED ABSTRACT

Carbohydrates, such as cellulose, hemicellulose, cellobiose, starch and sucrose are the most widely distributed sources of natural biomass, which hold impressive potential as alternatives to fossil fuels for sustainable production of fuels and chemicals. Current industrial processes to convert carbohydrates are typically catalyzed by either enzyme[1,2] or mineral acid catalyst[3,4]. The enzyme hydrolysis has no pollution and by-products, but is of low efficiency and high costs, while the latter process is efficient and technical maturely except that the acid recovery is difficult, the environmental pollution and the equipment corrosion is serious. Efficient conversion of carbohydrates to a variety of chemicals is important for a sustainable future.

In this work, sulphur dioxide ($SO_2$) dissolved in hot water was found to be an efficient catalyst for the selective conversion of carbohydrates (cellulose, xylan, starch, inulin and sucrose) to chemicals such as glucose and levulinic acid.

The effects of temperature, reaction time and $SO_2$ dosage on the conversion of cellulose are summarized in Table 1. The results reveal that cellulose can be effectively converted into chemicals, sequentially from cellulose to glucose, 5-hydroxymethylfurfural (HMF) and levulinic acid (LA) with the increase of temperature, reaction time and $SO_2$ dosage, which is the same with the products obtained from acid catalytic reactions.

---

* Corresponding author; Email: wzwu@mail.buct.edu.cn, Tel./Fax: +86 10 64427603.

**Table 1** Cellulose conversions and product yields under different conditions

| Entry | $T$ (°C) | $t$ (min) | $W_{SO2}$ (g/mL) | $C$ (%) | Yield of products (mol %) | | | | |
|---|---|---|---|---|---|---|---|---|---|
| | | | | | Glucose | Fructose | HMF | LA | Sum |
| 1 | 210 | 30 | 0.006 | 31.4 | 22.9 | 2.4 | 0.2 | - | 25.5 |
| 2 | 210 | 30 | 0.015 | 58.8 | 40.0 | - | 2.6 | - | 42.6 |
| 3 | 210 | 30 | 0.028 | 80.8 | 48.0 | - | 0.7 | 3.3 | 52.0 |
| 4[a] | 210 | 15 | 0.046 | 40.9 | 29.7 | 3.7 | 1.2 | - | 34.6 |
| 5[a] | 210 | 30 | 0.046 | 92.9 | 34.7 | - | 0.1 | 17.3 | 52.1 |
| 6 | 210 | 45 | 0.046 | 82.2 | 0.2 | - | - | 45.0 | 45.2 |
| 7[b] | 190 | 30 | 0.038 | 52.9 | 37.4 | 1.2 | - | 0.7 | 39.3 |
| 8 | 200 | 30 | 0.038 | 77.3 | 46.9 | - | - | 8.0 | 54.9 |
| 9[b] | 210 | 30 | 0.038 | 89.2 | 39.1 | - | - | 12.8 | 51.9 |

Reaction conditions: cellulose, 0.5 g; water, 15 mL; reactor volume, 25 mL; [a], values obtained by interpolation from Figure S1; [b], values obtained by interpolation from Figure S2; -, undetected.

As expected, the conversion of cellulose enhances considerably with $SO_2$ dosage (see Entries 1-3, 5 and 9 in Table 1). As we know, $SO_2$ in aqueous solution can form sulfurous acid and generate $H^+$, which acts as the catalyst for cellulose reaction. The more $SO_2$ is added into the aqueous solution, the more is concentration of generate $H^+$, which results in the increase of conversion rate with added $SO_2$. Besides, reaction time is crucial for the distribution of products (see Entries 4-6 in Table 1). When cellulose is converted at 210 °C for 15 min, its main products are glucose with small amounts of fructose, furfural, and HMF as intermediate products, and then they gradually turn into LA with a maximum yield of 45.0 % for 45 min. However, when the reaction time is 45 min, the cellulose conversion has a small reduction in the cases that the added $SO_2$ is more than 0.027 g/mL (Entry 6 in Table 1). This is caused by the formation of dark insoluble substances, called humin in literature.

Table 1 also shows that temperature has a great influence on the conversion of cellulose, yields of glucose and LA. With an increase of temperature, the conversion of cellulose increases greatly, the yield of glucose first increases and then decreases, and the yield of LA increases (Entry 7-9). Because the increase of temperature can enhance the cellulose reactions to produce glucose and also can accelerate the dehydration of glucose, leading to the growing of cellulose conversion and LA yield, but showing a maximum of glucose yield. Moreover, less $SO_2$ is needed to reach the highest glucose yield at a higher temperature. The reason is that high temperatures are favorable to break the hydrogen bonds existing in inter- and intra-molecular of cellulose, resulting in enhancing conversion rate.

The results above indicate that the conversion of cellulose and yields of products can be controlled by the amount of added $SO_2$, temperature and reaction time. Moreover, it is noted that the conversion of cellulose is always higher than the total yield of products, and their difference is enlarged with the increase of temperature, reaction time, and amounts of added $SO_2$. The conversion of each product from cellulose indicates that HMF is rather active in acid solution, and it can react with itself, LA or other intermediates to form polymeric substances, which are difficult to measure.

Based on the study of cellulose conversion above, we have also done some researches on the conversion of other carbohydrates catalyzed by $SO_2$ in hot water, such as xylan, starch, inulin and sucrose. They are polymers formed by different sugar units, such as glucose, fructose, xylose and arabinose. They are more amenable to be converted than cellulose since their degrees of polymerization are much lower. All of their conversions show the same rule as that of cellulose. At first, the main products are monosaccharide in accordance with their own structure. Then, the obtained monosaccharide is further converted into furan compounds and LA.

Importantly, we found that the catalyst $SO_2$ in product mixtures can be easily recovered by steam stripping.[5] Our results indicate the residual concentration of $SO_2$ in a product mixture via stripping time. Only in 5 min, the concentration of $SO_2$ in solution decreased from 6.73 g/L to 0.31 g/L. In 20 min, this value decreased to 0.01 g/L, and after 30 min, it is lower than 0.007 g/L (limit of detection). This can be understood from two aspects. First, increasing temperature reduces the solubility of $SO_2$ in aqueous solution. Second, the steady flow of steam takes away the $SO_2$ in gas phase, breaking the vapor-liquid equilibrium of $SO_2$, so the escape rate of $SO_2$ from liquid phase speed up. The steamed $SO_2$ from the product mixture was collected and verified by an iodometric method. In our work, the product mixtures of all experiments with $SO_2$ added were swept for 30 min using the same method before the products were analyzed by HPLC, and no $SO_2$ or $SO_3^{2-}$ was detected by HPLC analysis.

In conclusion, $SO_2$ can act as a catalyst to efficiently convert cellulose and other carbohydrates in high temperature water with a high glucose yield of 48.0 %

and a high LA yield of 45.0 %. $SO_2$ molecular and ionization $H^+$ act together as catalysts, leading to less corrosion to reactor than dilute sulfuric acid. The conversion of cellulose and the distribution of products are greatly influenced and can be controlled by temperature, reaction time and amount of $SO_2$. $SO_2$ in the product mixture can be easily recovered completely by steam stripping, which supplies an environmentally benign process for carbohydrates conversion to chemicals.

**KEYWORDS:** conversion, carbohydrates, hot water, recoverable catalyst, $SO_2$

## ACKNOWLEDGEMENT

This work was supported by the National Natural Science Foundation of China (21176020), the Beijing Natural Science Foundation (2082017), and the Program for New Century Excellent Talents in University (NCET-08-0710).

## References

1. F. B. Sun, H. Z. Chen (2008), "Enhanced enzymatic hydrolysis of wheat straw by aqueous glycerol pretreatment", *Bioresour. Technol.*, 99, pp. 6156–6161.
2. T. A. Nguyen, K. R. Kim, S. J. Han, H. Y. Cho, J. W. Kim (2010), "Pretreatment of rice straw with ammonia and ionic liquid for lignocellulose conversion to fermentable sugars", *Bioresour. Technol.*, 101, pp. 7432–7438.
3. A. Orozco, M. Ahmad, D. Rooney, G., Walker (2007), "Dilute acid hydrolysis of cellulose and cellulosic bio-waste using a microwave reactor system" *Trans IChemE, Part B, Process Saf. Environ. Prot.*, 85, pp. 446–449.
4. Y. Sun, L. Lin, C. H. Pang, H. B. Deng, H. Peng, J. Z. Li, B. H. He, S. J. Liu (2007), Hydrolysis of cotton fiber cellulose in formic acid. *Energy & Fuels*, 21, pp. 2386 –2389.
5. W. N. Liu, Y.C. Hou, W.Z. Wu, M.G. Niu, W.H. Wang (2007), "Conversion of wheat straw hemicellulose using $SO_2$ as a recoverable catalyst", *Bioresource Technology*, DOI: http://dx.doi.org/10.1016/j.biortech.2012.08.028.

# Extraction of value-added chemicals from fast pyrolysis oils

*Duo Wang, Dong-can Lv, Yun-quan Liu**
*School of Energy Research, Xiamen University, Xiamen 361005, China*

## Introduction

As a renewable energy source, bio-oils from fast pyrolysis of biomass have great potential to replacing petroleum. However, bio-oils are mixtures totally different from petroleum in regard to their physical and chemical properties. Therefore, direct application of bio-oils is very limited, and is being used only as a fuel for combustion in boilers today. To be used as transportation fuels (e.g. gasoline or diesel), however, bio-oils need to be upgraded via hydrotreating or hydroprocessing, which usually requires a lot of hydrogen due to high oxygen content in bio-oils, causing the process from bio-oils to transportation fuels non-competitive with that from petroleum in terms of production cost right now. Therefore, exploration of other methods that could expand applications of bio-oils while still cost-effective is needed. This work is intended to explore the methods of recovering value-added chemicals such as phenols, acetic acid, and/or acetates from bio-oil. Two kinds of extraction methods were studied.

## Experimental

Bio-oils from pine pyrolyzed at 500 °C were used for extraction. The raw bio-oils have the composition like Table 1. Before extraction, raw bio-oils were pretreated with vacuum distillation and the distillate of boiling point <160 °C was collected for extraction. Table 2 is the composition of bio-oil distillate compared with raw bio-oil. As you can see, with distillation the phenolic compounds in distillate are enriched.

After extraction, the samples from both aqueous and organic phase were analyzed using GC-MS.

**Table 1.** Typical composition of bio-oils made in our lab

| Name | wt% | Component |
|---|---|---|
| Acids | 10.32 | Formic acid |
| | | Acetic acid |
| Esters | 3.12 | Methyl acetate |
| | | Butyl acetate |
| Alcohols | 3.56 | Cyclopropylmethanol |
| | | Glycol |
| | | 2 - Furfuryl alcohol |
| Ketones | 4.69 | 1 - hydroxy-2 – butanone |
| | | 2 (5H) – furanone |
| Aldehydes | 7.61 | Formaldehyde |
| | | Acetaldehyde |

| | | |
|---|---|---|
| | | Butanediol |
| Furans | 7.88 | Furfural |
| | | Hydroxymethylfurfural |
| Phenolics | 15.50 | Phenol |
| | | 2 - methoxyphenol |
| | | 4 - methyl-2 - methoxyphenol |
| Others | 20.68 | 2,2-diethyl-3-methyl-oxazolidine |
| | | 2 - Amino - 1 – propanol |

**Table 2.** Phenolic compounds in the distillate are enriched

| Compounds | wt% | |
|---|---|---|
| | Raw Bio-oil | Distillate |
| Phenol | 0.66 | 1.17 |
| 2 – methyl phenol | 0.43 | 0.84 |
| 4 - methyl phenol | 0.36 | 0.61 |
| 2 - methoxyphenol | 4.23 | 9.96 |
| 4-methyl-2 - methoxyphenol | 4.56 | 7.50 |
| 4-ethyl-2 - methoxyphenol | 0.90 | 0.74 |
| Eugenol | 1.56 | 0.58 |
| Vanillin | 1.77 | 0.01 |
| 2-methoxy-4-(1-propenyl)-phenol | 0.80 | 0 |
| Total Phenols | 15.50 | 21.41 |

Two extraction processes were attempted in order to extract value-added chemicals from bio-oil: 1) extracting phenolic compounds using organic solvents; 2) recovering phenolic compounds and acetic acid simultaneously using complexing agents.

**1. Extraction of phenolic compounds using organic solvents**

For this method, the process schematic is shown in the schematic below.

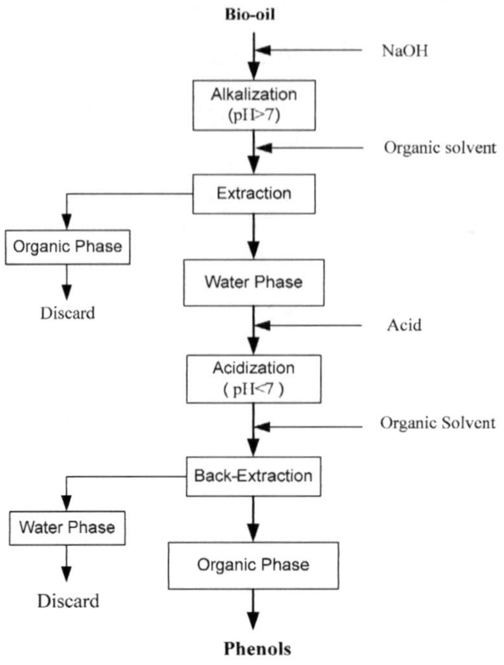


**Figure 1.** The Schematic of extracting phenols using organic solvents

## 2. Complexation extraction using metal ions

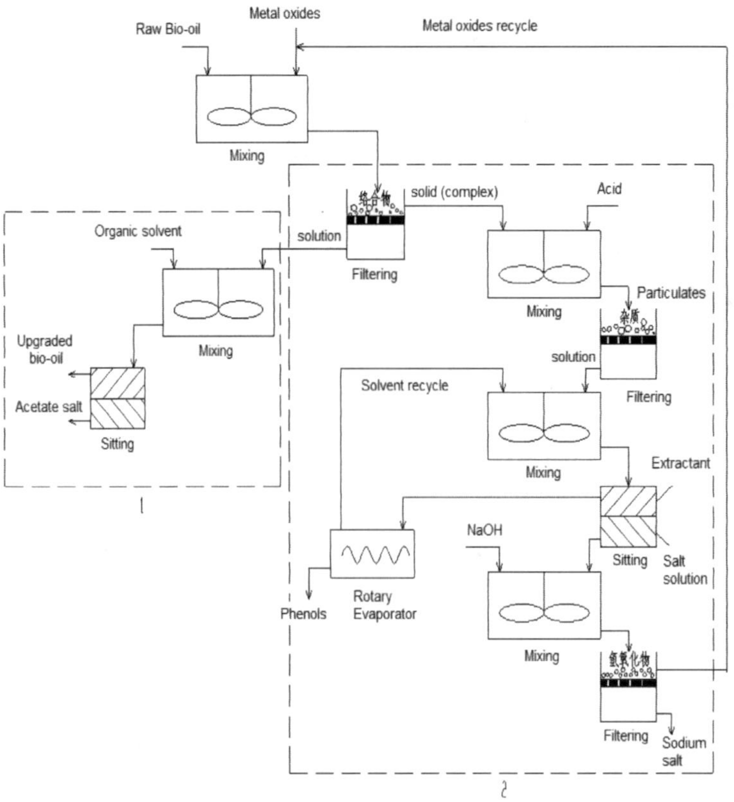


**Figure 2.** The schematic of extracting phenols using metal ion oxides

# Results & Discussion

## Method 1 - Extraction using organic solvents

### *1. Distribution of phenols in organic and aqueous phases for different solvents*

various solvents were tested to extract phenols from bio-oil, including dichloromethane (DCM), ethyl acetate (EA), butyl acetate (BA), n-hexane (n-Hex), cyclohexane (c-Hex), n-pentane (n-Pent), petroleum ether (PE), n-butanol (n-but), benzene (Ben), and Trioctanamine (TOA). The following are some of the preliminary results.

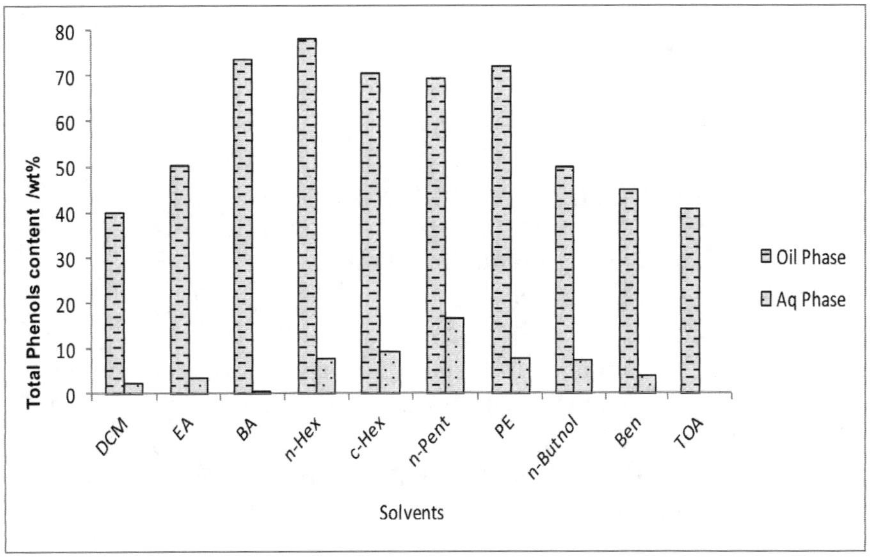


**Figure 3.** Distribution of phenols in organic and aqueous phases for different solvents

As you can see, butyl acetate is the best for extracting phenols from the bio-oil, as the organic phase (oil phase) contains much more phenols than in water phase after extraction.

### *2. Phenol recovering rate for different solvents*

Figure 4 shows the recovering rate of phenols with different solvents. As you can see, butyl acetate has the highest phenol recovering rate of ~92.3%, which means it is the best solvent for extract phenolic compounds from bio-oil.

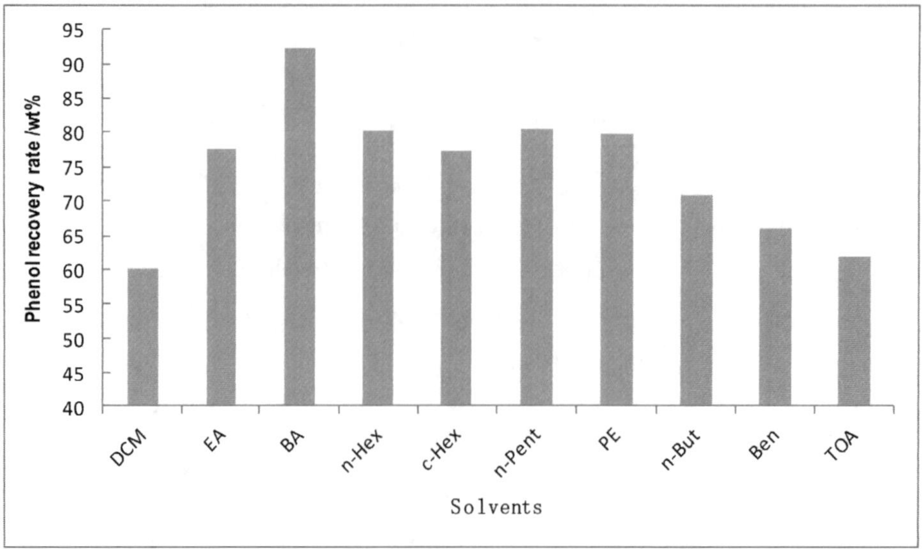


**Figure 4.** Phenol recovering rate for different solvents

## Method 2 - Complexation extraction using metal ions

Different metal ions, such as $Al^{3+}$, $Zn^{2+}$, $Fe^{3+}$, $Mg^{2+}$, $Ba^{2+}$, $Ca^{2+}$ were tested to form precipitations - complexation compounds of phenols to recover phenols from bio-oil. Both the metal oxides and metal hydroxides could be used for this method. Table 3 is the composition of phenolic mixture recovered using CaO as the complexing agent. As you can see, the recovered phenolic mixture contains more than 90% of phenolic compound, among which, 2 – methoxyphenol and 4-methyl-2 – methoxyphenol are the two kinds of phenols with the highest concentration.

**Table 3.** Composition of phenolic compounds obtained after complexation extraction

| Number | Retention time | Chemicals name | Formula | Area percent (%) |
|---|---|---|---|---|
| 1 | 3.412 | Butanedial | $C_4H_6O_2$ | 1.35 |
| 2 | 4.704 | Formamide, N-methoxy- | $C_2H_5NO_2$ | 1.15 |
| 3 | 7.433 | Butanoic acid, 4-hydroxy- | $C_4H_8O_3$ | 2.47 |
| 4 | 7.521 | 2(5H)-Furanone | $C_4H_4O_2$ | 2.3 |
| 5 | 10.472 | Phenol | $C_6H_6O$ | 3.37 |
| 6 | 13.406 | Phenol, 2-methyl- | $C_7H_8O$ | 3.15 |
| 7 | 14.222 | Phenol, 4-methyl- | $C_7H_8O$ | 1.67 |
| 8 | 14.667 | Phenol, 2-methoxy- | $C_7H_8O_2$ | 43.25 |
| 9 | 18.667 | Phenol, 2-methoxy-4-methyl- | $C_8H_{10}O_2$ | 29.82 |
| 10 | 20.423 | 2,3-Dimethoxytoluene | $C_9H_{12}O_2$ | 2.41 |
| 11 | 21.789 | Phenol, 4-ethyl-2-methoxy- | $C_9H_{12}O_2$ | 5.26 |
| 12 | 24.466 | Eugenol | $C_{10}H_{12}O_2$ | 2.66 |
| 13 | 24.804 | Phenol, 2-methoxy-4-propyl- | $C_{10}H_{14}O_2$ | 1.14 |

Figure 5 is the GC-MS chart of phenolic compounds extracted from bio-oil using CaO as the complexation agent. The peak numbers are corresponding to the Table 3.

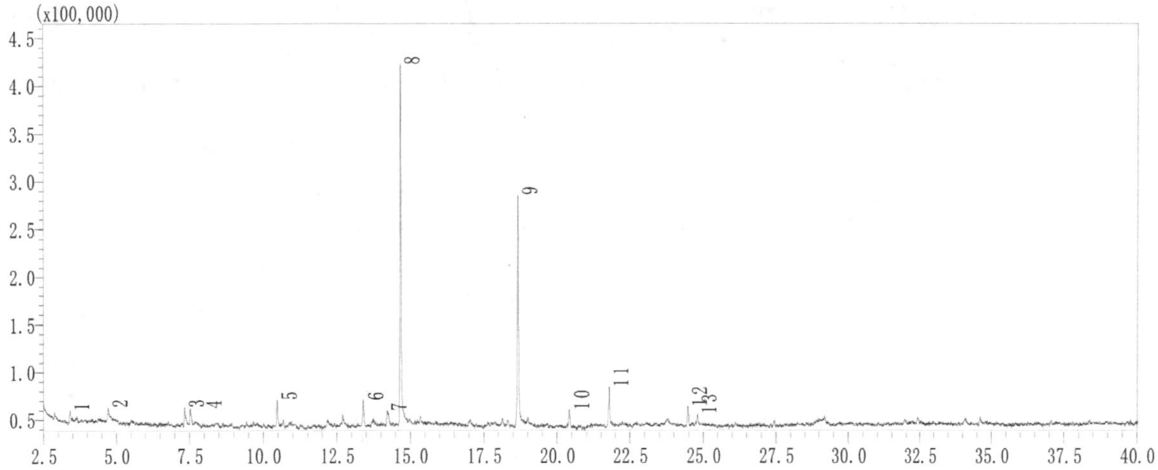

**Figure 5** GC-MS Chart of phenol mixture extracted

In addition, as you can see from Figure 2 when using complexing agent such as CaO or MgO for recovering phenols from bio-oil, acetic acid in bio-oil can also be extracted simultaneously because it can react with CaO or MgO to form acetate (see the Fig 2), which will be transferred to the aqueous phase. By adding acids such as HCl or $H_2SO_4$ to the aqueous phase, acetate will be converted into acetic acid. Then, acetic acid can be extracted with solvents such as butyl acetate. Our preliminary experimental results indicated that a recovering rate of ~80% for acetic acid is achieved.

## Summary

Besides using the regular extraction for recovering phenolic compounds from biomass pyrolysis oil, a novel method that could recover both phenolic compounds and acetic acid simultaneously was developed in this work. Further work will be continued to improve the recovering rate and to obtain phenols with much higher purity.

### References

1. King, C. J. Separation process based upon reversible chemical complexation. Rousseau R W. Handbook of Separation Process Technology, Chap 15. New York: John Wiley & Sons. 1987:760~774.

2. Chen, C. A., H. Pakdel, C. Roy. Separation of phenols from *Eucalyptus* wood tar[J].Biomass and Bioenergy.1997, 13( l/2):25-37.

3. Žilnik, L. F., A. Jazbinšek. Recovery of renewable phenolic fraction from pyrolysis oil. Separation and Purification Technology. 2012, 86:157-170.

4. Chum, H. L., S. K. Black. Process for fractionating fast-pyrolysis oils, and products derived therefrom. US Patent 4942269, Midwest Research Institute, 1990.

# (474e) Biochar Based Activated Carbon As Adsorbent to Recover Volatile Chemicals From Photobioreactor

**Zhengrong Gu**[1], Joseph Polin[2], Xiaomin Wang[2], Muthu Ramanathan[2] and Kasiviswanathan Muthukumarappan[3], (1)Agricultural and Biosystems Engineering, South Dakota State University, Brookings, SD, (2)South Dakota State University, Brookings, SD, (3)Center for Bioprocessing Research and Development and Department of Agricultural and Biosystems Engineering, South Dakota State University, Brookings, SD

Transformative concept is to establish an continuous *cyanobacterial platform* to convert $CO_2$ into energy dense fuels and organic chemicals. This will involve the following objectives: 1) Engineer *cyanobacterial* to produce and excrete linalool, 2) Develop an integrated, recirculating adsorption recovery system, which harvest volatile products from gas stripping from photobioreactor. The model product is linalool, a C10 alcohol with an energy density of 40 mj/kg, heat of vaporization of 0.19 mj/kg, and octane of 102.

At room temperature, 1 atm, linalool evaporated from spiked *cyanobacterial solution, which contains* linalool (0.4%wt at beginning) by air bubbling (100 ml/min), has been completely captured by activated carbon (from woody biochar after steam physical activation at 800 C for 20 minutes). After selective temperature swing desorption, 95% pure linalool was recovered as product. The adsorption dynamic and equilibrium was investigated. Simultaneouly, lifetime of activated carbon adsorbent was quantified.

This transformational technology will allow us to exploit two un- or under-utilized resources in the Midwest and Great Plains (wind and $CO_2$) by converting them into storable, transportable, energy-dense, and infrastructure-compatible renewable liquid fuels.

# (474f) Prediction of Weak Acid Toxicity in S. Cerevisiae Using Genome-Scale Models

**Patrick B. Hyland**, Serene Lock Sow Mun and Radhakrishnan Mahadevan, Department of Chemical Engineering and Applied Chemistry, University of Toronto, Toronto, ON, Canada

**Introduction**

Renewable and economically feasible production of biofuels and biochemical relies on the efficient use of lignocellulosic biomass as a feedstock. However, necessary pretreatment of the biomass generates an array of growth inhibitors such as weak acids, phenolics and furan derivatives that adversely affect the fermentation characteristics. The concentrations and relative amounts of each inhibitor in the hydrolysate vary among the sources of lignocellulosic biomass [1]. Acetic acid is a large contributor to weak acid inhibition in lignocellulosic hydrolysate and ranges in concentration from 1.6-4.4 g/L [1].

One mechanism by which weak acids inhibit cellular growth, known as the uncoupling theory, is a result of the diffusion of undissociated weak acids through the cytosolic membrane into the cytosol. Due to the near neutral pH of the cytosol, the weak acid dissociates into its conjugate base and proton, thereby decreasing cytosolic pH. In order to maintain pH neutrality, protons are pumped from the cytosol against the concentration gradient by the membrane bound ATP synthase at the cost of ATP, thus reducing the amount of ATP available for cell growth. Alternatively, the accumulation of anions in the cytosol has been suggested to play a role in growth inhibition.

Metabolic modeling has emerged as a powerful tool for enhancing our understanding of microorganisms and aiding in design of microbes for specific phenotypes. Previous studies have modeled the impact of weak acids on microorganisms using kinetic and thermodynamic approaches. The focus of such studies has largely been the toxic affects of weak acids in bacteria for food preservation purposes. However, the yeast *S. cerevisiae* is the most abundant organism used for industrial fermentations, as it is known to be more robust to inhibitory compounds present in hydrolysate. It would therefore be of interest to examine the affects of weak acids on the growth of an organism relevant to the biotechnology industry such as *S. cerevisiae*.

An alternative modeling approach, flux balance analysis (FBA), is a commonly used linear programming based, steady state framework that uses a stoichiometric model of the organism and a set of physicochemical and thermodynamic constraints to establish intracellular fluxes.

This work is a combined computational and experimental approach to examine the affects of weak acid toxicity in *S. cerevisiae*. We focused on the toxicity of acetic acid as it is an abundant inhibitor in lignocellulosic hydrolysate. It was hypothesized that the energy usage required to pump protons out of the cytosol via ATP-synthase could be represented using a genome-scale metabolic model. Specifically, we extended the iMM904 consensus model of *S. cerevisiae* to include the uncoupling mechanism of weak acid toxicity [2]. We implemented the uncoupling mechanism by adding a series of reactions to the iMM904 model that describe the inhibitory effect. Growth rate was optimized with glucose as the carbon source under varying concentrations of acetic acid and pH under both aerobic and anaerobic conditions. Predicted growth rates and product profiles of *S. cerevisiae* were compared to experimental results for aerobic conditions obtained in this study, as well as anaerobic conditions reported by Tahezadeh *et al* [3].

Batch cultivations of *S. cerevisiae* CEN.PK MAT A/a were conducted in glucose mineral medium under aerobic conditions in the presence of acetic acid ranging from 0-4 g/L. Throughout the cultivations, temperature and pH were maintained at 30°C and 5.5, respectively. Growth and metabolic concentrations were monitored throughout the batches by optical density and HPLC measurements.

**Results**

Growth of microorganisms such as *S. cerevisiae* in the presence of acetic acid is hindered due to the consumption of ATP necessary to maintain near-neutral cytosolic pH. Model predictions were consistent with the aerobic growth rates in the batch cultures with varying concentrations of acetic acid at pH 5.5, as shown in Figure 1. In this figure, the experimental growth rate is shown to decrease linearly ($R^2 = 0.93$) with increasing acetic acid concentration. Growth rates predicted by FBA at pH 5.5 follow this trend closely and predicted flux of metabolites were also in agreement with experimental data. Interestingly, these results suggest that the toxic affect of acetic acid may be largely attributed to the uncoupling mechanism, rather than anion accumulation.

Growth rates and metabolite profiles were also predicted for varying acetic acid concentration for anaerobic conditions, and compared to the experimental values reported by [3]. Model predictions of growth rate and ethanol yield in this case were also in good agreement with experimental values.

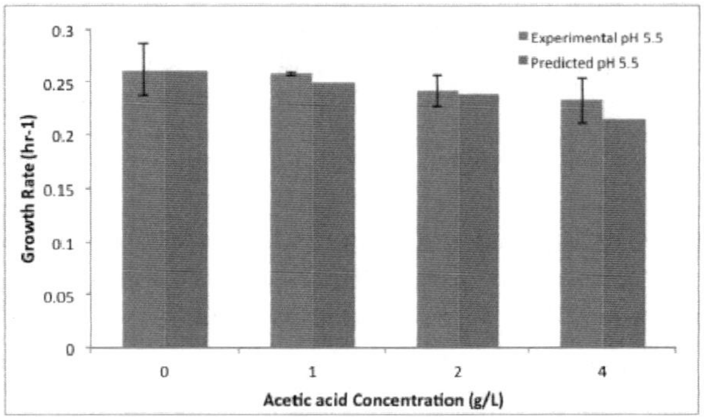


**Figure 1:** Experimental and predicted growth rates of *S. cerevisiae* in the presence of acetic acid. Error bars indicate the standard deviation of two or more batch cultivations.

## Conclusion

It was shown in this work that the genome-scale metabolic model is able to capture the growth associated energetic burden incurred by the presence of acetic acid. The consensus model of *S. cerevisiae* was extended to include the uncoupling mechanism of weak acid toxicity, whereby intracellular dissociation of acids generates protons that are subsequently exported at the expense of ATP. Model predictions of acetic acid toxicity agreed well with aerobic batch cultivation data of *S. cerevisiae* CEN.PK MAT A/a. The model was also shown to predict the toxicity of acetic acid during anaerobic growth.

We expect this work to find application in the optimization of industrial processes. By systematic evaluation of the model, the trade off between product yield, cellular growth rate and neutralization cost may be established, thereby improving the bioconversion of lignocellulosic hydrolysate to renewable chemicals.

## References

[1] Almeida JRM, Modig T, Petersson A, Hahn-Hagerdal B, Liden G, Gorwa-Grausland MF. (2007) **Increased tolerance and conversion of inhibitors in lignocellulosic hydrolysates by *Saccharomyces cerevisiae*.** *J Chem Technol Biotechnol*, **82**:340-349.

[2] Mo ML, Palsson BO, Herrgard MJ. (2009) **Connecting extracellular metabolomic measurements to intracellular flux states in yeast.** *BMC Syst Bio*, **3**:37.

[3] Taherzadeh MJ, Niklasson C, Liden G. (1997) **Acetic acid – friend or foe in anaerobic batch conversion of glucose to ethanol by *Saccharomyces cerevisiae*?** *Chem Eng Sci*, **52**(15), p.2653-2659.

# (488a) Optimal Design of LNG Plant for Cost Minimization

**Jeongwoo Jeon**, Kiwook Song, Jeongnam Kim and Chonghun Han, School of Chemical and Biological Engineering, Seoul National University, Seoul, South Korea

Natural gas (NG) is naturally generated gas mixture lying under the earth and collected from gas well directly. Natural gas emits little amount of pollutant when it burns and is buried around the world widely. Because of these properties, NG is receiving attention as low-carbon, eco-friendly alternative fuels. But most of NG reserves are far from the market physically and Natural gases have very large gas-to-liquid volume ratio. The liquefaction process that reduces volume of natural gases is essential for effective storage and transport. Through the liquefaction process, natural gas becomes liquefied natural gas (LNG). There are several kinds of natural gas liquefaction process. Among the liquefaction processes, propane pre-cooled mixed refrigerant (C3MR) process is the most widely used natural gas liquefaction process to date.

In this study, a steady-state simulation of the C3MR process was performed using ASPEN HYSYS™. And operating and capital cost of the process were calculated. The capital cost of the process was converted to equivalent annual cost in order to be compared with the operating cost. Then, the process was optimized to reduce total cost. Optimization formulation is developed to minimize the sum of operating cost and capital cost of the process. Temperature, pressure, and flow-rate of refrigerants were selected as optimization variables.

This research was supported by a grant from the LNG Plant R&D Center funded by the Ministry of Land, Transportation and Maritime Affairs (MLTM) of the Korean government.

**MITIGATION OF REFINERY PRE-HEAT TRAIN FOULING BY OPTIMISATION OF OPERATING CONDITIONS AND APPLICATION OF HEAT EXCHANGER TECHNOLOGIES**

*Marion RATEL, TOTAL Raffinage Chimie, France*
** Zoé ANXIONNAZ-MINVIELLE and Bernard VINET, CEA Grenoble, LITEN, France*
*Yash-Vardhan KAPOOR, TOTAL Raffinage Chimie, European Technology Centre, France*
*Luc SEMINEL, TECHNIP, France*
** email : zoe.minvielle@cea.fr*

## Extended abstract

### Introduction

Fouling is considered as one of the major unsolved problem in industrial heat exchangers. It generates additional resistances to heat transfer, a decrease of the thermal effectiveness and an increase of pressure drop in industrial units. In order to maintain stable crude inlet conditions upstream of the distillation column, the energy consumption is increased with detrimental environmental and economic consequences.

Hydrocarbon fouling is a complex phenomenon as several fouling mechanisms often act in parallel, interacting with one another. They can be broadly classified in two categories:

- Fouling due to the fluid physico-chemical properties. These fouling mechanisms such as precipitation or sedimentation are predominant at low temperatures (T<180°C).
- Fouling due to a change in fluid chemical composition. Such fouling mechanism includes chemical reaction and corrosion and is promoted at high temperature (T>180°C).

In a crude preheat train, fouling is mainly due to chemical reaction fouling. It is located downstream of the desalter where temperatures are the highest.

Chemical reaction fouling can be modelled by two simultaneously acting but opposing phenomena:

- A deposition term which depends on hydrocarbon velocity (Reynolds number), bulk and wall temperatures and hydrocarbon composition.
- A suppression term which depends on hydrocarbon velocity and heat transfer surface geometry (shear stress).

The work presented here is based on an innovative test loop specifically designed and commissioned to achieve the following objectives:

- Identify the main parameters involved in hydrocarbon fouling (crude oil and atmospheric distillation residue);
- Assess heat exchanger performance under fouling conditions;
- Investigate the impact of heat exchanger geometries on fouling tendency;
- Compare experimental results to fouling models prediction;
- Predict industrial heat exchanger fouling rates and improve the efficiency as well as the design of the pre-heat train.

### Experimental setup

The fouling rates encountered in the refinery are often relatively slow (on the order of few micrometers to few millimeters of deposit per month). Thus in order to hasten the process in a laboratory setup, in most of the fouling rigs described in the open literature, crude is heated by an external heating source.

This increases the wall temperature which accelerates the reaction kinetics and leads to increased fouling rates. Thus, main test sections are not representative of real heat exchanger operating

conditions. Furthermore, most of the test rigs use single tube or annular tube as the test section wherein fouling occurs only on one side [1-6]. Currently, no fouling test rig, that approached both hydrodynamic and thermal conditions of crude preheat train, was identified in the literature.
Thus, an innovative test loop was designed to study the hydrocarbon fouling in a pre-heat train of a refinery. This pilot plant allowed to investigate the fouling behaviour of various heat exchanger technologies. Tests were made in controlled conditions to mimic typical refinery operating conditions (nature of fluids, velocity, temperatures and fouling rate).
The pilot plant consists of five closed loops in cascade. Each loop is connected to the subsequent one with a heat exchanger. The test loop is illustrated in the figure 1.

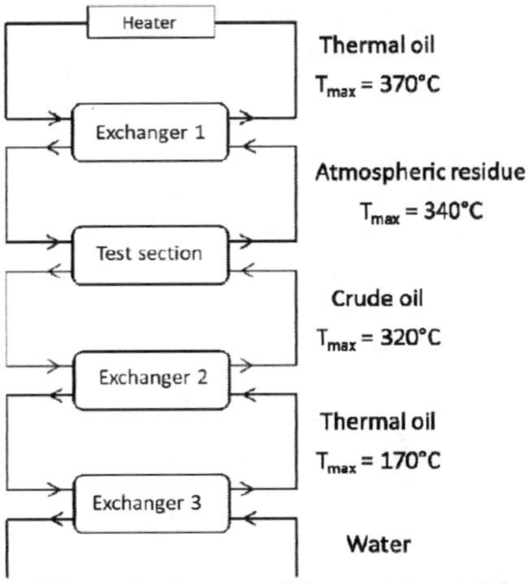

Figure 1: Schematic diagram of the experimental fouling rig

The pilot plant and the test sections were designed to mimic industrial heat exchangers geometries, operating conditions and fouling at both sides of the heat exchangers.
Indeed, fouling may occur in each heat exchanger in contact with crude oil or residue (exchanger 1 and 2 and test section). Fouling is monitored by recording the inlet and outlet temperatures and the pressure drops across the heat exchangers during one test.
Since the fluids are operated in a closed circuit, an inherent risk of depletion of fouling species initially present in the fluid had been identified. In order to mitigate the risk, a non–invasive sampling system was designed and samples were collected at regular intervals during the complete test campaign. Several analyses were conducted to monitor any variations in fluids chemical composition. In case of modification, a fresh feed is used.

## Test sections

### *Shell and tube heat exchanger*

The first studied technology was a shell and tube heat exchanger. It was designed to have the same heat flux, overall heat transfer coefficient, shear stress and construction material as an industrial heat exchanger. Two crude oils were tested in this heat exchanger. The operating conditions used for this test campaign were chosen to be representative of a refinery. They are listed in table 1.

| Operating parameters | Range |
|---|---|
| Crude temperature (°C) | 200-300 |
| Inside tubes velocity ($m \cdot s^{-1}$) | 1-2.3 |
| Inside tubes pressure (bar) | 25-30 |
| Overall heat transfer coefficient ($W \cdot m^2 \cdot K^{-1}$) | 500-600 |
| Duty per unit area ($kW \cdot m^{-2}$) | 12.5 |
| Tubes diameter (mm) | 19.9 |
| Test section diameter (mm) | 270 |
| Test section length (m) | 2.5 |

Table 1: Shell and tube overall dimensions and operating parameters

### *Cross-flow plate heat exchanger*

The second technology tested was a cross-flow plate heat exchanger. It was designed so as to transfer the same duty as did the shell and tube heat exchanger under the same operating conditions. In order to compare the two tests sections, tests were run with identical fluids and same pressures and temperature range. Furthermore, shear stresses were chosen to be representative of a refinery pre-heat train.

## Experimental results

### *Influence of operating conditions*

Fouling rates were calculated from temperature and flow rate measurements. Flow-meter as well as temperature sensors, with high precision (± 0.1°C), were set at the inlets and outlets of each heat exchanger.

Once fouling is initiated the temperature profile changes and global heat transfer coefficient decreases. This coefficient is assessed by:

$$U = \frac{Q}{A \cdot LMTD} \tag{1}$$

U ($W \cdot m^{-2} \cdot K^{-1}$) depends on the heat exchanger duty, Q (W), the heat transfer area, A ($m^2$) and the logarithmic mean temperature difference, LMTD (K).

Fouling resistance is defined as the drop in the inverse of overall heat transfer coefficient.

The fouling resistances of the three heat exchangers in which hydrocarbons circulated had been monitored during the test duration. The evolution of fouling resistances showed that both sides of the test section (atmospheric distillation residue and crude oil) encountered fouling. Moreover, it was confirmed by the detection of fouling deposits on the two sides of this heat exchanger while dismantling it.

The proof of concept of the fouling test rig was validated by measuring fouling rates close to literature and industrial values.

A fouling index ($I_f$) is used to compare experimental results. It is given by:

$$I_f = \frac{U_t}{U_{t=0}} \tag{2}$$

Where, $U_{t=0}$ is the overall heat transfer coefficient at the beginning of the test (clean heat exchanger) and $U_t$ is the overall heat transfer coefficient at time t.

Figure 2 represents the fouling index of one of the fouling tests.

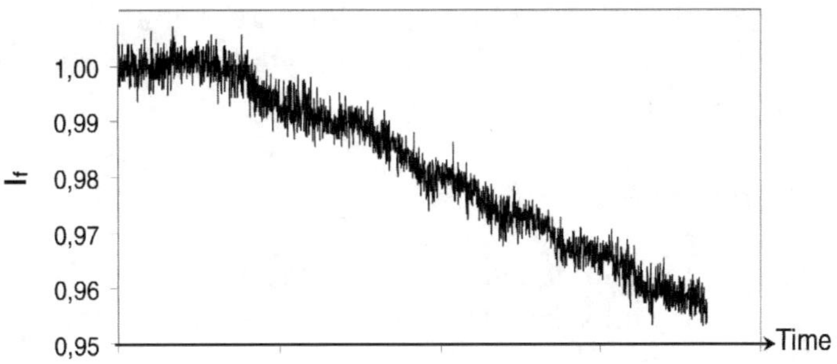


Figure 2: Fouling Index versus time

In figure 2, the drop in $I_f$ with time signifies a loss in thermal efficiency of the exchanger due to fouling on the heat exchanger surface.
Tests were carried at 3 different crude velocities keeping the temperature constant and vice versa. Each test lasted ten days.
Figure 3 represents the fouling index versus time for tests made at three different velocities (u1<u2<u3). The results are illustrated in the figure below.

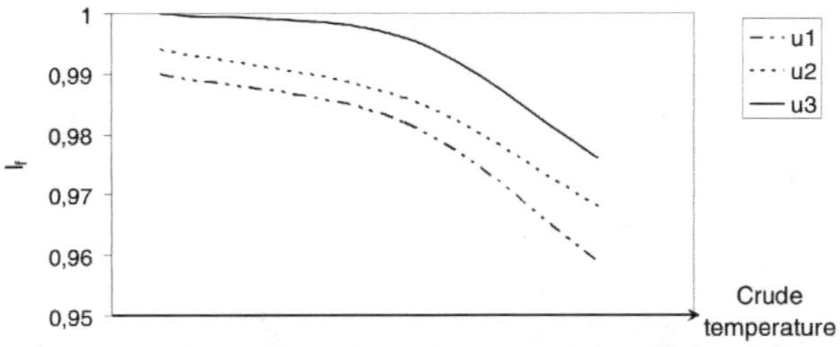


Figure 3: Effect of velocity and temperature on fouling index

For each temperature, results show that fouling index increases with an increase in crude velocity, indicating that fouling rate decreases with velocity. This can be attributed to the fact that when the velocity increases, the shear stress increases thus enhancing deposit suppression removal rates.
Furthermore, it was observed that fouling index decreases non-linearly with increasing bulk and film temperatures. An increase of the temperature promotes chemical reactions and thus fouling deposit rate.

### *Influence of crude composition*

Two crude oils were tested on the test loop, crude A and crude B. Crude A is a low fouling crude based on HLPS test (Hot Liquid Process Simulator) whereas crude B is a medium fouling crude. Figure 4 represents the fouling index of a test conducted with the two kind of crude at the same temperature and same velocity.

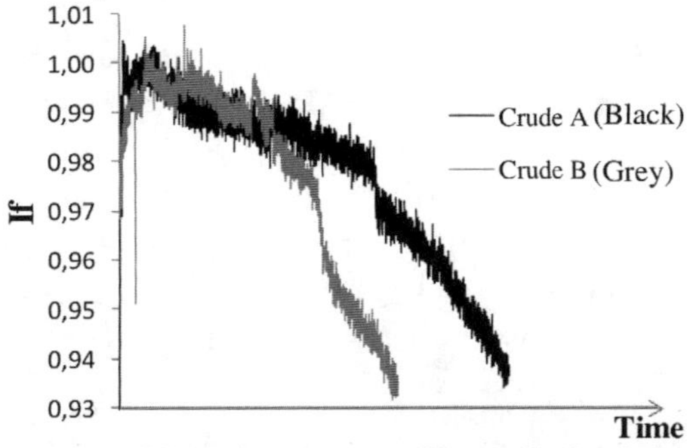


Figure 4: Effect of crude composition on fouling index

The same initial fouling index value was obtained with the two crudes. Nevertheless, fouling index of the crude B test decreases more quickly with time than that of the crude A test. This comparison confirms the effect of crude composition on fouling.

### *Influence of heat exchanger geometry*

The impact of heat exchanger technology was also studied. The two test sections were operated at same volumetric flow and shear stress. Figure 5 represents the fouling index for tests run at the same crude oil volumetric flow.

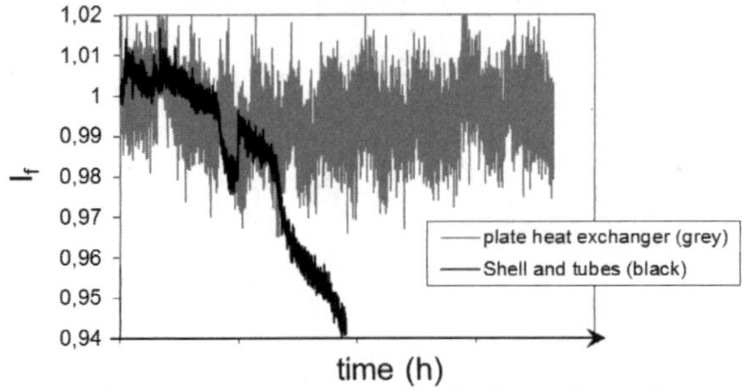


Figure 5: Influence of the heat exchanger geometry on fouling index at the same volumetric flow.

Figure 6 represents the fouling index for tests conducted at the same shear stress on the two test sections.

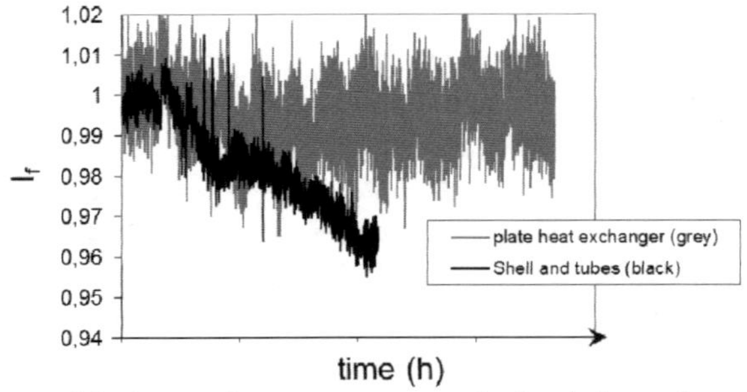


Figure 6: Influence of the heat exchanger geometry on fouling index at the same shear stress.

In figure 5 and 6, the fouling index of the shell and tube heat exchanger decreases with time whereas the fouling index of the plate heat exchanger remains almost constant. It seems that in a plate heat exchanger the fouling tendency is mitigated under the given operating conditions due to higher turbulence.

### *Use of fouling models*

Several fouling models have been established to predict the fouling rates based on operating temperatures and flow rates. Ebert and Panchal and its variant are among the most frequently used models.

The Ebert and Panchal model [7] is based on the fouling threshold concept which corresponds to theoretical zero fouling for a given operating temperature and velocity. From the model, typically at low temperatures and high velocities, fouling rates can be minimised. The fouling rate can be estimated by the difference of two opposing factors, a term of deposit and a term of suppression:

$$\frac{dR_f}{dt} = \alpha \cdot \mathrm{Re}^{\beta} \cdot \exp\left(\frac{-Ea}{\mathrm{R} \cdot \mathrm{T}_{\mathrm{fi}}}\right) - \gamma \cdot \tau \qquad (3)$$

Where $\alpha$, $\beta$ and $\gamma$ are constants, Ea is the activation energy ($J \cdot mol^{-1}$), $T_{fi}$ is the film temperature (°C) and $\tau$ the shear stress (Pa).

Later based on the above model, several authors presented variants with minor modifications.
Panchal *et al.* [8] proposed a modified version in 1999 by introducing the Prandtl number in the deposition term:

$$\frac{dR_f}{dt} = \alpha \cdot \mathrm{Re}^{\beta} \cdot \mathrm{Pr}^{-0.33} \cdot \exp\left(\frac{-Ea}{RT_{fi}}\right) - \gamma \cdot \tau \tag{4}$$

The use of the Prandtl number allows to take into account the variations of crude chemical properties with temperature.
In 2002, Polley *et al.* [9] based on the same hypothesis, proposed a modified version of the Ebert and Panchal model. The new equation was given as:

$$\frac{dR_f}{dt} = \alpha \cdot \mathrm{Re}^{-0.8} \cdot \mathrm{Pr}^{-0.33} \cdot \exp\left(\frac{-Ea}{RT_{\mathrm{surf}}}\right) - \gamma \cdot \mathrm{Re}^{0.8} \tag{5}$$

For this model, Polley *et. al.*, replaced the film temperature with the wall temperature in the deposition term. Furthermore, one of the parameters $(\beta)$, exponent to Reynolds's number, was also set to 0.8. In the suppression term, the shear stress term was replaced by Reynolds number to the power 0.8 in order to better represent turbulent flow.
In 2006, Nasr and Givi [10] used experimental results of Saleh *et al.* [11] to establish another variant of the threshold fouling model. Their correlation is valid for laminar flow (Re < 2100).

$$\frac{dR_f}{dt} = \alpha \cdot \mathrm{Re}^{-\beta} \cdot \exp\left(\frac{-Ea}{RT_{film}}\right) - \gamma \cdot \mathrm{Re}^{0.4} \tag{6}$$

For each experimental test run in our rig, theoretical film or wall temperatures were estimated from each model. Those temperatures were compared to our experimental data. The figure 7 shows, for each test, the percentage of difference between our experimental results and the fouling models predictions.

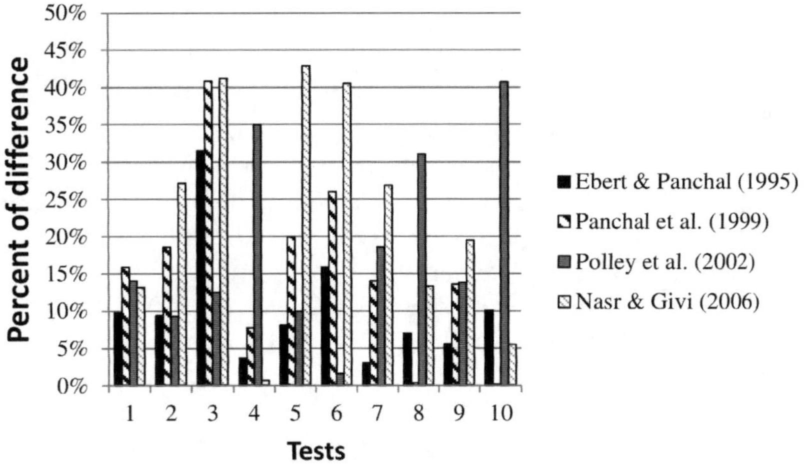

Figure 7: Comparison between experimental results and fouling models prediction.

Temperatures measured in the fouling rig were a reasonably good fit (except for test n°3) to those predicted by the Ebert & Panchal fouling models (1995 and 1999) and those observed at industrial scale. Nasr and Givi model prediction were close to experimental results for tests made at low velocity (tests n°1, 4, 8 and 10). As expected, their model predictions are far from the experimental results for high velocity tests since their correlation is valid for laminar regime.
Experimental results were also compared with experimental data found in the literature. The Panchal *et al.* empirical model is the most frequently used model to predict industrial fouling rates. The model parameters were fitted to reduce the difference between experimental and prediction results [12]. The figure 7 represents the model prediction before and after optimization (dotted line),

in an X-Y plot between surface temperature and $\tau \cdot Re^{0.66} \cdot Pr^{0.33}$. On this plot, empty markers are used for no fouling test and filled markers are used for fouling test.

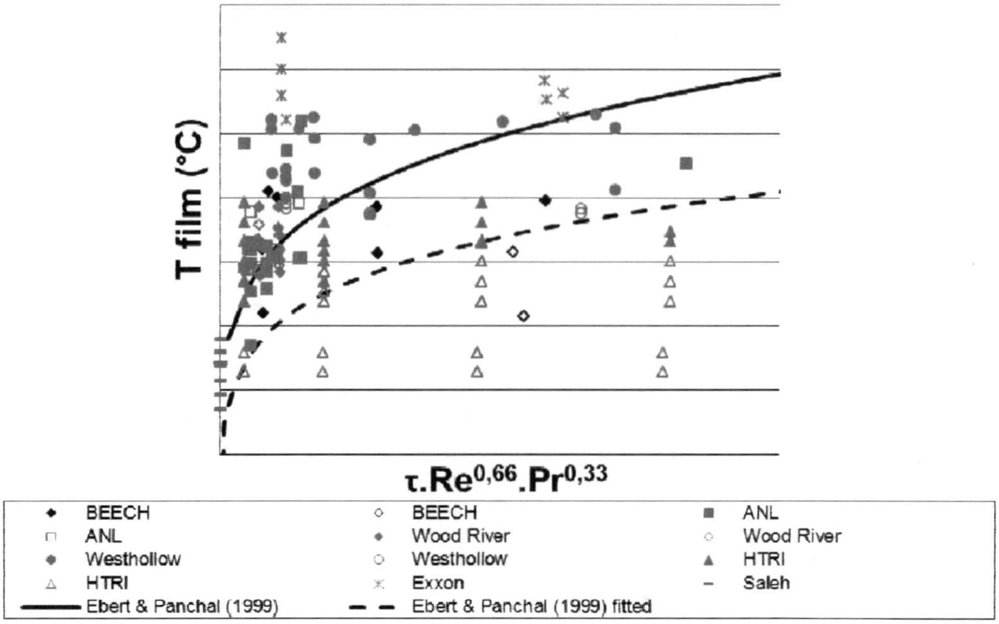


Figure 8: Models prediction on experimental data

For low $\tau \cdot Re^{0.66} \cdot Pr^{0.33}$, fouling results are scattered. Fouling seems to be non-repeatable. Nevertheless, high for $\tau \cdot Re^{0.66} \cdot Pr^{0.3}$, optimization allows reducing the difference between experimental and prediction results for most of the series data.

## Conclusion

A novel experimental fouling rig was designed and commissioned to investigate the fouling behaviour of various heat exchanger technologies. Firstly, the influence of various operating parameters was assessed in a standard shell and tube heat exchanger. The results highlighted that fouling rate decreases with increasing velocity and decreasing temperature. Furthermore, it was also shown that the use of a cross flow plate heat exchanger allowed reducing the fouling rates as compared to that in a shell and tube heat exchanger.

Moreover, Panchal *et al.* model parameters were optimised to reduce the difference between the model predictions and the literature and experimental fouling data.

**Keywords:** Fouling, crude oil, residue, shell and tubes heat exchanger, plate heat exchanger

## References


1. Dickakian G. B., Crude oil fouling control by a fouling analyzer, In National Heat Transfer Conference, Philadelphia, USA, August 6-9, 1989.

2. Watkinson A. P., Comparison of crude oil fouling using two different probes, In Heat Exchanger Fouling and Cleaning IV, Santa Fe, New Mexico, USA, May 18-21, 2003.

3. Srinivasan M., Watkinson A. P., Fouling of some Canadian crude oils, In Heat Exchanger Fouling and Cleaning IV, Santa Fe, New Mexico, USA, May 18-21, 2003.

4. Bennett C. A., Kistler R. S., Nangia K., Al-Ghawas W., Al-Hajji N., Al-Jemaz A., Observation of an isokinetic temperature and compensation effect for high temperature crude oil fouling, In Heat Exchanger Fouling and Cleaning VI, Tomar, Portugal, July 01-06, 2007.

5. Crittenden B. D., Kolaczkowski S. T., Takemoto T., Phillips D. Z., Crude oil fouling in a pilot-scale parallel tube apparatus, In Heat Exchanger Fouling and Cleaning VI, Tomar, Portugal, July 01-06, 2007.

6. Young A., Venditti S., Berrueco C., Yang M., Waters A., Davies H., Hill S., Millan M., Crittenden B. D., Characterisation of crude oils and their fouling deposits, In Heat Exchanger Fouling and Cleaning VII, Schladming, Austria, June 14-19, 2009.

7. Ebert W. A., Panchal C. B., Analysis of Exxon crude-oil-slip stream coking data, In Fouling mitigation of industrial heat exchangers, San Luis Ebispo, USA, June 18-23, 1995.

8. Ebert W. A., Panchal C. B., Liao C. F., Palen J. W., "Threshold conditions for crude oil fouling". In *Understanding Heat Exchanger Fouling And Its Mitigation.* Ed. Begell House, New York, (1999).

9. Polley G. T., Wilson D. I., Yeap B. L., Pugh S. J., Evaluation of laboratory crude oil threshold fouling data for application to refinery pre-heat trains, *Applied Thermal Engineering*. 22: 777 - 788 (2002).

10. Nasr M. R. J., Givi M. M., Modeling of crude oil fouling in preheat exchangers of refinery distillation units, *Applied Thermal Engineering*. 26: 1572 - 1577 (2006).

11. Saleh Z. S., Sheikholeslami R., Watkinson A. P., Fouling characteristics of a light Australian crude oil, *Heat Transfer Engineering*. 26: 15 - 22 (2005).

12. Bories M., Patureaux T., Preheat train crude distillation fouling propensity evaluation by Ebert and Panchal model, In Heat Exchanger Fouling and Cleaning IV, Santa Fe, New Mexico, USA, May 18-22, 2003.

# (488c) Enhanced Development and Control of Continuous Processes Using Real Time in Situ FTIR Analytics

Sean Cusack, Mettler Toledo AutoChem, Inc., Columbia, MD

Multiple reactions were studied using ATR-FTIR as a convenient, specific, inline monitoring technique to examine various aspects, challenges, and opportunities posed by flow chemistry: Dispersion in the column, rapid screening of reaction conditions, reactive intermediate identification, hazardous compound detection to protect the experimenter, microliter scales experiments. This initial study showed that the time it usually takes to optimize reaction conditions was shortened Due to the fast response time and structural information in each of the many reaction mixture spectra obtained, short-lived intermediates were observed *in situ*, which helped gain a better understanding of reaction mechanism. In a later investigation, formation of a pyrazole derivative, the use of ATR-FTIR enabled precise control of reagent streams addition in a complex multi-step reactions sequence. For the first time, it is therefore possible to control pumps to dispense reagents based upon the real time concentration of reaction intermediates in multi-step processes. As a result of this high level of control via inline monitoring, multi-step sequences were run more efficiently. For instance, in the event of a reaction failure, precious material would be saved since the lowered concentration of intermediate would be accounted for in real-time.

# (488d) Analysis of the Influence of Heat Flows in Optimal Design and Operation of Multicomponent Distillation Columns: Extractive Distillation Case Study

**Arley F. Nova**[1], Manuel A. Ramos[2] and Jorge Mario Gómez[2], (1)Chemical Engineering, Universidad de los Andes, Bogota, Colombia, (2)Chemical Engineering, Universidad de los Andes, Bogotá, Colombia

Distillation is a mayor separation system in chemical process industries. It uses heat supplied from a higher temperature level at the reboiler, and rejects almost equal amount of heat in the condenser at lower temperature levels. Therefore, it resembles a heat engine producing a separation work with a rather low efficiency[1]. Just in the U.S distillation of chemicals and petroleum uses about 53 percent of the total energy used for industrial separations and is the largest energy-consuming process in industry[2]. Besides the second-law efficiency of conventional distillation is very low, only around 5-20%. This means that distillation is associated with a high entropy production (exergy loss) and hence degradation of energy. To reduce the exergy wasted on it is recommended to alter design and operation of the distillation process. This is achieved by spreading the heat requirements over the whole length of a distillation column. Such design consideration is referred to as diabatic distillation.[3]

Previous studies of this kind of systems have been considered, with some contemplations about heat exchange between stages, such as heat exchange in all stages[4], systems in which each stage could have whether a condenser or a reboiler[5] and columns with sequential heat exchangers[6]. These studies compare entropy production of each of the proposed systems, against an adiabatic distillation column with same operating conditions, and only analyze binary systems (50% toluene-50% benzene in most cases).

In the present work the concept of diabatic distillation is applied into a extractive distillation system (extractive distillation of ethanol using glycerol as entrainer, an innovating distillation technique because the use of glycerol which recently has a high availability as byproduct in the production of biodiesel) in order to study how heat flows influence the design and operation of multicomponent distillation systems using Mixed Integer Nonlinear Programming (MINLP).

The mentioned distillation system is modeled applying the rigorous MESH equations (Mass balances, Equilibrium relationships, Summation of compositions and energy balance) into each separation stage within the column, taking into account a heat flow whether entering or leaving the stage. The model (assuming adiabatic operation) is compared with a simulation of the same distillation column in Aspen Plus V7.2 achieving accurate results. Then, optimization of the operating conditions was carried out for both diabatic and adiabatic extractive distillation column with an economic objective function, assuming heat flow as a binary variable which may exists or not at each stage for the diabatic case. The same energy consumption for adiabatic and diabatic systems was found, but in the diabatic system the heat is distributed all through the column reducing heat requirement for the reboiler. In this vein an exergy term is included in the model to attain a thermo - economic optimization of the mentioned case study, to identify the location, magnitude and source of the exergy destruction and energy loses in the system computing the associated cost with these terms.

From all the analysis about operating conditions of the case study, a methodology for the optimal design and operation of distillation systems including heat flows through the column is proposed, using MINLP in order to determine number of stages and feed locations of these particular separation systems.

**REFERENCES**

1. Demirel, Y., *Thermodynamic Analysis of Separation Systems.* Chemical ans Biomolecular Engineering Research and Publications, 2004(Thermal Mechanics).

2. Sciences, N.A.o., N.A.o. Engineering, and N.r. Council, *Real Prospects for Energy Efficiency in The United States*.

2010, Whashington, D.C.: The National Academies Press. 349.

3. Schaller, M., *Numerically Optimized Diabatic Distillation Collumns*, in *Natural sciences*2007, Technischen Universität Chemnitz: Chemnistz. p. 86.

4. De Koeijer, G., A. Røsjorde, and S. Kjelstrup, *Distribution of heat exchange in optimum diabatic distillation columns.* Energy, 2004. **29**(12-15 SPEC. ISS.): p. 2425-2440.

5. Chen, M.-C., K. Takeshita, and M. Ishida, *Computational Study on Binary Distillation of Heat-Driven Distillation System.* Industrial & Engineering Chemistry Research, 2005. **44**(24): p. 9156-9163.

6. Jimenez, E.S., et al., *Optimization of a diabatic distillation column with sequential heat exchangers.* Industrial and Engineering Chemistry Research, 2004. **43**(23): p. 7566-7571.

# A HYBRID SIMULATION-OPTIMIZATION APPROACH FOR THE DESIGN OF INTERNALLY HEAT-INTEGRATED DISTILLATION COLUMN

*Juan A. Reyes-Labarta*, Miguel A. Navarro, José A. Caballero*
*Department of Chemical Engineering, University of Alicante, Ap. Correos 99, Alicante 03080, Spain*
*(e-mail: ja.reyes@ua.es)*

## Introduction

Distillation is still one of the most important separation techniques, even though it is an expensive operation in terms of capital and energy costs. This fact with the major current environmental problems such as the global warming and depleting resources have led to growing interest in more efficient and sustainable chemical processes. In these sense, since the distillation columns normally present sections with different heat requirements (to vaporize a liquid in the bottom and to condensate a vapor in the top) as we can see in Figure 1, internally heat-integrated distillation columns (HIDiC) can be an efficient way of saving energy [1-3].

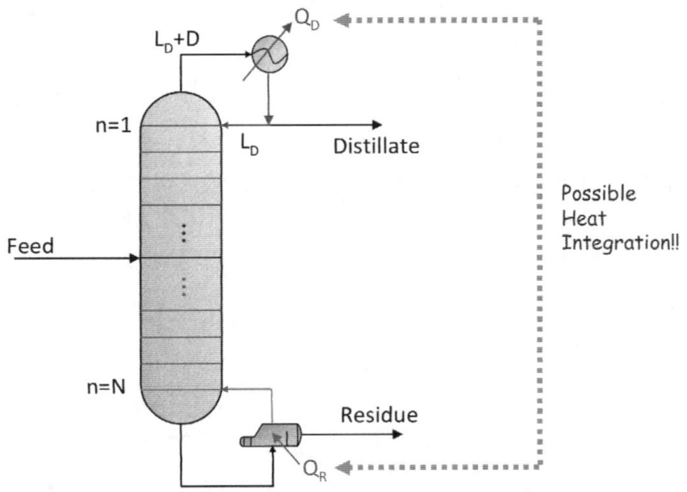


***Figure 1.** General configuration of a single distillation column.*

On the other hand, the presence of intermediate reboilers and intermediate condensers in the corresponding stripping or rectifying section of a distillation column, can improve its feasibility and efficiency. This fact is due to the changes of internal vapor and liquid flows that provokes the intermediate heat supply and/or heat removal, as it can be quickly analyzed by using the McCabe-Thiele method [4-6].

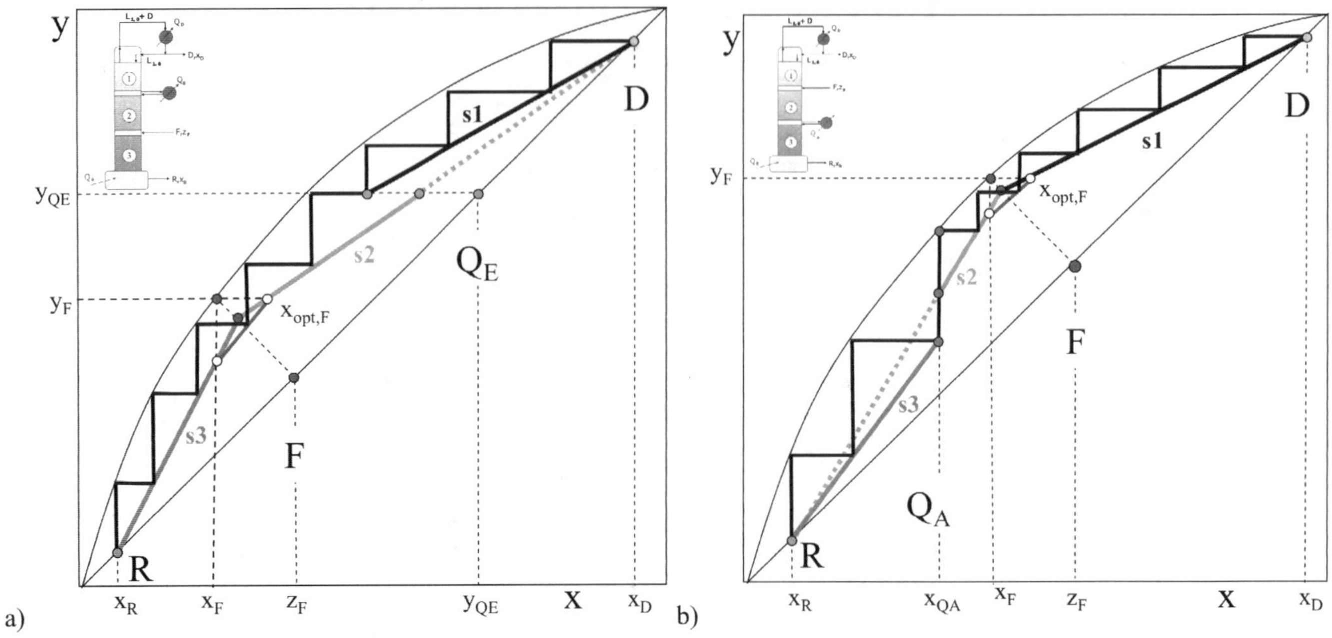


***Figure 2.*** *McCabe-Thiele diagram for a distillation column with 3 sectors due to the presence of two side streams: one two-phases feed stream (F) and one intermediate heat exchanger: a) intercondenser ($Q_E$) located at the rectifying section b) interreboiler ($Q_A$) located at the stripping section.*

HIDiC configurations are characterized by the existence of internal heat integration between the rectifying section and the stripping section. In order to allow the heat transfer from the rectifying to the stripping section, the temperature of the rectifying has to be increased with a compressor. Normally, the stripping-section pressure is the same as the feed pressure, while the rectifying-section pressure should be elevated enough that the different between the dew-point temperature of the distillate and the bubble point of the bottoms product were significant (larger than 10 ºC).

In this kind of designs, once we have defined the operating pressures, the main variables to get the best column configuration for a specified separation are: the number of theoretical stages of each section ($n_{RS}$, $n_{SS}$), reboiler duty ($Q_R$), distillate flow rate (D), number and location of the intermediate heat exchangers, overall internal heat transfer ($Q_{IHT}$) and compressor shaft work ($W_S$).

Due to the inclusion of very efficient simulation algorithms in commercial packages the design of distillation columns has been usually performed by successive simulations [7,8]. However, this trial-and-error procedure is very time consuming and of doubtful utility when trying to evaluate complex configurations or sequences of interrelated columns.

On the other hand, the use of new design methodologies based on superstructure models and mathematical programming (General Disjunctive Programming or Mixed-Integer Non-Linear Programming techniques) present difficult convergence due to the high non-linearity of the equations used (especially when rigorous equilibrium calculations are modeled). Additionally, these methodologies normally have a strong dependence of the initial values used to start the optimization calculations and need the utilization of good initialization procedures and tight bounds on the variables to convergence [9-11].

Thus, in the present paper, we present a hybrid simulation-optimization approach for the optimal and rigorous design of heat integrated distillation columns (HIDiC). This approach allows easily analyze and quantify the effect of the presence of intermediate heat exchangers over the overall heat duty, the total annualized cost or even over the environmental impact, by combining the capabilities of process simulation packages with optimization tools and Life Cycle Assessment (LCA) [12,13]. The number of trays and pressure of each section can be optimized in an outer loop providing the method a great flexibility. The commercial process simulator Hysys® has been used to implement the process and solve the rigorous VLE using the available thermodynamic models such as NRTL, and MatLab® to implement a Particle Swam Optimization (PSO) algorithm [14].

## System description

Figure 3 represents the general configuration of a HIDiC. The basic components are the rectifying section (RS), the stripping section (SS), the compressor (Comp), the reboiler (Reb), the condenser (cond), the expansion valve (EV) and the intermediate heat exchanger (IHE). An additional pre-heater (p-H) is also show in Figure 3, just before the compressor, to avoid the possibility of presence of a liquid phase in the compressor. This pre-heated is controller by the Hysys adjustment tool (A) that introduce the minimum amount of heat in order to have a saturated vapor in the stream leaving the compressor.

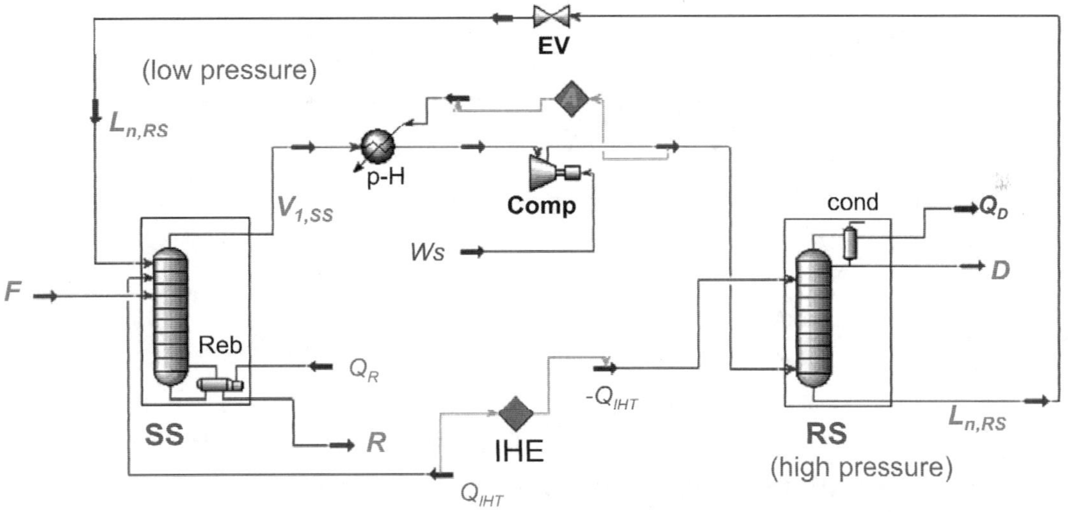


***Figure 3.*** *General configuration of an internally heat-integrated column (HIDiC).*

The system operation at two pressure levels is as follows. The feed steam (F) is introduced in the top of the stripping section. The distillate (D) and residue (R) streams are the products living its corresponding section. The vapor stream leaving the stripping section ($V_{1,SS}$) is compressed before to be introduced in the rectifying section. As commented before, the stream $V_{1,SS}$ could be preheated in order to avoid the presence of any liquid phase in the compressor. The liquid stream leaving the rectifying section ($L_{n,RS}$) is expanded before to be introduced in the stripping section. The overall internal heat transfer ($Q_{IHT}$) is exchanged between the two rectifying and stripping section by the IHE equipment.

## Objective functions

Two different objective functions have been used: the operating cost ($C_{op}$) and the total annualized cost, which includes the capital and operating cost.

$$TAC = C_c + C_{op} \tag{1}$$

The capital cost ($C_c$) includes the cost of the heat exchangers ($C_{HXS}$) and compressor ($C_{comp}$). The cost of the valves is neglected. The sum of the equipment cost is multiplied by the capital recovery factor, which is function of the interest rate ($ir$=10%) and the life time of the equipment ($n$=8 years).

$$C_c = (C_{HXS} + C_{comp}) \cdot \frac{ir \cdot (ir+1)^n}{(ir+1)^n - 1} \tag{2}$$

The cost of the heat exchangers is estimated by the correlation proposed by Kilizkan et al. [15]. The compressor cost is estimated using correlations from Turton et al. [16].

$$C_{HXS} = \sum_{k=hxs} (c_1 \cdot A_k + c_2) \tag{3}$$

$$C_{comp} = 5.8 \cdot 10^{[k1 + k2 \cdot \log(W_S) + k3 \cdot \log^2(W_S)]} \tag{4}$$

Note, that the heat exchangers area is calculated using the logarithmic mean temperature difference ($\Delta T_k^{lm}$), the overall heat transfer coefficient ($U_k$) and the heat duty ($Q_k$).

$$A_k = \frac{Q_k}{U_k \, \Delta T_k^{lm}} \tag{5}$$

The operating cost ($C_{op}$) includes the cost of the steam ($C_q$) used in the distillation column ($Q_R$), the cost of electricity ($C_e$) consumed by the compressor ($W_S$) and the cost of the cooling water ($C_w$).

$$C_{op} = (C_q \cdot Q_R + C_e \cdot W_S + C_w \cdot Q_D) \cdot t_{op} \tag{6}$$

A Multi-Objective Optimization (MOO) framework could also be stated by accounting simultaneously the minimization of the total annualized cost and the associated environmental impact of the system (ISO 14044. 2006). In this sense, just to mention that the Life Cycle Impact Assessment (LCIA) aims the evaluation of the significance of potential environmental impacts according to the life cycle inventory of emissions and feedstock requirements, and the corresponding set of damage factors [12,13]. The damage factors can be obtained from Ecoinvent® Database (e.g. Eco-Indicator 99 that is the sum of the 10 impact categories that are aggregated into one using normalization and weighting parameters [17]). The solution of this problem should be given by the corresponding set of Pareto optimal points (Pareto curve) that trade-off the economic and environmental performance.

## Results

Different binary and multicomponent systems have been studied using the suggested procedure to analyze the advantages of HIDiC configurations with a small number of heat exchanges:

- benzene + toluene
- propylene + propane
- methanol + water
- benzene + cyclohexane + toluene
- methanol + acetone + water
- methanol + ethyl-acetate + water
- propylene + i-butane + propane
- benzene + toluene + ethyl-benzene + styrene

In order to show an example of the advantages that the proposed approach offers, Figure 4 shows the results of a sensitivity analysis where, for a defined configuration of a HIDiC (i.e. concrete number of trays and pressures in both sections), the overall internal heat transfer ($Q_{IHT}$) has been varying from 0 to 2000 kW.

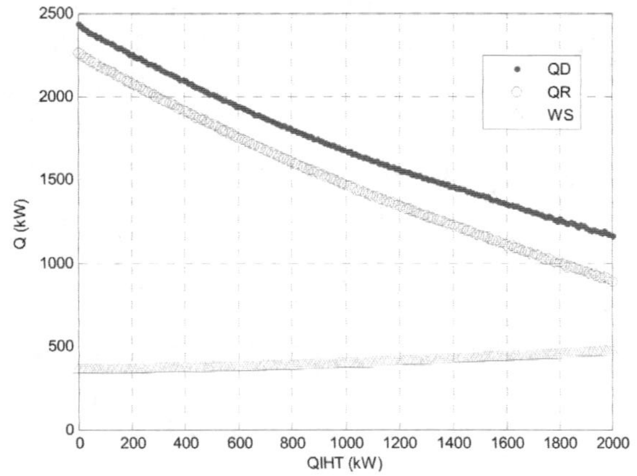


***Figure 4.*** *Condenser and reboiler duties and compressor shaft work vs overall internal heat transfer ($Q_{IHT}$).*

As we can observe in Figure 2, the existence of only one internal heat exchange between both sections can produce interesting reduction in the total energy demand and therefore significant cost saving, without being necessary heat exchange between each tray of both sections. This results are consistent with those proposed by different authors [2,18,19]. Additionally, to remark that internal heat integration could also be applied to distillation columns with no direct connections [1].

## Conclusions

A hybrid simulation-optimization design method has been suggested and applied to analyze the advantages and benefits of using internal heat exchangers in the distillation columns. The effect of the overall internal heat transfer on the energy requirements has studied for a defined HIDiC configuration. The proposed approach can also be applied to vapor recompression columns (VRC).

Obviously, the economic interest of the HIDiC configurations depends on the ratio between the cost of the electricity and the cost of the heating steam. However, environmentally friendly and energy efficient technologies should be introduced to provide sustainable processes that minimize the environmental impacts.

The main ideas obtained of the present analysis are the followings:

- HIDiC configurations with a small number of intermediate heat exchangers can present interesting benefits, providing savings around 20-30% depending on the components of the system, the flow rate and composition of the feed stream, and the product purities desired.
- In this sense, the possibility of heat exchange between stages that are not at the same high in both sections can also improve the efficiency of the configuration.
- The optimal location of the feed stream is generally the first stages of the stripping section.
- The pressure difference between both sections is also a key point of the process. The pressure of the rectifying section should be high enough to allow the heat transfer between both sections, but it should be as low as possible to minimize the compressor costs.
- Traditional HIDiC schemes (with heat transfer along the total length of each section) are more efficient with non-flat temperature profiles inside the column and low products purity. However, HIDiC configurations with a small number of intermediate heat exchangers improve the feasibility and efficiency of the configuration in this kind of separations.
- Besides, this procedure is based on the robustness of the commercial simulators, more work is needed in order to reduce the actual convergence problems and local optimal solutions (e.g. surrogate models, kriging interpolation [20]).

## Acknowledgements

The authors would like to acknowledge financial support from the Spanish Ministerio de Ciencias e Innovación (PPQ, CTQ2009-14420-C02-02).

## References

1. Nakaiwa M, Huang MK, Endo A, Ohmori T, Akiya T, Takamatsu T. Internally heat integrated distillation columns: A review. Chem. Eng. Res. Des. 2003;81:162-177.
2. Shenvi AA, Herron DM, Agrawal R. Energy efficiency limitations of the conventional heat integrated distillation column (HIDiC) configuration for binary distillation. Ind. Eng. Chem. Res. 2011;50:119-130.

3. Ho TJ, Huang CT, Lee LS, Chen CT. Extended Ponchon-Savarit method for graphically analyzing and designing internally heat-integrated distillation columns. Ind. Eng. Chem. Res. 2010;49:350-358.
4. Huang K, Matsuda K, Iwakabe K, Takamatsu T, Nakaiwa M. Graphical Synthesis of an Internally Heat-Integrated Distillation Column. J. Chem. Eng. Jpn. 2006, 39, 703-708.
5. Marcilla A, Serrano MD, Reyes-Labarta JA. Analysis of the Connecting Zone in the McCabe-Thiele Method Covering Multiple Feeds, Products and Heat Transfer Stages. AIChE J. 2012 (in revision).
6. Marcilla A, Serrano MD, Reyes-Labarta JA. Review and Extension of the McCabe-Thiele Method. Open Academic Repository of the University of Alicante (RUA). http://hdl.handle.net/10045/23195.
7. Douglas JM. Conceptual Design of Chemical Processes. McGraw-Hill. 1988.
8. Doherty MF, Malone MF. Conceptual Design of Distillation Systems. McGraw-Hill. 2001.
9. Caballero JA, Milán-Yañez D, Grossmann IE. Rigorous Design of Distillation Columns: Integration of Disjunctive Programming and Process Simulators. Ind. Eng. Chem. Res. 2005;44(17):6760-6775.
10. Reyes-Labarta JA, Grossmann IE. Disjunctive Optimisation Design Models for Complex Liquid-Liquid Multistage Extractors. AIChE J. 2001; 47(10): 2243-2252.
11. Reyes-Labarta JA, Caballero JA, Marcilla A. A Novel Hybrid Simulation-Optimization Approach for the Optimal Design of Multicomponent Distillation Columns. Computer Aided Chemical Engineering. 2012, 30, 1257-1261.
12. Guillén-Gosálbez G, Caballero JA, Jiménez L. Application of Life Cycle Assessment to the Structural Optimization of Process Flowsheets. Ind. Eng. Chem. Res. 2008;47(3):777-789.
13. Brunet R, Reyes-Labarta JA, Boer D, Guillén-Gosálbez G, Jiménez L. Combined Simulation-Optimization Methodology for the Design of Environmental Conscious Absorption Systems. Computers and Chemical Engineering 2012, 46: 205-216.
14. Ruiz-Femenia R, Caballero JA. A Particle Swarm Optimization for Solving Nlp/Minlp Process Synthesis Problems.Computer Aided Chemical Engineering. 2010, 28, 1877-1882.
15. Kizilkan O, Sencan A, Kalogirou SA. Thermoeconomic optimization of a LiBr absorption refrigeration system. Chemical Engineering and Processing: Process Intensification 2007; 46:1376-1384.
16. Turton R, Bailie RC, Whiting WB, Shaeiwitz JA. Analysis Synthesis and Design of Chemical Processes. 2003, New York: McGraw-Hill.
17. PRe-Consultants (2000). The Eco-indicator 99A damage oriented method for life cycle impact assessment. methodology report and manual for designers. Technical Report, PR'e Consultants Amersfoort, The Netherlands.
18. Harwardt A, Kraemer K, Marquardt W. Identifying optimal mixture properties for HIDiC application. Proceedings Distillation and Absorption 2010, 55-60.
19. Suphanit B. Optimal heat distribution in the internally heat-integrated distillation column (HIDiC). Energy 2011;36:4171-4181.
20. Caballero JA, Grossmann IE. An Algorithm for the Use of Surrogate Models in Molecular Flowsheet Optimization. AIChE J. 2008;54(10):2633-2650.

# (488f) Design of Reactive Distillation Process for the Production of n-Butyl Levulinate

Tzu-Hsuan Peng[1], Cheng-Liang Chen[2] and **I-Lung Chien**[2], (1)Department of Chemical Engineering, National Taiwan University, Taipei, Taiwan, (2)Department of Chemical Engineering, National Taiwan University, Taipei, Taiwan

Butyl Levulinate (Levulinic acid butyl ester, LABE) is one of the potential fuel additives due to its good characteristics such as high octane number, high oxygen content, and low water solubility. This chemical can be produced by a reversible esterification reaction of n-butanol and levulinic acid (LA). Both raw materials of this reaction can come from bio-conversions. For example, n-butanol can be produced by ABE (Acetone-Butanol-Ethanol) fermentation process, and LA can be produced from lignocellulosics through acid-catalyzed dehydration and hydrolysis of hexose sugars.

The traditional process for manufacturing LABE requires two-step reaction. First, to convert Levulinic acid to angelica lactone and then followed by reaction of angelica lactone with alcohol. However, the two-step reaction may result in complex process flowsheet and high operating cost. Direct esterification is an alternative way to produce LABE, However, the low equilibrium conversion limitation results in similar complex process flowsheet. This work aims to use reactive distillation (RD) technology to simplify the process by breaking the equilibrium conversion of this reversible reaction.

In the RD process, high-purity LABE is designed as bottom product while top vapor approaching the minimum-boiling azeotrope of n-butanol and water. With top vapor condensed in a decanter and an additional flash drum, high-purity water product at 99.9 mol% can be withdrawn from the liquid draw of the flash drum. The vapor stream of the flash drum is condensed and recycled back to the decanter. Aspen plus simulation shows that an economical design flowsheet with high overall conversion of 99.5% can be achieved with stoichiometric feed of two reactants in the proposed RD process.

# (488g) Application of Plasma Electrolysis Technology in Chlor Alkali Production Sector

**Fakhrian Abqari**, Chemical Engineering, University of Indonesia, Depok, Indonesia

Chlor Alkali sector is one of the most important sectors in chemical industry. But, this sector consumed the electricity energy the most because the electrolysis process. Plasma electrolysis is a method that can be applied in chlor alkali production process and can reduce the energy consumption several times. Plasma technology involves the formation of sparks of electricity by passing an electric current through a gas in a process called disconnection of electricity (electrical breakdown). Because of the electrical resistance throughout the system, it will produced a number of significant amounts of heat, which takes electrons from molecules of gas produces a stream of ionized gas, or plasma (Gomez et al, 2009). In previous researches, in hydrogen production from water, plasma electrolysis can reduce the energy consumption more than conventional electrolysis. In this paper, plasma electrolysis technology will be applied in chlor alkali electrolysis process. The plasma electrolysis in chlor alkali production will form the active radical species of NaCl and $H_2O$ which then make the productivity increases. Products resulted will be measured and compared to the result from conventional electrolysis. Finally, the energy consumption used will be calculated and the efficiency will be analyzed.

# (523a) Teaching Product with Process Design

Warren D. Seider, Chemical and Biomolecular Engineering, The University of Pennsylvania, Philadelphia, PA

This presentation addresses the challenges in teaching product design with process design. For the simplest chemical products, i.e., basic chemicals, which are often manufactured as commodities, the design focus is usually on manufacturing efficiency. Other chemicals involve special properties to meet customer needs. For these chemical products, the focus is on matching the latest technologies with the *voice of the customer*; i.e., on product design.

As the field of chemical engineering diversifies, college graduates increasingly become engaged in product design. Classical process design, which meshes nicely with most core chemical engineering courses, normally provides students with few, if any, product design experiences and perspectives. For product design, it's often important to find chemicals and chemical mixtures having desired properties and performance. Often small quantities of specialty chemicals are involved; e.g., pharmaceuticals, requiring the design of multipurpose batch chemical processes. Often to satisfy customer needs, product design focuses on the 3-dimensional product structure; e.g., LCDs, labs-on-a-chip, colloidal structure, ..., and their more complex process designs.

In this presentation, approaches are suggested for introducing product design topics while teaching a classical process design course. Subjects added include the product- and technology-development framework, the innovation map (linking new technologies with customer needs), the Stage-Gate™ Product-Development process (Cooper, 2002, 2004), and molecular-structure design; i.e., the identification of molecules that satisfy property specifications for new chemical products. These and related subjects are interspersed while our students solve process design homework exercises. Included are an example of configured-consumer product design; e.g., a lab-on-a-chip for high-throughput screening of kinase inhibitors, and an industrial product; e.g., a thin-glass substrate for liquid-crystal displays (LCDs).

While some groups of 3-4 students solve classical process design projects, others solve product design projects; e.g., to produce encapsulated drugs with controlled release profiles, and to model physical vapor deposition of thin films on integrated circuit substrates (using kinetic Monte-Carlo methods). While our entire senior class is not trained to be expert in each technology platform, each student group learns enough basics from its faculty advisor to work effectively on its product design.

# (523b) The Teaching of Capstone Design

David L. Silverstein, University of Kentucky, Paducah, KY, Lisa G. Bullard, Chemical and Biomolecular Engineering, North Carolina State University, Raleigh, NC, Warren D. Seider, Chemical and Biomolecular Engineering, The University of Pennsylvania, Philadelphia, PA and Margot A.-S. Vigeant, Chemical Engineering, Bucknell University, Lewisburg, PA

The authors will present the results of the fourth survey in the resumption of the Chemical Engineering Education Special Projects Committee series of annual surveys on how chemical engineering courses are taught. The surveys are now conducted under the auspices of the Education Division of the Institute. This year's survey covers capstone design, and seeks to determine what concepts are being taught, who is teaching the course(s), and what tools are being used. The role of industry in preparing instructors and in developing projects and case studies is also studied. Teaching methods and novel approaches are, as always, focus elements of this year's survey. Some comparison with previous studies on the subject will be included.

Following the presentation of the survey results, a panel consisting of selected participants in the survey will lead a discussion with the audience covering common issues in teaching capstone design. Ideas and approaches discussed during the session will be incorporated into the final survey report.

# (533a) Drug Product Model Design, Development, Qualification, and Maintenance in a QbD Framework

**Hayden Thomas**, Formulation Development, Vertex, Cambridge, MA

Model development and implementation is a core principle to employing a QbD approach in drug product design and process scale-up. The advantages in taking this approach are realized through the ease of visualization or mathematical expression of data sets for greater process understanding, which leads to clear decision making and results in superior product quality. There are several model types that can be developed and used at every stage in the QbD process. A favored approach is to begin with the end in mind by outlining up front a clear plan toward model design and development for the drug product process, considering how each model will be used, limitations, assumptions, qualification, and maintenance. The experimental approach to the development of each model should consider the type of model (mechanistic or empirical), selected variables, scale dependency and relationship to the overall drug product process. Once a model has been developed, steps are needed toward implementation through qualification of the model and evaluating uncertainty in predictions. Finally, for models selected for use in commercial manufacture, a maintenance plan is needed to support product life cycle, allowing model updates as needed. In this presentation, representative examples will be provided using these principles that have been used to support drug product filings.

# (533b) Prediction of Post-Relaxation Radial Tensile Strength of Pharmaceutical Tablets Using At-Line Collected near Infrared Data

Sameer Talwar[1,2], Benoit Igne[2], Zhenqi Shi[2], Carl A. Anderson[2] and James K. Drennen III[2], (1)Graduate School of Pharmaceutical Sciences, Duquesne University, Pittsburgh, PA, (2)School of Pharmacy, Duquesne University, Pittsburgh, PA

The prediction of relaxed radial tensile strength (RTS) of pharmaceutical tablets in real-time using near infrared (NIR) spectroscopic data collected at-line was investigated. Radial tensile strength is a critical quality attribute (CQA) of tablets which may correlate to drug product performance *in vivo*. However; post-compaction, the tablets undergo axial and radial relaxation, which influences dosage form characteristics such as disintegration time, dissolution time, and mechanical strength. It is the RTS of completely relaxed tablets that is of interest, as it is a completely relaxed tablet that will be administered to a patient. The existing work relating NIR spectra to RTS relies on the use of data collected from relaxed tablets (off-line NIR data) to predict the relaxed (off-line) RTS. However, the NIR analysis of tablets in a manufacturing setting, as desired in the current Quality by Design philosophy and accomplished in a modern real-time release environment, necessitates that tablet quality attributes are determined immediately following compression and prior to complete relaxation.

A six component system was compressed with variability in excipient ratio, disintegrant level, and compression force (total of sixteen design points). Eighteen tablets were collected per design point and scanned at-line using a commercial NIR spectrometer immediately after compression. Tablets were left to relax for two weeks. Their physical dimensions were then measured and they were crushed to calculate RTS values. A multiple linear regression (MLR) was used to predict RTS of relaxed tablets from at-line spectra. In a first step, partial least squares models were developed to predict variables with a significant relationship to RTS (tablet density, tablet volume). Outputs from the PLS models were then used along with spectral slope and formulation parameters as input in an MLR to predict RTS. Radial tensile strength predictions on an independent test set had an error of 0.25 MPa with a $R^2$ of 0.8. The MLR model had calibration and cross validation error between 0.2 – 0.3 MPa. Updating spectral information with physical features and chemical information through MLR allowed for predicting RTS of relaxed tablets using at-line information while compensating for the effect of tablet relaxation over time. This approach has the potential to be implemented on-line since all the information needed by the MLR model can be calculated from the same near infrared spectrum. Automating spectral collection will allow the real-time prediction of relaxed RTS at the tablet press level.

# (533c) Quantitative Analysis of Crystalline Materials in Binary Compacts Using Time Domain Spectra of THz Spectroscopy and Density Dependence

**Md. Anik Alam**, James K. Drennen III and Carl A. Anderson, Mylan School of Pharmacy, Duquesne University, Pittsburgh, PA

Polymorph and pseudo-polymorph systems play a major role in the pharmaceutical industry because of their potential to impact the performance of finished dosage forms. Quantitative analysis of polymorph and pseudo-polymorph is becoming increasingly important for solid oral dosage forms and transdermal products. A rapid, non-destructive analytical tool is required to reduce the time and cost for quantitative analysis. Terahertz Time Domain Spectra has emerged as a potential analytical tool for rapid and non-destructive analysis of these systems. Terahertz Time Domain Spectra provides high specificity for polymorphic study due to the ability to probe inter-molecular and phonon lattice vibration directly into the crystal structure.

The feasibility of Terahertz Time Domain Spectra for quantitative analysis of anhydrous theophylline and the monohydrate in compacts was studied. This study included the analysis of density dependence in quantification accuracy. The effect of density on the Terahertz Time Domain Spectra was demonstrated. An improvement in model performance was demonstrated when density was undertaken into consideration.

# (533d) Bulk and Micro-Scale Characterization of Continuous Powder Blending Integrated with an Multi-Point in-Line NIR As a PAT Tool

Juan G. Osorio, William E. Engisch Jr. and Fernando J. Muzzio, Chemical and Biochemical Engineering, Rutgers, The State University of New Jersey, Piscataway, NJ

Continuous manufacturing of solid pharmaceutical products has been an area of much research in the past few years. Most efforts have focused in characterizing the bulk properties and efficiency of continuous powder mixers as a function of operational and design parameters. The present study summarizes the bulk characterization of a new Glatt continuous powder (GCG-70) blender using experimental techniques previously reported [1,2]. While bulk characterization of a resulting blend is crucial to describe the mixing efficiency of a continuous mixer, it is not enough to determine the final quality of a blend, which can then be related to the performance in succeeding processes. A new approach to study final blends at the micro-scale using hyperspectral imaging to further understand phenomena such as API agglomeration and de-agglomeration in continuous mixing is reported here. Further along, comparison results from the integration of an in-line NIR as a PAT tool to monitor the concentration and variability of key components in final blends are described. Both approaches presented here were useful to further understand the continuous mixing process as a function of the design, operational and material properties parameters.

References:

[1] P.M. Portillo, M.G. Ierapetritou, F.J. Muzzio, Characterization of continuous convective powder mixing processes, Powder Technology 182 (2008) 368-378.

[2] A.U. Vanarase, F.J. Muzzio, Effect of operating conditions and design parameters in a continuous powder mixer, Powder Technology 208 (2011) 26-36

# (533e) Effect of Lubricant Mixing Time and Concentration, Compaction Speed and Compression Force On the Dissolution Profile of a Control Release Formulation

**Sarang Oka**[1], Pallavi Pawar[2], Fani Boukouvala[2] and Fernando J. Muzzio[1], (1)Department of Chemical and Biochemical Engineering, Rutgers, The State University of New Jersey, Piscataway, NJ, (2)Chemical and Biochemical Engineering, Rutgers University, Piscataway, NJ

Effect of lubricant mixing time and concentration, compaction speed and compression force on the dissolution profile of a control release formulation

*Sarang Oka, Pallavi Pawar, Fani Boukouvala, Fernando Muzzio*

Rutgers University

Segregation or de-mixing of granular matter has been studied in great depth and in the pharmaceutical industry continues to be a routine nuisance. In the current project, significant variability was observed in the dissolution profile between tablets made from a multiparticulate formulation. The formulation in discussion is a typical formulation consisting of an API in the form of granulated pellets, filler, Magnesium Stearate and a rapid disintegrant. The particle size of the pellets is 600 μm. Tablet dissolution is performed in a standard USP dissolution apparatus-2. The dissolution profile demands no more than 20% of the drug to be dissolved in 60 minutes, no more than 35% in 120 minutes, between 20-45% in 480 minutes and no less than 80% in 1440 minutes.

The standard deviation in the value of the amount of drug dissolved in one hour was found to be unacceptably high by regulatory standards. The observation was attributed to segregation since the final formulation exhibits a significant tendency to demix.

In order to enable a QbD approach for solving this problem, preliminary studies were conducted to examine the effect of several variables on the dissolution profile of the tablets. It was hypothesized that the amount of lubricant and the amount of shear in lubricated mixture could have a significant impact on the observed response. Compression speed and compression force were also considered to be potentially important.

The work focused on conducting a preliminary set of designed experiments to test the relative importance of multiple variables on a lab scale, perform statistical analysis to determine significant parameters affecting dissolution and consequently develop a more concise Design of Experiment to determine the design space at the manufacturing scale. The original formulation contains 0.2% MgSt. The study was a full factorial DOE with 0.2%, 0.4% and 0.6% MgSt mixed for 50, 100 and 150 revolutions that result in 10 blends (the $10^{th}$ blend employs 0.2% MgSt mixed for 300 revs). Mixing was performed in a conical twin-axial mixer at 30 rpm. The blend was sampled to test mixing homogeneity. Mixing homogeneity was quantified by performing offline NIR measurements on extracted samples. The blend is finally compacted in a tablet simulator at 3 speeds and 3 punch depths (compaction force) resulting in 9 compaction conditions per batch leading to 90 tested conditions in all.

Analysis of variance tests are performed on the ensuing data set. Statistical significance of compaction force, compaction speed, concentration of lubricant and mixing time of lubricant on percent drug dissolved in 60 minutes is tested. The compaction speed, lubricant concentration and lubricant mixing time are all found to be statistically significant parameters which affect the amount of drug dissolved in 60 minutes. Coupled interactions of the above parameters are found to be significant while three way interactions were found to be insignificant.

The dominant effects were the compression speed and the amount of lubricant (Magnesium Stearate in this case) in the formulation., leading to a large interactive effect with the compression force. At low levels of Magnesium stearate, an increase in compression force increased the dissolution rate, perhaps due to increased pellet breakage. On the other hand, at higher levels of Magnesium stearate, the rate of dissolution decreased with an increase in force, due to increased tablet consolidation.

# QbD Case Studies: An Integrated Multivariate Approach to Drug Product and Process Development


Presenting Author: Goldi Kaul (Ph.D.)

(Authors: Jun Huang, Ramarao Chatlapalli, Krishnendu Ghosh and Arwinder Nagi)


AICHE 2012

*(Paper ID number: 253849)*



## Introduction to QbD- Approach and Advantages in Product and Process Development.

1. QbD approach provides Links b/w API, excipients and drug product
2. Design space, PAT , QRM and control strategy concepts
3. Regulatory Wins: Site and scale change flexibility
4. Faster approvals and minimal post approval issues
5. Visibility with global regulatory authorities

# Case Study-I

## Problem Statement:

To explain the unexpected effect of a surfactant Poloxamer 188 on High drug Load, HSWG IR-tablet dissolution

**Approach:** Synergistic DoE effects, RSM , optimization and Multivariate data analysis

## Objectives:

1. To explain tablet dissolution slowdown in HSWG IR-tablets with increase in poloxamer-188 in a combination DoE with the following design variables:
   a. Poloxamer amount
   b. Granulating water amount
   c. Binder addition rate

2. To elucidate the effects if any of the process variables on tablet dissolution
   Preliminary dissolution spec.: Q NLT 70% in 30 minutes in 0.1 N HCl

9/21/2012 3

# What causes Poloxamer mediated dissolution slowdown?

- Poloxamer causes tablet DT slowdown mediated dissolution slowdown
- MOA of tablet DT slowdown:
  - Poloxamer mediated increased hydrophilicity of tablet matrix-exhausts early the disintegrant action of CPV
  - Poloxamer facilitates binder action of Povidone during HSWG by lowering granule IFT and ST of binder solution
    - » Evidence- Noisy /lumpy granulation (Run 2)
    - » High power consumption (Run 2)
  - Acts as a surrogate binder itself – Additional binding effect to Povidone.
  - Mechanism verified by no effect of poloxamer on CCS formulation

9/21/2012 4

# Summary and Conclusions

- Poloxamer retards tablet dissolution while promoting granule dissolution by its effect on tablet DT
- MVDA complementary tool to DoE effects analyses explains above effect:
  - High poloxamer runs have coarser granulations,
  - Poloxamer does have a surrogate/inherent binding effect
- Synergistic use of DoE effects and MVDA provided essential clues into the MOA of poloxamer in tablets

9/21/2012 Publication: *"Quality by Design Case Study: Investigation of the role of poloxamer in immediate release tablets by experimental design and multivariate data analysis"*- AAPS Pharm. Sci. Tech. (2011), Vol 12, No.4, 1064-1076

# Case Study II

## Problem Statement:

HS Wet granulated product- Poor Blend flow ("Funnel flow and Rat-holing") observed at small scale process DoE batches. Blend flow was identified per QRA exercise and earlier experiments as being dependent on the following CPPs –

(i) Water Amount
(ii) Wet Massing time (WMT) and
(iii) Lubrication time (LT)

## Objective/s:

1. Optimize blend flow using synergistic approach of DoE effects, RSM, optimization and MVDA
2. Develop a Control Strategy using the above 3 CPPs as design factors
3. Understand competing Blend flow and tablet dissolution

9/21/2012 6

## Roadmap: Design to Knowledge Space by DOE, RSA and Optimization

**DOE, and run experiments and obtain response variables** → **DOE effect analysis to study significance of factors on responses** → **Response surface modeling and Optimization to establish design space** → **Run confirmatory batches within design space**

| Run # | Lube Time (MIN) | WMT (MIN) | Water (G) | CTAB@ 30min | FFC |
|---|---|---|---|---|---|
| 1a | 3.0 | 1.0 | 268.7 | 88.6 | 7.4 |
| 2a | 2.0 | 4.0 | 306.3 | 76.4 | 11.0 |
| 3a | 3.0 | 5.0 | 268.7 | 83.6 | 7.6 |
| 4a | 1.0 | 3.0 | 268.7 | 87.1 | 7.3 |
| 5a | 5.0 | 3.0 | 268.7 | 85.0 | 6.0 |
| 6a | 3.0 | 3.0 | 287.5 | 80.0 | 9.1 |
| 7a | 4.0 | 4.0 | 306.3 | 83.7 | 4.8 |
| 8a | 3.0 | 3.0 | 287.5 | 79.9 | 8.9 |
| 9a | 3.0 | 3.0 | 287.5 | 81.5 | 8.2 |
| 10a | 2.0 | 2.0 | 306.3 | 86.0 | 7.2 |
| 11a | 4.0 | 2.0 | 306.3 | 86.9 | 7.2 |
| 12a | 3.0 | 3.0 | 325.0 | 82.7 | 9.7 |
| 13a | 3.0 | 3.0 | 250.0 | 82.7 | 10.5 |

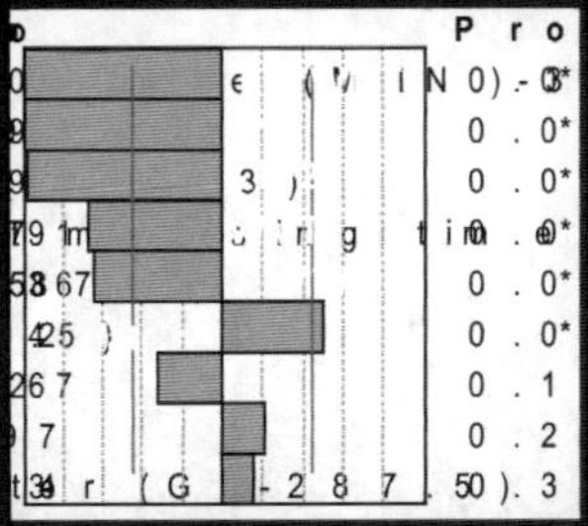

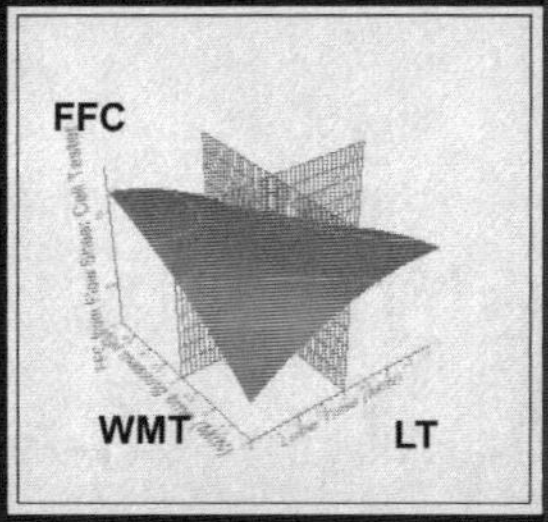


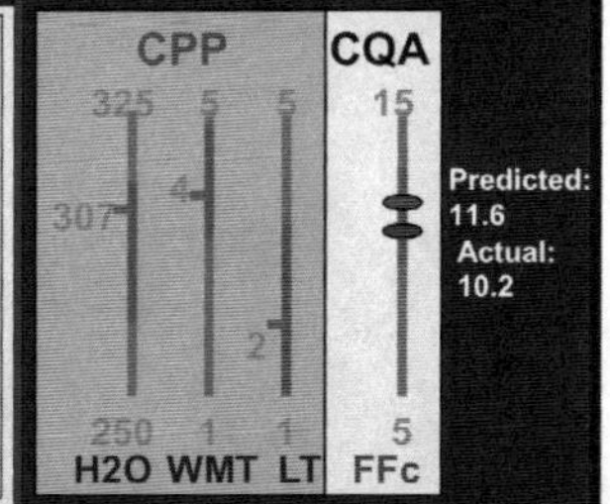


**Knowledge space**

DOE ranges of design variables were determined by trial runs (Sep 2008)

**DOE Effect Analysis**

Sort significance of modeling terms

**Design space**

An optimal region represented by a combination of variables that ensure desired CQAs

**Confirmation**

Run confirmatory batch using an optimal solution to verify design space (Dec 2008)



## Summary and Conclusions

- Case studies exemplifies the application of QbD principles and tools to drug product and process development.
- Identified conditions/reasons for poor and good flow of tablet blend, optimized flow and established a design space
- Blend flow found to be improved in confirmatory runs by use of process parameters suggested by DoE effects analysis- High water and High WMT with low Lube times identified as being optimal for blend flow
- Control Strategy put in place for Blend flow, to ensure product always in DS region
- Dissolution data analysis- By MVDA, helped identify several variables responsible for dissolution differences and relationships b/w blend flow and dissolution. Compromise may have to be reached b/w the two.

Publication: *"Quality by Design Case Study: An Integrated Multivariate Approach to Drug Product and Process Development"*- Int. J. Pharm. (2009), doi:10.1016/j.ijpharm.2009.07.031

# Acknowledgements

- L-Wyeth and Pfizer Drug product Co-dev team
- Chemometrician: Jun Huang
- Statisticians: Andrew McKeen and Brent Harrington
- Contract Manufacturing Facility.

# (533g) Automation and Control of Drug-On-Demand Technology

**Laura Hirshfield**, Arun Giridhar, Gintaras V. Reklaitis, Michael T. Harris and Venkat Venkatasubramanian, Chemical Engineering, Purdue University, West Lafayette, IN

Stimulated by initiatives such as PAT and QbD, the pharmaceutical industry has been focusing on the development of innovative, efficient manufacturing methods. The concept of "mini-manufacturing" of drugs, as opposed to mass production of drugs, has shown promise in this area. We describe one such mini-manufacturing process, "drug-on-demand," which uses drop-on-demand printhead technology to deposit active pharmaceutical ingredients (API) onto edible substrates. We use a high precision positive displacement pump to deposit polymer-drug melts. Besides minimizing a need for mass production of all drugs, this method allows for the layering of different APIs and the creation of individual dosing forms, in which the amount of API can be varied depending on the patient.

In this work, we focus on the automation, control and Exceptional Events Management (EEM) strategy for the drug-on-demand pilot facility done as part of the Engineering Research Center for Structured Organic Particulate Systems. A proper control strategy is necessary so that the system can run at optimal setpoints and adjust in real-time, allowing for an efficient manufacturing process and a precise on-specification drug product. The control strategy for our “drug-on-demand” setup allows us to execute an automated, optimized, and controlled print cycle while closely monitoring drop size, drug morphology, and drop deposition pattern. The implementation of EEM allows for detection, diagnosis, and mitigation of abnormal events that occur outside of the control space to ensure a more efficient, productive process. In this session, we summarize our achievements and current research in this area.

# (534a) Exergy-Loss-Minimization Based Process Synthesis for Cascade Refrigerant System in Ethylene Plants

**Ha Dinh**, Jian Zhang and Qiang Xu, Dan F. Smith Department of Chemical Engineering, Lamar University, Beaumont, TX

Refrigeration system holds an important role in most chemical/petrochemical processes. The traditional cascade refrigeration system (CRS) used in ethylene plants includes multiple refrigerants and multiple temperature and pressure levels. Each component subsystem includes a compression section, a condensing section and several temperature levels with sub-coolers, expansion valves, evaporators and flash drums. In this study, at first, in-depth thermodynamic analysis for a single-loop refrigeration system is carried out to identify exergy loss in heat exchanger network with non-isothermal phase changes, especially at sub-cooler and evaporator in the refrigerant system. The relationship between compressor work consumption and exergy change from one pressure level to another is also comprehensively demonstrated. The achieved methodology is applied in a cascade refrigeration system used in ethylene plants to explore system energy loss below ambient temperature.

The procedure of process synthesis involves three consecutive steps: Thermodynamic analysis (for both single-loop and cascade refrigeration systems), Synthesis model development and solving, and solution validation and evaluation. Rigorous steady-state model of a propylene cycle in the cascade system is firstly constructed and employed to generate Exergy-Temperature (B-T) chart in order to analyze their thermodynamics and energy distributions. An exergy-based mixed-integer linear programming (MINLP) model is developed to minimize the overall exergy loss of the system, where multiple refrigerants with multiple recycling loops are simultaneously addressed and all of cooling demands are fulfilled. The optimal solution is subsequently examined by conducting rigorous simulations to check their feasibility and consistency. Economic evaluation is also carried out from the calculated capital cost of new equipments and annual operating cost. The efficacy of the developed methodology is demonstrated by a case study of a cascade propylene refrigeration system used in an ethylene plant.

# (534b) Development of the Optimal Flowsheet of Isomerization of Light Gasoline Fractions

**Nickolay V. Kuzichkin**, Department of resource saving, Institute of Chem. Technology, Saint Penersburg, Russia

The two-tiered approach to the synthesis of the optimal structure of the process of isomerization of light gasoline fractions is proposed.

Problem optimal choosing of reactors and separation's system with a minimal use of auxiliary heating and cooling is solved at the low level, and on the top-level is carried synthesis of heat exchangers network.The volume of reactor, reflux ratio of the temperature and pressure in the distillation columns are the independent variables

The minimum amount of heat consumed by the auxiliary heating and cooling agents can be determined from the temperature - enthalpy diagram (TED).
The maximum possible heat recovery, and hence the minimal costs associated with the use of auxiliary fluids, obtained from TED are the lower boundary of the synthesis problem of heat exchange networks. Synthesis of heat exchangers network is executed by means of combinatoric method with aid estimated functions.

The developed method of choosing the optimal structure is universal and can be used to calculate construction and retrofit installations refinery

# (534c) A Systematic Framework for Synthesis and Design of Multi-Scale Processing Networks Using Incremental-Based Solution Strategy

Alberto Quaglia[1], Maxim Shishmarev[2], Bent Sarup[3], Gürkan Sin[2] and Rafiqul Gani[1], (1)CAPEC, Department of Chemical and Biochemical Engineering, Technical University of Denmark, Kgs. Lyngby, Denmark, (2)Department of Chemical and Biochemical Engineering, Technical University of Denmark, Kgs. Lyngby, Denmark, (3)Vegetable Oil Technology Business Unit, Alfa Laval Copenhagen A/S, Søborg, Denmark

**A systematic framework for synthesis and design of multi-scale processing networks using incremental-based solution strategy**

A.Quaglia,[1] B. Sarup,[2] G. Sin,*[1] R. Gani[1]

[1]CAPEC, Department of Chemical and Biochemical Engineering, Technical University of Denmark, DK-2800 Kgs. Lyngby, Denmark; email: aq@kt.dtu.dk. *email:gsi@kt.dtu.dk.

[2]Vegetable Oil Technology Business Unit, Alfa Laval Copenhagen A/S, 2860, Denmark

For processing companies, the synthesis and design of optimal industrial operations includes all the strategic and tactical decisions such as the selection of raw materials and products portfolio, as well as the synthesis and design of the processing network and the optimization of the material flows through it.

Recent development in process system engineering (PSE) focused on a class of multi-scale models (from meso to mega-scale), formulating them as Mixed Integer Non Linear Programming (MINLP) problems (Grossmann, 2005). As many contributions in the literature demonstrate, problems of this class are certainly non-trivial, where typical factors contributing to the increase in problem complexity are number of process alternatives considered, size and complexity of the process models used to describe the process alternatives and presence of recycles, bypasses and parallel processing. Several methods and solution strategies (e.g. model reformulation, problem decomposition, variable initialization and bounding) have been successfully proposed for the solution of complex network synthesis problems. These methods and tools are often tailor-made for the specific problem to be solved, and therefore their selection and implementation is not trivial and requires expert judgment.

In this contribution, we propose a systematic framework for the formulation and solution of processing network synthesis problem, and illustrate its application through the solution of industrial case studies, developed in collaboration with Alfa Laval. The key characteristic of the framework is to manage the complexity of the multi-scale problem through a systematic approach, in order to allow simple, efficient and robust formulation and solution of the problem. In the first step of the framework, the user is guided through the formulation of the problem. A superstructure containing all possible alternatives is generated, the objective function is selected, and all relevant data and models are collected, reconciliated and systematized in a database structure. All available engineering and commercial insights are translated in logical constraints aiming at excluding redundant solutions from the search space, that is, the superstructure. In the second step the formulated MINLP problem is analyzed to identify and suggest the most appropriate solution strategy.

One of the strategies employed in the framework is the incremental decomposition-based solution, the complexity of the formulated problem is decomposed, and a series of sub-problems with incremental levels of complexity is automatically generated. The generated sub-problems are then solved in a sequence, where a lower level sub-problem solution provides initialization and bounds to the higher level, until the full problem solution is obtained. Moreover, sensitivity analysis is used at each step, in order to prioritize the use of complex (non linear) models to describe only processes and phenomena which have a large impact on the overall objective, so to obtain a good tradeoff between model complexity and quality of the solution.

The proposed framework is applied to the formulation and solution of industrial process synthesis problems, in the field of edible oil production and refinement, wastewater treatment and biorefinery.Through the presentation of these industrial case studies, the features of the proposed framework are highlighted, and the complexity related to industrial application of process synthesis (large number of process alternatives, complex and non-linear models, presence of recycles, bypass and parallel processes etc.) is discussed.

**References**

A.Quaglia, B.Sarup, G.Sin, R.Gani, 2012, Integrated business and engineering framework for synthesis and design of enterprise-wide processing networks, Computers & Chemical Engineering (38) 213-223.

I.E. Grossmann, 2005, Enterprise-wide Optimization: A New Frontier in Process Systems Engineering, AIChE Journal, 51, 1846-1857.

# (534d) Design of a Reactive Distillation Column: Comparison Between Trays and Packings

**Paola Bastidas**[1], Julio C. Vargas[1], Alvaro Orjuela[2], Iván D. Gil[1] and Gerardo Rodriguez[1], (1)Departamento de Ingeniería Química y Ambiental, Universidad Nacional de Colombia, Bogotá, Colombia, (2)Departamento de Ingeniería Química y Ambiental, Universidad Nacional de Colombia, Bogota, Colombia

**Design of a reactive distillation column: comparison between trays and packings**

Paola Bastidas, Julio Vargas, Álvaro Orjuela, Iván Gil, Gerardo Rodríguez

Grupo de Procesos Químicos y Bioquímicos. Departamento de Ingeniería Química y Ambiental, Universidad Nacional de Colombia-Sede Bogotá, Cr. 30 45 – 03, Bogotá, Colombia

During the last decade, there has been an increasing application of reactive distillation technology in industry, due to the economic benefits that it offers. For example, the production rate and the conversion are improved by the constant removal of the generated products, and the capital costs are reduced since less units operations are required. This growing interest demands deeper study, in order to have more efficient processes.

The key to have efficient separation-reaction process is the proper design and selection of the hardware to be implemented inside the column. The current process simulators are powerful tools that allow comparing different types of internals, that is, to study the effect of the equipment geometry to achieve given separation and reaction grades. That is the case of the Aspen Plus software, where rigorous design of reactive distillation columns is possible through the non-equilibrium (NEQ) model, which has in to account the effect of heat and mass transfer velocities in the global and phase balances, and assumes that the equilibrium is reached only at the interphase.

In this work, the design of a reactive distillation column to produce isoamyl acetate is done with a NEQ model approach. Comparative studies between using trays and packings are developed through process simulations using Aspen Plus®software. In both cases, NRTL-HOC thermodynamic model is used to describe the phase equilibrium, and an equilibrium kinetic model using Amberlyst 35 as catalyst is implemented to describe the reaction behavior. A stream containing acetic acid, amyl alcohol, amyl acetate and water is feed in the lower zone of the reactive section. At the same time an excess of amyl alcohol is fed at the top of that section.

For the same column configuration, better results are obtained with packings than with trays. When using packings as internals, the isoamyl acetate purity is higher than with trays, as well as the global conversion. According to the global heat transfer rate, the feed streams vaporize when enter to the column, with a stronger tendency when using packings than when using trays. This change of energy in the reactive zone promotes the reaction in the desired direction, due to its endothermic nature. Because of the vaporization is weaker in the case of the tray column, the temperature profile goes down in this zone, making that one of the reactants be in the phase vapor rather than in the liquid vapor, which leads to a lower conversion than in the case of the packed column.

It was encountered also, that in the reactive zone were the higher tray efficiency and the lower HETP value, for the tray and packed column, respectively. That is, the reactive zone is the most efficient zone, mainly because that is the zone with bigger differences between the vapor and liquid temperatures.

# A State-Space-Based MIDO Formulation for Batch Distillation Process Synthesis

*Shan JIN, Hong-Guang DONG and Xin-Wen GUO, School of Chemical Engineering, Dalian University of Technology, Dalian, China*

## Introduction

Batch distillation is widely used in fine chemical and pharmaceutical industries, for the purification or recovery of high value-added products. The transient nature of the batch distillation process provides great flexibilities in both various operating policies and alternative complex configurations. Such unique characteristics help further explore the economic potential and process efficiency through the optimal synthesis of batch distillation process. However, the dynamic behavior and complicated economical trade-offs make a huge challenge for the synthesis problem.

Over the past decades, the optimal design and operation of batch distillation have received significant attentions (Logsdon, Diwekar and Biegler, 1990; Mujtaba and Macchietto, 1996; Sharif, Shah and Pantelides, 1998; Low and Sørensen, 2003). The previous works have provided some general insights on the comparative performances of various batch distillation systems and indicated that the interdependent process design and operating control variables should be optimized simultaneously. Furthermore, different configuration of batch distillation column was also taken into account as an optimization parameter with the method of a disjunctive model consisting a given set of equipment modules and the corresponding logical relationships provided by the design context (Oldenburg, Marquardt and Leineweber, 2003; Low and Sørensen, 2005).

With the development of manufacture and control in chemical process, more efficient and energy saving techniques are applied in batch distillation system. As a result, the number of possible batch distillation column configurations and complex operating strategies proposed in the literature continue to increase, such as the side withdrawals, the adaptive vapor recompression structures, thermally linked streams and other complex designs. So there is an urgent need to propose a completed superstructure of batch distillation system that encompasses all possible configurations and operating strategies to reduce the energy consumption and capital investment, and eventually achieve the profit maximization. The incompletion of the superstructure may preclude a series of design alternatives, where the optimal solution may actually lie.

In addition, because the batch distillation is a dynamic process, the separation difficulty also varies with the processing time, especially for multitask separation, the corresponding structure of the batch distillation system should be optimally changed in different time intervals. For example, two separation tasks with different relative volatilities, the optimal number of trays will certainly change during the different mixtures are processed. Even for a single separation duty, at the beginning it is relatively easier to get high purity of distillate, so less trays and reflux ratio are needed. As the process goes on, the separation difficulty is added, the number of trays and reflux ratio should be increased to accomplish the task. So there is plenty of room for the column configuration being optimized in different time intervals to make a better trade-off between capital and operating costs, which aims to maximize the total profit. Even more, due to the flexibility of batch distillation process, the temporary idle trays in certain time interval can even be used in other separation tasks, which can not only make the best of equipment resource, but also get extra profits. However, the constant vapor load and the fixed configuration on total time horizon are always adopted in most of previous works, and this limit may result in the compromise of the operation strategy and the design of configuration instead of the perfect match between each other in every time interval, which will increase the total cost.

In this work, we expect to propose a systematic method based on mixed integer dynamic optimization (MIDO) formulation for synthesizing the batch distillation process on both space and time dimension, which means that not only more novel advanced column configurations and operation strategies are taken into account, but also both of them can be optimized in every time interval. An improved superstructure based on state space representation (Bagajewicz and Manousiouthakis, 1992), a framework that takes all streams splitting and mixing possibilities into consideration, is constructed to achieve any potential optimal process. In the next section, the batch distillation process synthesis problem is presented including the problem definition, superstructure representation, and model formulation followed by objective function establishment. Finally, Binary and multicomponent separation case studies will be presented to illustrate the validity and advantages of the proposed approach.

## The Batch Distillation Process Synthesis Problem

### *Problem Definition*

The objective in this synthesis problem is to determine the most economical batch distillation process, which is capable of fulfilling all separation requirements. All the optimal parameters including configuration, design and operation strategy are achieved simultaneously. The problem is stated as follows:

*Given*: feed mixture to be separated with known batch size and composition; a set of product specifications (purities of key components in main products, amount, etc.); available production time.

*Determine*: For each time interval, the optimal configuration (position of the vessels, connections between trays and vessels), design (number of trays and vessels; vapor loads) and the operating strategy (liquid/vapor ratio, etc.).

*To maximize*: the profit.

### *Superstructure Representation*

The state space based superstructure has been successfully used for the synthesis of continuous distillation process to reduce both the utility and capital costs (Zou et al., 2012), and we take the unique characteristics of batch distillation process (dynamic behavior and tray-vessel configuration) into account to develop the superstructure.

To be specific, the superstructure of batch distillation system is viewed as a system of two interconnected parts (see Figure 1). One is referred to as the distribution network (DN), in which all mixers, splitters and the connections between them are embedded. The other is the so-called process operator (PO), which is further divided into two kinds of equilibrium stages, i.e. the vessels and the trays, with intermediate condenser and/or reboiler. From the superstructure we can see that the initial feed stream can be charged into arbitrary number of vessels according to its property, such as the number of components and each composition in the feed stream. Then the streams out of the vessels are delivered to DN for distribution to arbitrary trays for further mass and heat exchanges, and similarly, the trays are fully connected to each other through splitters and mixers on DN and also have opportunities to distribute streams to arbitrary vessels. Finally, when the required purities are satisfied, each product can be collected in certain vessels or withdrawn from certain trays.

Under this framework, different stream-connecting structures between trays and vessels are used to generate all kinds of feasible configurations and operating strategies. Moreover all the connecting streams are fluctuating over time, as a result the connecting structures have time dimension, which means that the operation strategies and corresponding configurations can be both optimized in different time intervals.

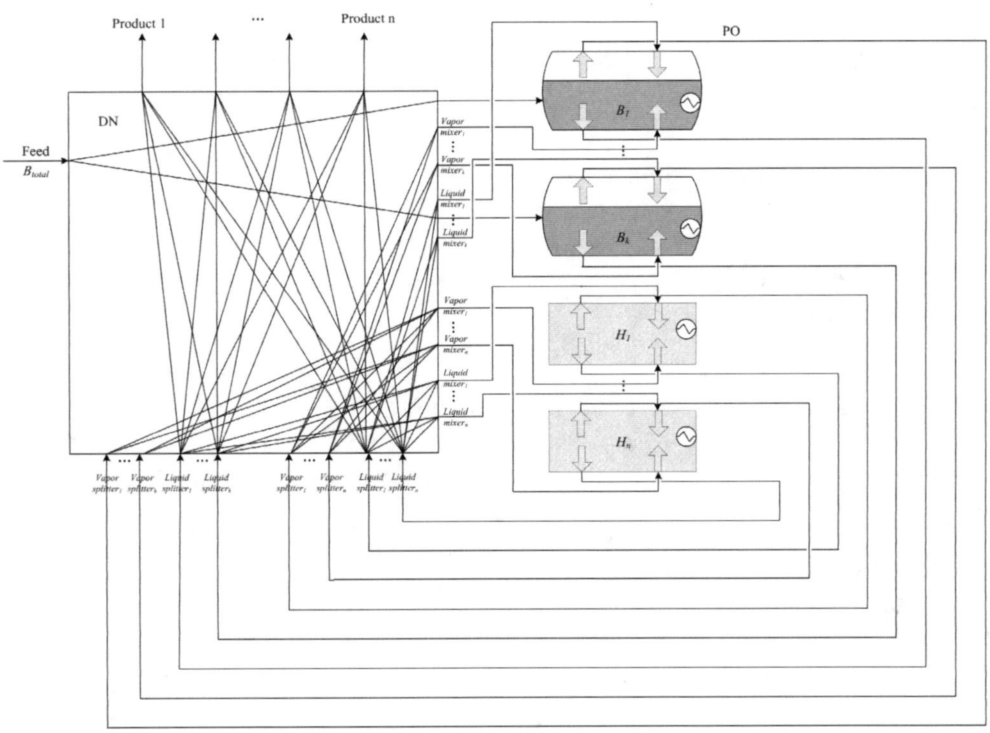


**Figure 1.** The improved state-space superstructure of batch distillation system

***Model Formulation***

In this work we have employed a detailed dynamic model considering Mass balance, Equilibrium relation, Summation check and Heat balance (MESH) equations for each of equilibrium stage, all mixers and splitters. Reasonable assumptions are also adopted here, such as constant molar liquid holdup on the trays, negligible vapor holdup, total condenser, constant pressure, perfect mixing and adiabatic trays. Thus the overall synthesis problem can be formulated as a mixed integer dynamic optimization (MIDO). The major mathematical model can be described by the following equations:

*Total mass balance for feed:*

$$B_{total} = \sum_{b=1}^{k} B_b \qquad \forall b \in NB \tag{1}$$

*PO: representation of mass and heat exchanges process on equilibrium stage. (Take tray i as example, the formulations of vessels have the similar forms.)*

*Mass balance for component j of tray i:*

$$\frac{d(H_i x_{i,j})}{dt} = L_{i,j}^{in} + V_{i,j}^{in} - V_{i,j}^{out} - L_{i,j}^{out} \qquad \forall i \in NT, \forall j \in C \tag{2}$$

Where $B_{total}$, $B_b$, $H_i$, $L$, $V$, $x_{i,j}$ are the total feed, vessel holdup, tray holdup, liquid flowrate, vapor flowrate, and liquid composition, respectively.

*Heat balance for tray i:*

$$\frac{d(H_i h(H_i))}{dt} = L_i^{in} \times h\left(L_i^{in}\right) + V_i^{in} \times h\left(V_i^{in}\right) - V_i^{out} \times h\left(V_i^{out}\right) - L_i^{out} \times h\left(L_i^{out}\right) + Q_{R,i} + Q_{C,i} \qquad \forall i \in NT \tag{3}$$

Where $h$ represents the specific heat content, and it is the function of composition of the mixture and tray temperature. $Q_{R,i}$, $Q_{C,i}$ denote the intermediate reboiler and condenser heat duty.

*DN: determination of stream-connecting structures. (For each time point)*
*Mass balance for component j of mixer i:*

$$L_{i,j}^{in} = \sum_{i'=1}^{n}(L_{i',i} \times x_{i',j}) + \sum_{b=1}^{k}(L_{b,i} \times x_{b,j}) \quad \forall i, i' \in NT, \forall b \in NB, \forall j \in C \tag{4}$$

$$V_{i,j}^{in} = \sum_{i'=1}^{n}(V_{i',i} \times y_{i',j}) + \sum_{b=1}^{k}(V_{b,i} \times y_{b,j}) \quad \forall i, i' \in NT, \forall b \in NB, \forall j \in C \tag{5}$$

*Mass balance for component j of splitter i:*

$$L_{i,j}^{out} = \sum_{i'=1}^{n}(L_{i,i'} \times x_{i,j}) + \sum_{b=1}^{k}(L_{i,b} \times x_{i,j}) + D_i \times x_{i,j} \qquad \forall i, i' \in NT, \forall b \in NB, \forall j \in C \tag{6}$$

$$V_{i,j}^{out} = \sum_{i'=1}^{n}(V_{i,i'} \times y_{i,j}) + \sum_{b=1}^{k}(V_{i,b} \times y_{i,j}) \qquad \forall i, i' \in NT, \forall b \in NB, \forall j \in C \tag{7}$$

Where $L_{i',i}$, $V_{b,i}$, $L_{i,b}$, $V_{i,i'}$, $D_i$, $y_{i,j}$ are the flowrate of liquid stream from tray *i'* to tray *i*, the flowrate of vapor stream from vessel *b* to tray *i*, the flowrate of liquid stream from tray *i* to vessel *b*, the flowrate of vapor stream from tray *i* to tray *i'*, flowrate of product stream from tray *i* and vapor composition, respectively.

From the DN formulation, we can notice that $L_{i,b}$, $V_{i,b}$, $L_{b,i}$, $V_{b,i}$ are used to represent the connecting relations between trays and vessels, and $L_{i',i}$, $V_{i',i}$, $L_{i,i'}$, $V_{i,i'}$ express the connecting relations between different trays. Through the analysis of all these variables on different time intervals, we can get the connecting structures of the batch distillation system, which means that the optimal configurations and operation strategies are obtained on the total time horizon.

***Objective Function Establishment***

In this work, we formulate the objective function to contain the sales revenue, operating costs (including costs of cold and hot utilities) and equipment costs (including costs of distillation columns and heat exchangers) and we convert these costs to an hourly based. The objective function has the following form (Logsdon, Diwekar and Biegler, 1990):

$$\psi = \frac{DP_r - B_0 C_0}{t_f + t_s} - \frac{T.A.C.}{Hrs} \tag{8}$$

The function of $T.A.C.$ is adopted from Zou (2012) as follows:

$$T.A.C. = f(N_T, D_{tray,i}, A_{R,i}, A_{C,i}, A_{R,b}, A_{C,b}) + C_w(\sum_{i\in NT} Q_{C,b} + \sum_{i\in NT} Q_{C,i}) + C_S(\sum_{i\in NT} Q_{R,b} + \sum_{i\in NT} Q_{R,i}) \tag{9}$$

For complex columns systems, the cost of a column was normalized on a "per tray" basis in Zou (2012), this strategy is also adopted in this work in order to make the best of every operation unit in every time interval.

The function of the cost of single tray is based on Warren D. Seider (2004)

$$CT = (193.04 + 22.72 D_{tray,i} + 60.38 D_{tray,i}^{\ 2}) \times 1.2 \times 1.5 \tag{10}$$

Then the $f(N_T, D_{tray,i}, A_{R,i}, A_{C,i}, A_{R,b}, A_{C,b})$ can be written as:

Invest cost= $\sum_{i\in NT} (193.04 + 2.49 V_i^{out\,0.5} + 1.304 V_i^{out}) +$

$$500(\sum_{i\in NT} (\frac{Q_{R,i}}{7000(T_s - T_i)})^{0.8} + \sum_{i\in NT} (\frac{Q_{C,i}}{3000(T_i - T_w)})^{0.8} + \sum_{b\in NB} (\frac{Q_{R,b}}{7000(T_s - T_b)})^{0.8} + \sum_{b\in NB} (\frac{Q_{C,b}}{3000(T_b - T_w)})^{0.8}) \tag{11}$$

Where $V_i^{out}$ denotes the vapor load of tray *i*, $T_i, T_b, T_s, T_w$ represent the temperature of tray *i*, vessel *b*, heating and cooling utilities respectively.

The mathematical model of complex batch distillation system has been formulated as a mixed integer dynamic optimization (MIDO) problem with a set of differential-algebraic equations (DAE). Then the integer variables are eliminated by means of continuous variables directly. To be specific, the binary variables assigned to each tray and heat exchanger with a value of "0–1", denoting the logic of existence or absence, can be also indicated with the value of nonzero or zero of the continuous variables respectively, like the vapor flowrate of certain tray or the heat duty of certain exchanger. The central difference method is used to discretize the problem into nonlinear program (NLP) in this work. The NLP model is solved using the CONOPT solver in GAMS environment with the random initial values and perturbation strategy to increase global search ability.

## Case Studies

In following sections, two synthesis case studies involving binary and multicomponent separation process are presented.

***Example 1***

The case of binary mixture separation of benzene and toluene is studied. In this example, the setup time and hold up of the trays are neglected, and we expect to synthesize the optimal batch distillation process to satisfy the separation requirements. The given data is presented in Table 1.

**Table 1.** Given data for example 1

| | |
|---|---|
| Column Specifications | |
| Upper bounds on the number of stages, $N_T$ | 11 |
| Upper bounds on the vapor boilup rate, $V_i^{out}$ *(kmol/h)* | 2 |
| Operation pressure, *P (kPa)* | 100 |
| Available annual production time, *Hrs (h/year)* | 8000 |
| | |
| Utility available | |
| Cooling water: 300K *($/GJ)* | 0.7 |
| Steam: 421K *($/GJ)* | 3 |
| | |
| Initial charge | |
| Feed, $B_b$ *(kmol)* | 8 |
| Feed composition, Benzene/Toluene $x_{b,1}/x_{b,2}$ *(mole fraction)* | 0.58/0.42 |
| | |
| Product specification | |
| Lower bounds on average Product purity of benzene $x_D$ *(mole fraction)* | 0.90 |
| Product yield of benzene per batch, *D (kmol)* | 4 |
| Top product price $P_{r1}$ *($/kmol)* | 34 |
| Bottom product price $P_{r2}$ *($/kmol)* | 34 |
| Raw material cost $C_0$ *($/kmol)* | 2 |

***Case 1 permanent configuration model***

First, permanent configuration model is employed, which means that the configuration of batch distillation column is unchangeable in total time horizon. The configuration, batch time, vapor flow, and distillate rate of the batch distillation system are optimized simultaneously to maximize the profit (See Table 2).

The result shows that under this scenario, the profit realized by regular column is about 61% more than that by inverted column, and this is mainly because the batch time of the regular column is much shorter than that of the inverted column, as a result the product revenue achieved by the regular column is greater than that of the inverted column.

**Table 2.** Summary of the result for permanent configuration model

| Configuration | $t_b$ <br> *h* | *D* <br> *kmol/b* | *N*t | *OC* <br> *$/h* | *CC* <br> *$/h* | *P* <br> *$/h* |
|---|---|---|---|---|---|---|
| *regular* | 5 | 4 | 11 | 0.213 | 0.486 | 50.501 |
| *inverse* | 8 | 4 | 11 | 0.214 | 0.488 | 31.298 |

***Case 2 changeable configuration model***

Then the changeable configuration model, meaning that the configuration of batch distillation column can optimally change in each time interval, is adopted to illustrate its advantage. In this case, we consider the condition of fixed product demand and strict product quality specifications to avoid unlimited production of the products (Mujtaba and Macchietto, 2004). We assume that the batch time and the amount of distillate per batch are fixed with 6 hours and 4 kmol respectively, so the total revenue is fixed and the objective function is transformed into minimizing the total annual cost (TAC). The results are represented in figure 2 and Table 3.

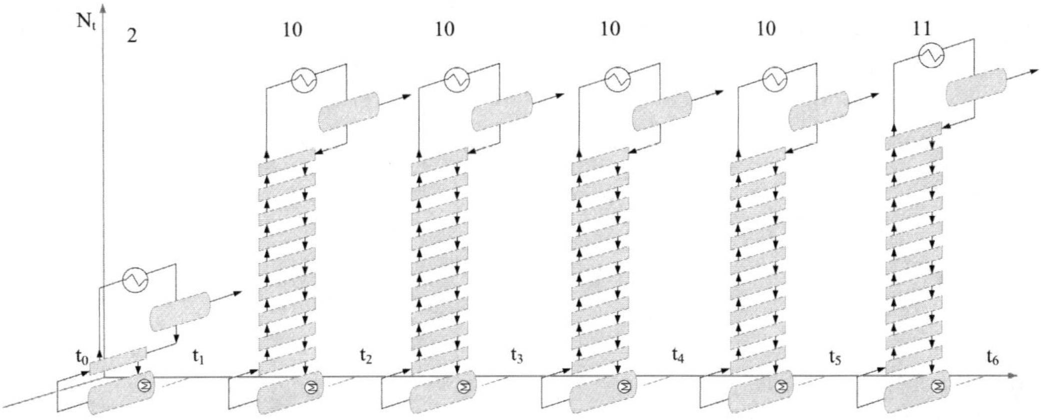


**Figure 2.** Optimal batch distillation process of changeable configuration model

**Table 3.** Summary of the result for changeable configuration model

| Configuration | $t_b$ *h* | *D* *kmol/b* | *N*t Each time interval | *AOC* *$/yr* | *ACC* *$/yr* | *TAC* *$/yr* |
|---|---|---|---|---|---|---|
| permanent | 6 | 4 | 11,11,11,11,11,11 | 1522.7 | 3854.7 | 5377.3 |
| changeable | 6 | 4 | 2,10,10,10,10,11 | 1408 | 3816 | 5224 |

The result shows that the number of trays is gradually increasing with the processing time in the changeable configuration model, which is in response to the increasing difficulty of the separation task as the process goes on. So there is a multistage optimization of both capital and operating costs in every time interval instead of compromising the operation strategy due to the fixed design of configuration in entire batch time. Thus, a 3% lower TAC is achieved in the changeable configuration model.

***Example 2***

In this section, the synthesis of optimal batch distillation process is investigated for the case of ternary mixture separation of benzene, toluene and ethylbenzene. In this example, the reflux of both condenser and reboiler is given changes to connect to arbitrary trays in different time intervals in order to adjust the liquid/vapor ratio on each equilibrium stage in real time. We expect to fully exploit the performance of the batch distillation system to synthesize an optimal process to maximize the amount of distillate of a specified concentration for a specified batch time. The given data is presented in Table 4.

Figure 3 shows that the liquid/vapor ratio on each equilibrium stage is optimized in each time interval to achieve the maximum amount of distillate. And the maximum amounts of each product are $D_{benzene}=4kmol$, $D_{toluene}=3.9kmol$ respectively, and the composition profile of each component in liquid holdup in condenser is represented in Figure 4.

**Table 4.** Given data for example 2

| | |
|---|---|
| Column Specifications | |
| Number of stages, $N_T$ | 11 |
| Upper bounds on the vapor boilup rate, $V_i^{out}$ *(kmol/h)* | 2 |
| Operation pressure, *P (kPa)* | 100 |
| Tray holdup, *M(kmol)* | 0.3 |
| Condenser holdup, $M_1$*(kmol)* | 0.5 |
| Processing time, $Hrs\,(h)$ | 50 |
| Initial charge | |
| Feed, $B_b$ *(kmol)* | 8 |
| Feed composition, Benzene/Toluene/Ethylbenzene $x_{b,1}/x_{b,2}/x_{b,3}$ *(mole fraction)* | 0.4/0.4/0.2 |
| Product specification | |
| Lower bounds on average product purity of Benzene/Toluene $x_{D,1}/x_{D,2}$ *(mole fraction)* | 0.95/0.80 |

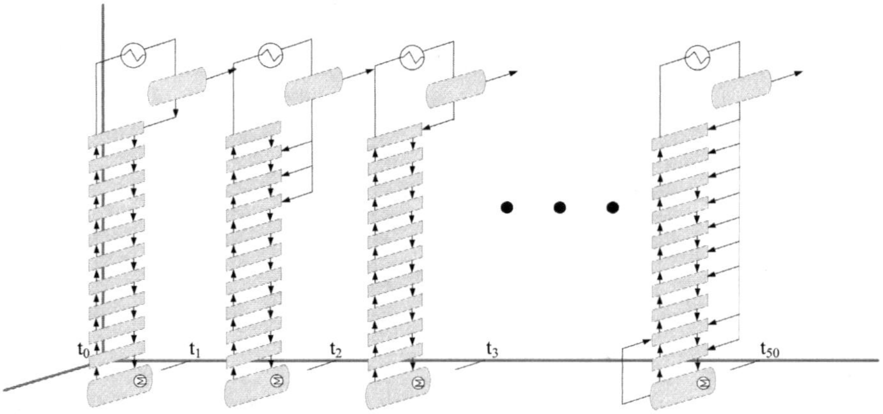


**Figure 3.** Optimal batch distillation process of ternary components separation task

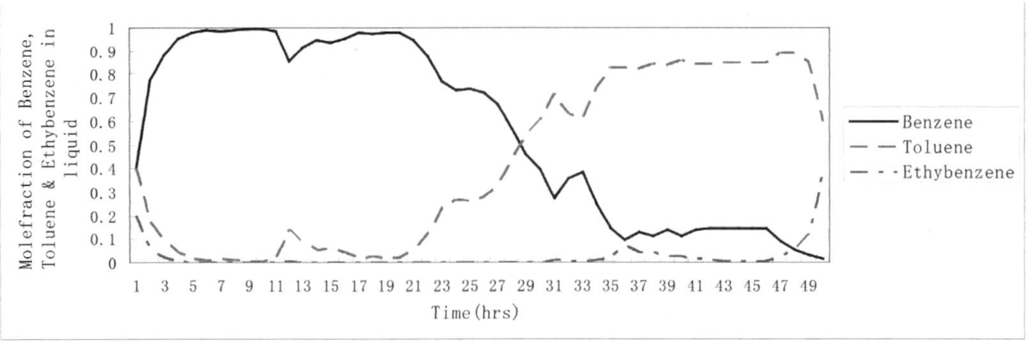


**Figure 4.**Composition profile in liquid holdup in condenser

## Conclusion

In this work, a systematic method based on mixed integer dynamic optimization (MIDO) formulation for synthesizing the batch distillation process on both space and time dimension is proposed. An improved superstructure based on state space representation is introduced to encompass all potential optimal process.

The case studies indicate that under this framework, optimal configuration, design and operating policies of the batch distillation system are solved simultaneously. Moreover, the advantage of changeable configuration model has been highlighted. From the comparison analysis of the example results, not only energy saving but also capital investment reduction has been brought in our novel conceptual designs. Additionally, a novel operating strategy to fully exploit and adjust the performance of batch distillation system in real time is represented. Due to complexity of considering all potential optimal design and operating strategy of multicomponent separation, more work on optimizing multicomponent batch distillation system, including both design and operating strategy, will be implemented in the future.

## Acknowledgment

This work is supported by the National Natural Science Foundation of China, under Grant no.21276029.

## References

1. Logsdon JS, Diwekar UM, Biegler LT. On the simultaneous optimal design and operation of batch distillation columns. *Transactions of IChemE*. 1990; 68: 434-444.
2. Mujtaba IM, Macchietto S. Simultaneous optimization of design and operation of multicomponent batch distillation column-single and multiple separation duties. *J of Process Control*. 1996; 6: 27-36.
3. Sharif M, Shah N, Pantelides CC. On the design of multicomponent batch distillation columns. *Comp & Chem Eng*. 1998; S22: 69-76.
4. Low KH, Sørensen E. Simultaneous optimal design and operation of multivessel batch distillation. *AIChE J*. 2003; 49: 2564-2576.
5. Oldenburg J, Marquardt W, Heinz D, Leineweber DB. Mixed-logic dynamic optimization applied to batch distillation process design. *AIChE J*. 2003; 49: 2900-2917.
6. Low KH, Sørensen E. Simultaneous optimal configuration, design and operation of batch distillation. *AIChE J*. 2005; 51: 1700-1713.
7. Bagajewicz M, Manousiouthakis V. Mass/heat-exchange network representation of distillation networks. *AIChE J*. 1992; 38: 1769-1800.
8. Zou X, Cui YH, Dong HG, Wang JQ, Grossmann IE. Optimal design of complex distillation system for multicomponent zeotropic separations. *Chem Eng Sci*. 2012; 75: 133-143.
9. Seider W, Seader J, Lewin D. *Product and process design principles - synthesis, analysis and evaluation*. 2nd ed. New York: Wiley; 2004.
10. Miladi MM, Mujtaba IM. Optimization of design and operation policies of binary batch distillation with fixed product demand. *Comp & Chem Eng*. 2004; 28: 2377-2390.

# (534f) Design and Synthesis of Energy Integrated Multipurpose Batch Plants Using a Robust Scheduling Platform

**Esmael Reshid Seid** and Thokozani Majozi, Chemical Engineering, University of Pretoria, Pretoria, South Africa

In recent years batch plants have attracted more attention due to their flexibility to produce different products using the same facility. Batch manufacturing is typically used in the pharmaceutical, polymer, food and speciality chemical industries, because it provides the necessary flexibility to accommodate various production requirements using the same processing facility. However, these very important features introduce extra degrees of complexity into the design, synthesis and operation of such plants. Utility requirements in these plants can be a major contributor to overall running costs. Heating and cooling are unavoidable aspects of many chemical processing facilities, with operations where heat is generated and others where heat is required. It is because of this occurrence that heat integration becomes a possibility for reducing utility requirements. Energy savings have often been neglected in the design and synthesis of batch processes in the past and so large percentage savings are possible.

The contribution in this paper is developing a comprehensive optimization technique for design and synthesis of heat integrated batch plants. In batch chemical processes it is highly imperative to optimize the schedule before design. The recent scheduling technique for multipurpose batch plants by Seid and Majozi (2012) is used as a platform. Computational studies are presented to illustrate the effectiveness of the proposed model. The result obtained shows that the utility design should be part of the same model for economic saving.

# (534g) Global Optimization of Nonconvex MINLP Problems by Domain and Image Partitioning with Applications to Heat Exchanger Networks

**Sung Young Kim**, Debora Faria and Miguel Bagajewicz, Chemical Engineering, University of Oklahoma, Norman, OK

Global Optimization of Nonconvex MINLP Problems by Domain and Image Partitioning with Applications to Heat Exchanger Networks

Sung Young Kim, Debora Faria and *Miguel Bagajewicz

*School of Chemical Engineering and Material Science, University of Oklahoma*

*100 East Boyd Street, T-335 – Norman, OK 73019-0628 USA*

** Corresponding Author*

Abstract

We propose a new method to obtain the global optimum of full nonconvex MINLP problems. The method is based on partitioning the domain and image of nonconvex functions. The procedure we propose uses an MILP lower bound constructed using domain/image partitioning. Then a newly developed bound contraction procedure is applied and compared to branch and bound as well as branch and bound with bound contraction at each node. To illustrate the method we focus on the heat exchanger network stage-wise model. Results show a robust behaviour where the solution does not need initial values as local MINLP model like Dicopt need. In the presentation we will compare the performance of all options.

# (546a) An Integrated Methodology for Design of Tailor-Made Blended Products: Biofuels and Bio-Based Lubricants

**Nor Alafiza Yunus**, CAPEC, Department of Chemical and Biochemical Engineering, Technical University of Denmark, Kongens Lyngby, Denmark, Krist V. Gernaey, Department of Chemical and Biochemical Engineering, Technical University of Denmark (DTU), Kgs. Lyngby, Denmark, John M. Woodley, Department of Chemical and Biochemical Engineering, Technical University of Denmark, Lyngby, Denmark and Rafiqul Gani, CAPEC, Department of Chemical and Biochemical Engineering, Technical University of Denmark (DTU), Kgs. Lyngby, Denmark

The demand for bio-based products has been growing in recent years due to increasing concerns about the environment, resource depletion, and/or sustainability. Bio-based products can be designed by blending of conventional materials (e.g. fossil fuels) with other chemicals produced from renewable resources, such as biomass, reducing thereby the fossil fuel consumption and amount of pollutants released into the environment, and increasing the product safety. A tailor-made bio-based blended product is a mixture of several chemicals and designed to match specific product attributes. The most challenging task and critical step in tailor-made bio-product design is to identify and select the appropriate chemicals to be used as building blocks for the blend. At the initial stage, searching for the blend chemicals (additives) is uncertain and involves a very large search space. Therefore, it is important to have a method to systematically and efficiently reduce the search space within which the solution to the blending problem may be found. Once the candidate chemicals have been identified, their optimal compositions in the blend are determined, while still matching the target properties.

In this study, an integrated methodology for design of tailor-made bio-based blended products is presented. The methodology is able to find the composition of optimal blended bio-based products, and consists of two stages; The first stage is product design, where a computer-aided methodology is implemented to identify and evaluate the most promising blend candidates. Then, in the second stage, the blend attributes are validated experimentally or with rigorous simulation models. The first stage consists of four tasks. The first task is to define the design problem, where the product needs are identified, translated into target properties and given target values. In the second task, the required target property models are retrieved from a property model library developed specifically for bio-based blended products. The property model library stores the pure and mixture property models with their parameters. Also, a chemical database, which is one of the important tools in this integrated methodology, has been developed. This database stores the pure compounds which originate from renewable and non-renewable sources with their properties that may be used as building blocks in the blend design. In the third task, the blend problem is solved using a blend design algorithm to find the blends that best match the design targets. In this way, the amount of bio-based chemicals in the blend can be controlled and adjusted. This algorithm employs a decomposition based solution strategy to screen for the feasible blend candidates, thus reducing the search space in a systematic way. The result is a set of blends that match the desired targets, the compositions, values of the target properties and information about their miscibility. Finally, the mixture target property values are verified by rigorous models for the properties and mixtures that require it. Here, collection of data in the database helps in verifying the predicted blend properties.

The application of this systematic methodology is highlighted through case studies related to the design of sustainable and better performing blended gasoline and lubricants with bio-based chemicals. The objective of the case studies is to identify suitable bio-based chemicals as well as the blended products where they can be used.

# (546b) Process Synthesis and Technoeconomic Evaluation of a Catalytic Biomass-to-Fuels Strategy

**Sercan Murat Sen**[1], David Martin Alonso[1], Stephanie G. Wettstein[1], Elif I. Gurbuz[1], Carlos A. Henao[2], James A. Dumesic[1] and Christos T. Maravelias[3], (1)Chemical and Biological Engineering, University of Wisconsin-Madison, Madison, WI, (2)Chemical and Biological Engineering, University of Wisconsin - Madison, Madison, WI, (3)Department of Chemical and Biological Engineering; DOE Great Lakes Bioenergy Research Center, University of Wisconsin-Madison, Madison, WI

Lignocellulosic biomass, an alternative for petroleum, is a sustainable source of carbon for producing chemicals and fuels in an environmentally friendly way. Conversion of lignocellulosic biomass to valuable intermediates is a critical step to develop effective biomass-to-biofuel strategies. Levulinic acid (LA) is one of these platform chemicals that can be produced from lignocellulosic biomass and transformed into liquid fuels, fuel additives and even other specialty chemicals. In this respect, we developed an LA-based catalytic strategy, in which an alkylphenol solvent is used for extraction of intermediates from the sulfuric acid solution, to convert lignocellulosic biomass into liquid hydrocarbon fuels. Then, we compared the alkylphenol strategy with a previously reported butyl acetate strategy in terms of cost-effectiveness.

In the butyl acetate strategy, first, the hemicellulose fraction of the biomass is solubilized and removed using dilute acid pretreatment. Pretreated biomass, which contains cellulose and lignin, is sent to a cellulose deconstruction unit, in which cellulose is converted to levulinic acid and formic acid with 55% yield in the presence of sulfuric acid, whereas the remaining cellulose is degraded to humins as a result of undesired polymerization reactions. After cellulose deconstruction, solid residues (lignin and humins) are filtered from the product stream, then mixed with the hemicellulose fraction that was removed by dilute acid pretreatment, and sent to a boiler/turbogenerator to produce heat and electricity covering the utility requirements of the process. Excess electricity is sold to the grid. Then, LA is converted to $\gamma$-valerolactone (GVL) over a RuRe/C catalyst in the presence of sulfuric acid. GVL is extracted from the sulfuric acid and GVL aqueous solution using butyl acetate solvent and sulfuric acid is recycled back to the cellulose deconstruction reactor. Finally, GVL is separated from butyl acetate by distillation and purified GVL is converted to butene and to butene oligomers.

The potential shortcoming of the butyl acetate strategy is the use of an expensive RuRe/C catalyst to prevent the negative effects of sulfuric acid in the catalytic processing. To address this shortcoming, an alkylphenol (2-sec-butyl phenol (SBP)) solvent is used for extraction of LA, which is produced in low concentration, from sulfuric acid solution. This strategy also allows for the aqueous phase containing sulfuric acid to be recycled back to the cellulose deconstruction reactor efficiently. Another advantage of alkylphenols is their high boiling points, so that extraction at elevated temperatures, i.e. at cellulose deconstruction temperature, is possible and leads to potential energy and equipment cost savings. The LA obtained from pretreated biomass is selectively reduced to GVL (while the FA is decomposed to $H_2$ and $CO_2$) over a RuSn/C catalyst in the presence of SBP solvent without hydrogenating the solvent. After the GVL production, SBP and GVL can be separated by distillation recovering GVL as a top product. The GVL would then be processed to liquid fuels. This sulfuric acid management strategy provides downstream catalytic processing at high yields in the absence of sulfuric acid and with a cheaper catalyst to convert LA into GVL.

After developing a model for the alkylphenol-based separation strategy, we investigated the economic impact of integrating this strategy with the original process. To determine the economic potential of these strategies, we carried out detailed process simulation studies (based on experimental results for all associated reactions) and capital/operational cost calculations. After optimizing the process configurations of the alkylphenol strategy, the minimum selling price (MSP) for butene oligomers is around $4.70 per gallon of gasoline equivalents (GGE). Also, we identified that the MSP is sensitive to changes in the catalyst lifetime. For a catalyst lifetime of less than 6 months, the alkylphenol strategy results in better MSP than that can be obtained from the butyl acetate strategy. Finally, a comparison of the alkylphenol strategy with a lignocellulosic ethanol production facility, which has an MSP equal to $5.13/GGE, reveals the attractiveness of this LA-based catalytic approach.

# (546c) Use of Octanol-Water Partition Coefficient and Toxicity Prediction in Design of Ionic Liquids for *in Situ* Fermentation Extraction

**John White**[1], Brock C. Roughton[2], Kyle V. Camarda[1] and Aaron M. Scurto[1], (1)Chemical and Petroleum Engineering, University of Kansas, Lawrence, KS, (2)Bioengineering Graduate Program, University of Kansas, Lawrence, KS

Fermentation is widely used to produce fine and bulk chemicals Due to toxicity of the product towards the producing microorganisms, overall product concentrations are often dilute. After fermentation, downstream separation is required which often has high energy inputs. *In situ* product removal with an immiscible solvent, such as certain ionic liquids (ILs),offers a mechanism by which product could be continually removed during production, reducing the effects of toxicity and product inhibition while also decreasing downstream separation demands. The IL solvent should have good partitioning of the product, but must have low toxicity to the microorganism to allow in situ processing. Environmental and health concerns should be addressed to ensure sustainability. Ionic liquids, consisting of cations and anions, are highly tunable towards fermentation extractions. For instance, they can be designed with hydrophobicity, low to no vapor pressure, low toxicity, etc. Hundreds of ionic liquids have been synthesized with up to $10^{14}$ possible combinations. The use of molecular descriptors within property prediction models provides an efficient method of searching the possible chemical space and selecting/designing candidate ionic liquids for a given extraction.

Predictive models for octanol-water partition coefficient and toxicity for various model organisms (*Escherichia coli*, *Vibrio Fischeri*, and rat) are presented. Experimental data collected from previously published literature and presented here have been used to develop the models. Ionic liquid descriptors were calculated using E-Dragon (http://www.vcclab.org/lab/edragon/). Descriptor selection was performed using an exhaustive search to find descriptors that provided a good fit to the experimental data (high $R^2$) while avoiding overfitting (represented by Mallow's $C_p$ statistic). Using the selected descriptors, linear regression was used to generate the predictive property models. Leave-one-out cross-validation (LOOCV) was used to evaluate the predictive power of the models. Once the models were finalized, they were used in a computational molecular design (CMD) framework to optimally design ionic liquids for a given extraction. The optimization problem was formulated as a mixed-integer nonlinear program (MINLP) where a molecule structure is determined which minimizes the difference between target properties and the predicted properties of the solution molecule. A stochastic method, Tabu search, was used to identify locally optimal solutions to the optimization problem. The overall framework has been used previously by the authors in the design of ionic liquids for use in refrigeration. The target properties included low octanol-water partition coefficient, low *E. coli* toxicity, low *V. Fischeri* toxicity (indicative of aqueous environmental safety) and low rat toxicity (indicative of human toxicity). The results indicate that several designed ionic liquid candidates have desirable properties for solvent extraction from fermentation reactions. Further property prediction tools will allow more refined design of ionic liquids for more specified fermentation applications.

# (546d) Bio-Based Gasoline Additives: Product and Chemistry Selection Through Network Generation and Optimization

**W. Alex Marvin**, Srinivas Rangarajan and Prodromos Daoutidis, Chemical Engineering and Materials Science, University of Minnesota, Minneapolis, MN

Several biomass-derived oxygenates, such as butanol and dimethylfuran (DMF), have been proposed as alternatives to ethanol for blending into current gasoline fuels. Each of these oxygenates can be synthesized from bio-based platform chemicals such as 5-hydroxymethylfurfural (HMF), acetone, ethanol, etc. in multiple ways. Consequently, there are several options for choosing the product (gasoline additive) and the synthesis route. We propose a strategy for optimal product and process chemistry selection by combining rule-based reaction network generation and mixed-integer linear programming (MILP).

Our strategy is composed of two steps. First, we use a reaction network generation tool, RING [1], developed in our group to construct a network of all possible reactions that are involved in converting bio-based platform compounds to potential gasoline additives. RING takes in as input reaction rules that determine the chemistries being explored. In this case, we include a broad spectrum of heterogeneous acid, base, and metal catalyzed chemistries as reaction rules.

Second, we formulate an MILP to determine the optimal vector of steady-state fluxes through the reaction network that minimizes catalyst use and reactor capital, and corresponds to an oxygenate blend that satisfies the requirements for gasoline additives. The objective function incorporates estimated reaction kinetics for each reaction based on the corresponding reaction rule. Due to the detailed chemistry of the reaction network being embedded into the MILP, the optimal vector determines (1) an oxygenate blend compatible with gasoline and (2) synthesis routes to produce each oxygenate from bio-based platform chemicals. Alternative synthesis routes and oxygenate blends are identified that have competitive objective functions.

This two-step strategy proposed here is generic and scalable for any product-cum-chemistry selection problem. The set of reaction rules, rate estimation, objective function and constraints formulation, and thermochemical analysis of the optimal synthesis routes will be presented and discussed.

[1] Rangarajan, S.; Bhan, A.; Daoutidis, P. Rule-based generation of thermochemical routes to biomass conversion, *Industrial & Engineering Chemistry Research* **2010**, *49* (21), 10459-10470

# (546e) On the Systematic Synthesis Screening and Integration of Real-Life Biorefineries

Marinella Tsakalova, Athanassios Nikolakopoulos and Antonis Kokossis, School of Chemical Engineering, National Technical University of Athens, Athens, Greece

Biomass technologies have been inundated with new ideas and innovations the industrial world and the labs posing new challenges to chemical engineering. Individual bio-based paths, production of a single chemical product or fuel have gradually taken a significant position in markets. Given the plethora of degrees of freedom related to different feedstocks, paths and portfolio of products, the area seems to be chaotic. Opposing to the vast amount of options and possibilities, systems technology has on offer powerful methods to optimize designs and systematize improvements of biomass processing systems. Systematic screening leads to a holistic approach of the problem far from the applications that focus on individual processes or plants.

This paper introduces a methodology to address the systematic integration of biorefineries supporting decisions for selection and the integration of processing paths. The implementation takes the form of a synthesis model that is generalized with process and intermediates as appropriate to the case. For the estimation of cost the method uses regression models based on the thermodynamics of the process. The investment cost is quickly and roughly calculated based on energy losses resulting from the difference of the Lower Heating Value (LHV) of the inlet and outlet flows of a process.

The methodology is applied to 82 different chemistries that lead to the production of more than 80 intermediate and final products. The biomass is decomposed to three main streams of C5 sugars, C6 sugars and lignin. Those intermediates are the precursors for bio-based products and fuels such as polyamides, PVC, polypropylene, ethanol, aromatic polyols, resins, vanillin etc. that can be their petrochemical counterparts. The model produces a rank of products identifying the optimal integrated flows. The optimization model results to an optimal solution of coproduction of xylonic acid polyamide and lignin based castor oil. An impressive observation coming out from the results is that high market price of the product do not offer tops in the ranking lists. Several case studies are presented additionally to illustrate the wide framework of the methodology. The case studies concern the effect of the market demand and prices, the impact of the supply chain – transport cost, risk factors and LCA aspects.

# Integrated Pathway Optimization for Biorefineries

Pascual Eduardo Murillo-Alvarado[a], Ismael Leonardo Raso[a], José María Ponce-Ortega[a], Medardo Serna-González[a], and Mahmoud El-Halwagi[b]

[a] *Chemical Engineering Department, Universidad Michoacana de San Nicolás de Hidalgo, Morelia, Mich., 58030, México.*

[b] *Chemical Engineering Department, Texas A&M University, 3122 College Station, TX, USA, 77843-3122.*

## Abstract

This work proposes a new systematic approach to find the optimal pathway of a biorefinery considering simultaneous the selection of the products, feedstocks and processing steps, the model is based on a disjunctive programming formulation. The proposed model also allows selecting several products and feedstocks in the optimal solution. Also is possible considering the simultaneous maximization of the net profit and the minimization of the greenhouse gas emissions. The applicability of the proposed methodology is shown through a case study for the optimal pathway of a biorefinery and several scenarios have been analyzed. Results show that the combination of different feedstocks and products allows compensating simultaneously the economic and environmental objectives.

## Introduction

Along of the years the main energy sources are the petrochemical fuels; however, the depletion of the proved oil reserves as well as the significant quantities of greenhouse gas (GHG) emissions associated to the use of these fossil fuels have promoted the searching of new and cleaner energy sources. For this reason, biofuels offer the potential to solve this problem. The biomass processing is performing in an industrial facility called biorefinery, where biomass is processed through physical, chemical or/and biological technologies to obtain biofuels in addition to other products as can be seen in **Figure 1**. However, several factors must be considered for the installation of a biorefinery like the selection of the feedstocks, the pre-processing stages, processing technologies and the manufactured products in addition to biofuels for any specific case. Different types of feedstocks can be considered, and the processing routes depend on the selected biomass because of the constituted chemical components is different, which may include lignocellulose, sugars, starch and lipids. According to the content of the main components is possible to determine the optimal processing routes for obtaining the desired products.

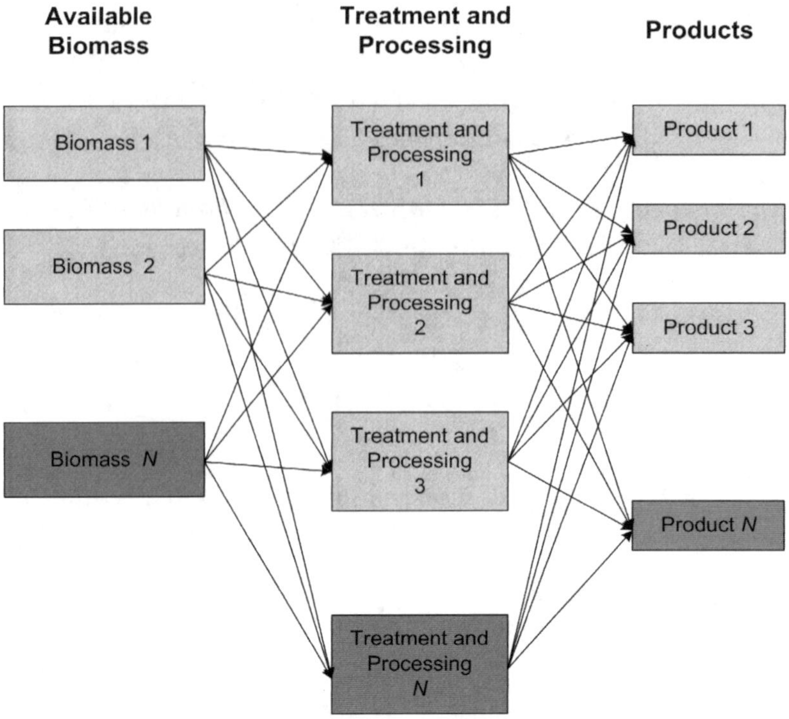


**Figure 1.** Biorefinery configurations.

In this context several works have been reported to highlight the importance of biorefineries; for example, Huber et al. (2006), Saxena et al. (2008, 2009) and Goyal et al. (2008) presented reviews of the processing of the biomass to obtain several products. However, these works did not present systematic procedures to obtain the optimal route of biomass processing. Recently, Pham and El-Halwagi, (2011) and Ponce-Ortega et al. (2012) presented a systematic approach and a formal optimization model based on disjunctive programming for the automatic optimization to determine the optimal pathway for a biorefinery configuration, respectively. Previous works did not consider the optimal selection of the type of feedstock and product, and the possibility to manufacture several types of products and feedstoks. For this reason, this paper proposes new mathematical programming model to take into account the selection of several feedstocks and products according with the availability and demand in the optimization of the pathway for a biorefinery. The proposed model considers simultaneously maximizing the total net profit and minimizing the associated net GHG emissions in the entire life cycle.

## Superstructure Formulation

The problem here is to identify the interconnections between feedstocks and products through the available chemical routes. The approach first consists in the forward synthesis of biomass, where all potential intermediate products and their corresponding conversion technologies are identified. The second stage consists in backward generated product; where for a given final product, the required preceding compounds as well as their corresponding technologies are determined. The final step for the superstructure generation is the interception of two intermediate compounds obtained in the forward and backward steps identifying the set of conversion technologies required to connect these compounds. Using this approach, a superstructure like the one shown in **Figure 2** is obtained for each specific problem (i.e., given available feedstocks and desired products).

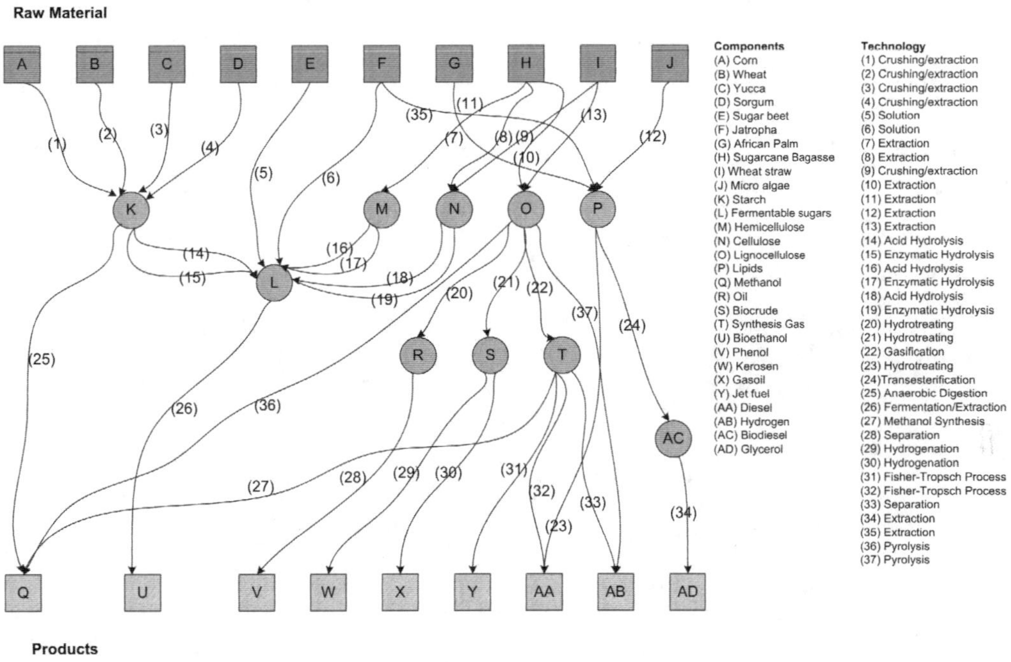


**Figure 2**. Example for the options of biorefinary pathway configuration.

Notice in previous figure that there are several possibilities for the biorefinery pathway, and that is not obvious to identify the pathway with the lowest cost, lowest GHG emissions and minimum processing steps; in addition, the solution with the minimum cost may not correspond to the solution with the lowest environmental impact or the minimum GHG emissions (and remember that this is one of the purposes for the use of biofuels).

## Optimization formulation

Based on the generated superstructure, first the following scripts are defined, $i$ represents any compound preceding the compound $j$ in the superstructure. To yield any compound $j$, several preceding $i$ can be used. The script $t$ is used to represent any chemical route that connects any pair of compounds.

The following logical relationship states that when the compound $j$ is selected to appear in the optimal pathway, then one technology $t_j$ to produce it must be used. This is modeled using binary variables as follows:

$$y_j = \sum_{t_j} z_{j,t_j}, \; \forall j \in J, j \neq feedstock$$

When one technology $t_j$ is used to produce $j$ from $i$, then there is required an additional constraint to state that the preceding compound $i$ must exist and this is modeled as follows:

$$\left(1 - z_{j,t_j}\right) + y_i \geq 1, \; \forall j \in J, j \neq feedstock, \forall t_j \in T_j$$

When the chemical route $t_j$ is selected, then this technology has associated a conversion factor $\alpha_{i,t_j}\left(\mathbf{O}_{t_j}\right)$ to yield $j$, and the corresponding cost $Cost_j$ and environmental impact $EI_j$ have to be determined properly. On the other hand, when the component $j$ is not present in the optimal pathway, then the associated flowrate, cost and environmental impact must be zero. This situation is modeled through the following disjunctive formulation:

$$\begin{bmatrix} Y_j \\ M_{f_j}^{\min} \leq f_j \leq M_{f_j}^{\max} \\ Cost_j = M_{Cost_j} \\ EI_j = M_{EI_j} \end{bmatrix} \vee \begin{bmatrix} \neg Y_j \\ f_j = 0 \\ Cost_j = 0 \\ EI_j = 0 \end{bmatrix}, \; \forall j \in J, j = feedstock$$

$$\bigvee_{t_j \in T_j} \begin{bmatrix} Z_{j,t_j} \\ f_j = \alpha_{i,t_j}\left(\mathbf{O}_{t_j}\right) f_i, \; i \text{ preceds } j \text{ through } t_j \\ Cost_j = c_{t_j}\left(\mathbf{O}_{t_j}\right) f_j \\ EI_j = ei_j\left(\mathbf{O}_{t_j}\right) f_j \end{bmatrix}, \; \forall j \in J, j \neq feedstock$$

Previous disjunction can be reformulated as an algebraic problem as follows. First the Boolean variables $Y_j$ and $Z_{j,t_j}$ are reformulated as binary variables $y_j$ and $z_{j,t_j}$. This means that when the Boolean variable is true, the associated binary variable must be one and the equations inside the disjunction must be applied; on the on the hand, when the Boolean variable is false, the associated binary variable must be zero and the relationships inside do not apply. Therefore, the objective function consists in the simultaneous maximization of the total profit and the minimization of the environmental impact taking into account the number of processing steps. This objective function is formulated as follows:

$$TotCost = H_Y \sum_{j \in J} Cost_j \qquad Sales = H_Y \sum_{j \in j_{products}} \text{Price}_j \times f_j$$

$$Profit = Sales - TotCost$$

The total environmental impact for the pathway is calculated as follows:

$$TotEI = \sum_{j \in J} EI_j$$

**Results**

To show the applicability of the proposed model, a case study is used to determine the optimal configuration of a biorefinery attending the economical and environmental aspects and considering the availability of feedstocks and demands of products of Mexico. The available feedstocks, unit costs and unit GHG emissions considered in this case study are showed in **Table 1**, which represents the most important.

**Table 1.** Data for the feedstocks for the case study.

| Raw Material | Available (Kg/Day) | Cost (USD/kg) | Gas Emission (Kg) |
|---|---|---|---|
| Corn | 80 000 000 | 0.0878 | -0.3835 |
| Wheat | 35 000 000 | 0.0912 | -0.4165 |
| Yucca | 1 000 000 | 0.0827 | -0.2465 |
| Sorghum | 25 000 000 | 0.0914 | -0.2191 |
| Sugar beet | 1 500 000 | 0.096 | -0.2164 |
| Jatropha | 7 500 000 | 0.0931 | -0.4273 |
| African Palm | 1 800 000 | 0.08205 | -0.3561 |
| Sugarcane Bagasse | 27 500 000 | 0.0907 | -0.4082 |
| Wheat Straw | 2 000 000 | 0.0769 | -0.2794 |
| Micro Algae | 1 250 000 | 0.0825 | -0.1878 |

The model considers the demand of products, price of sale and GHG emissions generated; which is shown in **Table 2**.

**Table 2**. Data for the products for the case study.

| Products | Demand (Kg/Day) | Cost Sale (USD/kg) | Emission (kg $CO_2$/kg product) |
|---|---|---|---|
| Methanol | 90 000 | 0.87 | 3.239 |
| Bioethanol | 150 000 | 0.523 | 1.375 |
| Phenol | 45 000 | 0.594 | 2.095 |
| Kerosene | 43 000 | 0.634 | 2.808 |
| Gasoil | 40 000 | 0.692 | 2.289 |
| Jet Fuel | 120 000 | 0.769 | 2.458 |
| Biodiesel | 175 000 | 0.79 | 2.346 |
| Hydrogen | 30 000 | 0.818 | 3.239 |
| Glycerol | 15 000 | 0.668 | 0.1 |

This case consists of 158 continuous variables, 66 binary variables and 409 constraints; and this model was coded in the GAMS software where the solver CPLEX was used to solve it. The optimal solution selects only one product (biodiesel), where 546,190 kg of micro algae per day to produce 175,000 kg of diesel per day are required. The optimal solution has a net profit of US$17,638,614.74/year. The associated net GHG emissions are $1.8139 \times 10^8$ kg $CO_2$ eq./year that represent the $CO_2$ captured in the growth of the micro algae, the $CO_2$ produced in the processing steps and the $CO_2$ produced in the combustion of the biodiesel.

For the case when is needed to satisfy the bioethanol demand (150,000 kg/day; this is called the scenario B), the optimal values for the net profit and the associated GHG emissions are 1,170,988.55 US$/year and 2.732x108 kg/year, respectively. **Figure 3** shows the Pareto curve for the objectives Profit versus GHG emissions for this scenario.

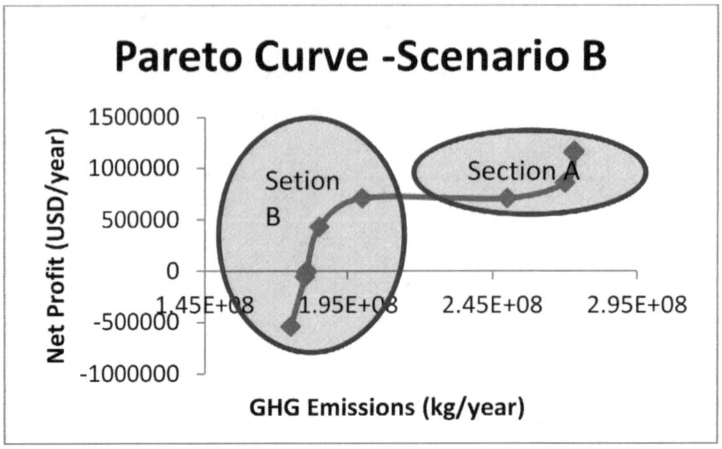


**Figure 3**. Pareto curve for Profit versus GHG emissions for scenario B.

The corresponding optimal pathway configuration is shown in **Figure 4**, where two sections are identified, one corresponding to high profits and high GHG emissions (section A) and the other one corresponding to negative profits and low GHG emissions (section B).

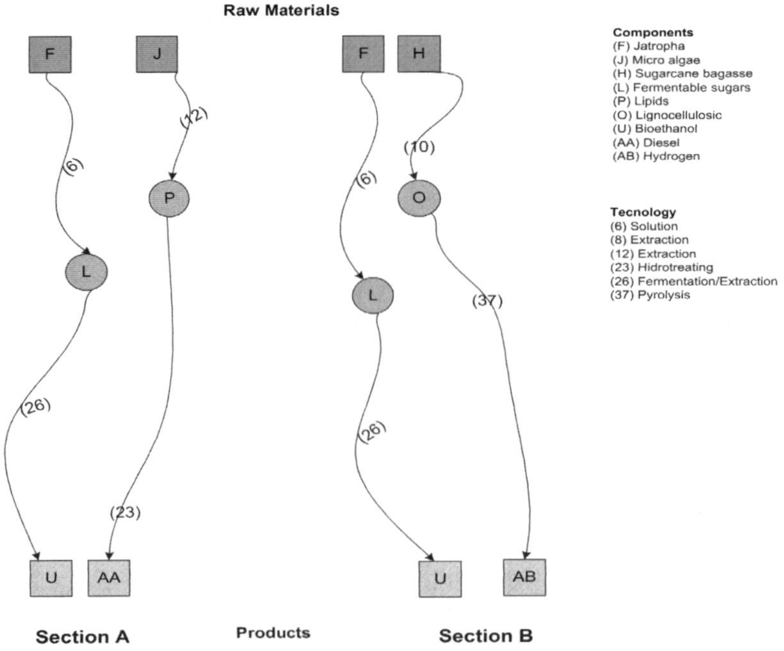


**Figure 4**. Optimal pathway for section A and B for scenario B.

## Conclusions

This paper presents a systematic approach for the optimal pathway configuration for a biorefinery based on a distinctive programming formulation. The proposed model takes into account the simultaneous selection of the raw materials, products and processing steps. Also, the proposed model considers the simultaneous maximization of the net profit and the minimization for the associated GHG emissions taking into account the processing steps. The results obtained are presented through Pareto curves to identify the tradeoffs of the considered objectives. In addition, the proposed approach allows selecting several feedstocks and manufactured products. Finally, the proposed approach is applied to a case study of Mexico, and several scenarios have been identified to account the tradeoffs of the proposed objectives.

## References


- Goyal H.B.; Seal D.; Saxena R.C. Bio-fuels from thermochemical conversion of renewable resources: A review. *Renewable and Sustainable Energy Reviews*. **2008**, 12 (2), 504-517.
- Huber G.W.; Iborra S.; Corma A. Synthesis of transportation fuels from biomass: Chemistry, catalysts, and engineering. *Chemical Reviews*. **2006**, 106 (9), 4044-4098.
- Pham V.; El-Halwagi M.M. Process synthesis and optimization of biorefinery configurations. *AIChE Journal*, **2011**, 58 (4), 1212-1221..
- Ponce-Ortega J. M.; Pham V.; El-Halwagi M.M.; El-Baz A. A disjunctive programming formulation for the optimal design of biorefinery configurations. *Industrial and Engineering Chemistry Research*, **2012**, 51(8): 3381-3400.
- Saxena, R.C.; Adhikari, D.K.; Goyal, H.B. Biomass-based energy fuel through biochemical routes: A review. *Renewable and Sustainable Energy Reviews*. **2009**, 13(1), 167-178.
- Saxena, R.C.; Seal, D.; Kumar, S.; Goyal, H.B. Thermo-chemical routes for hydrogen rich gas from biomass: A review. *Renewable and Sustainable Energy Reviews*. **2008**, 12 (3), 1909-1927.

# (546g) Optimization of Biochemical Conversion Process of Lignocellulosic Biomass to Sugars by Integrating the Kinetic Models

Vasanth Natarajan, Department of Chemical and Biomolecular Engineering, National University of Singapore, Singapore, Singapore and Iftekar A. Karimi, Dept of Chemical & Biomolecular Engineering, National University of Singapore, Singapore, Singapore

Over the years, interest in new ways of producing biofuels is increasing rapidly in response to rising prices of gasoline, depletion of fossil fuels and the non-renewable nature of crude oil. Bioethanol, the first generation biofuel is a highly established renewable fuel today which comes from sugar or grain (starch). The production of second generation biofuel from cheaper, abundant lignocellulosic feedstocks (LCF) emerged as a potential alternative feedstock to highly competitive food based raw materials. For lignocellulosic biofuels to become a reality, research in the development of cost and energy effective biochemical conversion of biomass to sugars are highly needed. With a lot of improvements in the experimental studies of various pretreatment, saccharification and fermentation processes, a significant improvement from a modeling and optimization perspective is required.

In the literature, studies have been done on kinetic modeling [1], statistical and non-statistical techniques [2, 3] for process design and optimization. Until the present, most of the modeling based works focus on the optimization of size reduction process [4], pretreament process, saccharification and fermentation individually. For predicting the behaviour of the entire conversion system it is essential to integrate existing models for efficient process optimization [5]. In this work, we propose a model based optimization approach for the biochemical conversion of biomass to sugars by considering the overall conversion process. Our major objective of this work is to minimize the amount of energy required for the conversion process by determining the optimal design variables. Here, we integrate the kinetic models of dilute acid pretreatment, enzymatic hydrolysis and fermentation process. Also, we consider the physical (size reduction) and chemical aspects (diffusion models) of the biomass for a more detailed study of the above system. The above problem has been formulated in GAMS including the mass balance equations, kinetic equations, and the equations involved in size reduction and diffusion processes.

Finally, we present the results based on the minimal amount of energy consumed by the entire conversion process of lignocellulosic biomass to sugars. We have considered cornstover as our raw material due to its vast availability of experimental and theoretical data. This study shows the need for sophisticated models that considers the integrated models of the conversion process and also, the physical and chemical aspects of the biomass which is the novelty and the contribution of this work.

1. Esteghlalian, A., et al., *Modeling and optimization of the dilute-sulfuric-acid pretreatment of corn stover, poplar and switchgrass.* Bioresource Technology, 1997. 59(2-3): p. 129-136.

2. Ramirez, M.T. and E.S. Fraga, *Mathematical Modelling of Feed Pretreatment for Bioethanol Production*, 2009. p. 1299-1304.

3. Vargas Betancur, G.J. and N. Pereira Jr, *Sugar cane bagasse as feedstock for second generation ethanol production. Part I: Diluted acid pretreatment optimization.* Electronic Journal of Biotechnology, 2010. 13(3): p. 1-9.

4. Sayed Ali Hosseini, R.L., Sergei Kucherenko, and Nilay Shah, *Multiscale Moldeling of Hydrothermal Pretreatment: From Hemicellulose Hydrolysis to Biomass Size Optimization.* Energy and Fuels, 2010. 24: p. 8.

5. Ziyu Wang, J.X., and Jay J. Cheng, *Modeling Biochemical Conversion of Lignocellulosic Materials for Sugar Production: A Review.* BioResources, 2011. 6(4): p. 25.

# (659a) Simplified Life Cycle Analysis As a Route to Fast Decision Making in Product and Process Development for Sustainability

**Santiago Faucher**, Technologies, Hatch, Mississauga, ON, Canada

The accounting of cradle-to-grave life cycle inventories for specific products often requires a substantial investment in time and resources. For this reason, few products are sold with an accompanying inventory of their life cycle impact on the environment. This lack of a baseline and the onerous effort required to obtain it can often lead to confusion in the development of new processes or products designed to be more sustainable. The new product or process is often designed based on the inventor's baseline knowledge and perception of what is more sustainable. After the new product or process has been shown to achieve the desired performance attributes, a full life cycle inventory of the new product and the baseline product is then commissioned as it is now seen as more justified. Ideally, the life cycle impact of the new product developed is shown to be lower than the original product it is intended to replace.

This approach can be successful if the inventor understands intimately the effect of both material and process selection on the life cycle inventory and can predict the results of a life cycle assessment study. However, in most cases the products and processes are complicated and comprised of many parts making such estimations difficult. Thus the risk of designing a new product that is less sustainable than the baseline product it is intended to replace can be high. To address this, simplified life cycle analyses can be applied in the product/process design phase. While this approach does not map out the usage phase of the product, it can capture the substantial differences that process and material selection may have on the life cycle impact of the new product. With the use of simplified life cycle analysis prior to experimental product or process development, the life cycle impacts of several new hypothetical design options can be compared without investing considerable time in experimental development and later full life cycle analyses. This approach increases the speed and efficiency of product and process development for sustainability.

# (659b) Process Development for a Higher Quality and Cost Optimized Monomer

Ahmed A. Youssef, SABIC, Innovative Plastics, Mt. Vernon, IN and Mohan Khadilkar, SABIC Innovative Plastics, Mt. Vernon, IN

This study addresses a process improvement that involves the development of a new treatment system for a purge (tar) stream resulting as a by-product in the condensation reaction of phenol with a ketone to produce a bisphenol monomer at one of SABIC's sites. The conventional tar recovery section is cumbersome and includes a multiplicity of vacuum distillation units.

This work discusses the installation of an acid catalyzed cracking process where the monomer is mainly converted to its original constituent (phenol) and the impurities are separated out.

Such improvement yields higher final product purity as a result of a decrease in the total impurities recycled into the main system. It also presents a cost effective process driven by the amelioration in the raw material usage.

The transformation of the recovery section empowers the manufacturing unit at SABIC (Innovative Plastics business) to operate with a reduction in the purge stream by ~ 30%.

# (659c) Carbon Footprint Analysis of Dairy Feed From a Mill in Michigan, USA

**Felix Adom**[1], Greg Thoma[2] and David R. Shonnard[1], (1)Chemical Engineering, Michigan Technological University & Sustainable Futures Institute, Houghton, MI, (2)Department of Chemical Engineering, University of Arkansas, Fayetteville, AR

A carbon footprint analysis was conducted for a single dairy feed mill located in Michigan USA with the aim of developing a preliminary assessment of dairy feed mill operations. The goal was to develop life cycle assessment methodology applicable to the animal feed mill industry and also to gain understanding of the relative importance of milled dairy feed inputs on the greenhouse gas (GHG) emissions of the outputs of the mill. In the final analysis, GHGs emitted from feed mill operations on the basis of one kilogram of dairy feed output from the mill (kg $CO_2$ eq $kg^{-1}$ of milled dairy feed) was estimated. Scenarios were investigated to consider the prevalence of different feed ingredient inputs likely to represent different U.S. mill locations.

The functional unit used for this study was 1 kg of milled dairy feed at its exit moisture content (an average feed formulation for dairy animal nutrition). Inputs and activities identified in this cradle-to-mill-to-dairy farm analysis included; feed ingredients, onsite energy, and transportation of feed inputs to the milling site and transportation mill output to dairy farms. Sources of Life cycle inventory (LCI) input data include; peer reviewed journal articles, purchase history documents of the milling facility, internet searches (ISI, Google scholar, ProQuest, etc), and a site visit. Using Ecoinvent™ unit processes in SimaPro, an open IO model and other peer reviewed articles, the cradle-to-mill-to-dairy farm (receiving) gate analysis was estimated using two allocation methods (economic and mass).

Feed mill GHG emissions were estimated to be 0.62 and 0.93 kg $CO_2$ eq (equivalent) $kg^{-1}$ of milled dairy feed for economic and mass allocation, respectively. The highest emissions were due to the feed ingredient inputs which contributed 73-82% towards the carbon footprint depending on the allocation method. Energy and transportation impacts together contributed between 8-12%. Scenario analyses were include to simulate different mill locations where dominant feed inputs change compared to the MI location, for example to a prevalence of dry distiller grains and solids for mills located in Iowa and a prevalence of oats for a mill located in North Dakota.

# (659d) Urban Mining: A Process and Technology Review

Daniel Brosig, Hatch Ltd., Mississauga, ON, Canada

The recovery of non-ferrous metals (e.g., copper, aluminum, lead, nickel and precious metals) from urban waste can be economically attractive and represents a more sustainable alternative to the extraction of metals from naturally-occurring deposits. A brief overview of the economic and environmental performance behind the metal recovery processes from urban waste is provided to shed some light on the nature of the challenge and opportunity. Although in principle it could appear to be a simple concept, in practice the mining of "urban ore" is accompanied by a complex host of process and technology problems that span from logistical issues around the separation of the target waste to metallurgical recovery inefficiencies. Currently practiced urban mining processes and the associated state-of-the-art technological solutions are reviewed in this presentation. A conceptual roadmap is presented to explore the development of the processes and technologies that may one day aid in making urban mining ubiquitous around the world.

# (659e) Sizing of a Packed Reactive Distillation Column for Isoamyl Butyrate Production Using an Equilibrium Model

Jairo A. Duran[1], Karen A. Cañon-Rubio[1], Alejandro Lopez Gomez[2], Alvaro Orjuela[1] and Gerardo Rodriguez[3], (1)Department of Chemical and Environmental Engineering, Universidad Nacional de Colombia, Bogota, Colombia, (2)Departament of Chemical and Environmental Engineering, Universidad Nacional de Colombia, Bogota, Colombia, (3)Departamento de Ingeniería Química y Ambiental, Universidad Nacional de Colombia, Bogotá, Colombia

**Sizing of a Packed Reactive Distillation Column for Isoamyl Butyrate Production Using an Equilibrium Model**

*Jairo Duran, Karen Cañon, Alejandro López, Álvaro Orjuela*

Grupo de Procesos Químicos y Bioquímicos. Department of Chemical and Environmental Engineering. Universidad Nacional de Colombia. Bogotá. Colombia

Reactive distillation has been considered as an alternative for sustainable process in the production of green chemicals because their capabilities of reaching high conversion in equilibrium reactions, flexibility and relatively low costs. Several studies have pointed the importance of taking into account high-quality thermodynamic and kinetic parameters in order to obtain reliable estimations on the performance of the integrated reaction-separation operation. Among different biobased esters, isoamyl butyrate that is used as green solvent and food additive can be obtained from renewable raw materials. For instance, fermentation broths from fuel ethanol production contain up to 2.5% (volume basis) of amyl alcohols which are currently blended with fuel ethanol or used as fuel for steam boilers. As a potential upgrading alternative, these alcohols can be esterified with different acids using a reactive distillation scheme to improve reaction conversions and energy savings with respect to the traditional esterification processes.

In this work, kinetic and phase equilibrium parameters from experimental data were used to simulate and size a reactive distillation column. NRTL and Hayden O'Connell models from regressed experimental data were used in phase equilibrium calculations. Kinetics was measured in a batch reactor using Amberlyst® 35 as catalyst. A full factorial design was established to carry out the kinetic tests. Experimental data were fitted adequately with a pseudo-homogeneous model. In addition, autocatalytic kinetics was measured using an identical methodology. The module RADFRAC of Aspen Plus® was used in modeling of reactive distillation. A basis of 200.000 kg per year of isoamyl butyrate (food grade, >99% weight) was considered in sizing and sensibility analyses. Amberlyst® 35 contained in KATAPAK SP-11® was considered in the simulation. MELLAPAK-250Y® was used in non catalytic packing sections, where autocatalytic kinetics was considered. Column height was calculated assuming an HETP of 0.5 meters, as reported by several authors.

According with simulation results a conversion of butyric acid higher than 96% was possible. Nevertheless, including a pre-reactor in the scheme, a conversion higher than 99.1% was observed. Best performance of operation was found when using a mole excess around 1.5 of isoamyl alcohol to butyric acid in the system.

**Keywords:** reactive distillation, esterification, catalytic packing, ion exchange resin, flooding factor.

# (659f) Analysis of Alternative Processes for Anhydrous Ethanol Production, an Economic and Environmental Approach

Michel Kahwaji Janho[1], Jorge E. Gatica[2], María Rosa Hernández[3], Fernando Daniel Mele[4] and Mauricio Colombo[4], (1)Chemical And Biomedical Engineering, Cleveland State University, Cleveland, OH, (2)Chemical and Biomedical Engineering, Cleveland State University, Cleveland, OH, (3)Chemistry and Chemical Engineering, Universidad Nacional de Tucuman, Tucuman, Argentina, (4)Ingenieria de Procesos y Gestion Industrial, Universidad Nacional de Tucuman, Tucuman, Argentina

The production of sugar is one of the oldest agro-industrial activities. It includes the production of sugar cane, sugar and dairy products: paper, alcohol, energy, etc. In addition to the production of sugar for domestic and international markets, one of the activities that is growing in importance is the production of alcohol from molasses. Energetic integration has also resulted in bringing power generation from bagasse as a secondary activity in modern sugar-cane processing industries.

A new chapter for bio-energy products associated with the manufacture of sugar has gained relevance as blending gasoline with ethanol is becoming part of energetic policies in most industrialized countries. Indeed, energy products from renewable sources are receiving increasing attention, and this has generated a renewed stimulus to examine alternative processes for bio-ethanol production. It has become common for sugar mills to have complementary distilleries with production capacities ranging from 50 to 400 liters per day of non-anhydrous alcohol.

At sugar plants, steam is generated in boilers using as fuel bagasse-cellulosic fiber waste obtained crushing the sugar cane- and it is used to give useful heat and shaft work to the plant, but it can also be used to generate electricity with export opportunities to the electrical network. The great number of process alternatives outlines a serious decision making problem in order to take advantage of the resources.

At steady state conditions the amount of bagasse obtained is more than enough to supply the demands of useful heat and energy. Thus these industries can advantage of the emergence of a new potential market, as they have an increased potential for generating a surplus of electricity for sale to the grid.

In this paper, a classic bio-ethanol distillery is adopted as case-study and simulated using AspenPlus® and UniSim ® commercial simulators. Alternative process flow sheets of varying complexity are analyzed aiming to reduce power consumption and effluent generation.

The process simulation allows a systematic analysis that leads to a good combination of all the values of process variables. In particular, since they demand significant energy influx, an important subsystem for analyzing is the separation of alcohol-water mixtures, There are also different technologies for this dehydration. In this paper we consider azeotropic distillation and hybrid schemes with membrane separation. The development of a simulation module for calculating the membranes performance, typically not included in commercial process simulators,

The consumed electric energy and steam depends on the adopted scheme and they have, depending to their source or generation policy, an associated cost and environmental impact. Among the energy systems that deserve special attention, cogeneration in the sugar industry must be pointed out. This part of the process makes efficient use of a common fuel for the generation of heat and power. As a result, the environmental impact and costs associated with the use of electricity and useful heat change appreciably, such as anhydrous alcohol production schemes can be achieved with a lesser environmental at the expense of decreasing the economic benefit which would result from selling electricity to the grid.

Cogeneration is not new in the sugar-cane processing industry. Indeed, typically co-generation refers to the generation of heat as the main objective and the generation of electrical and mechanical power as secondary objectives. To date, however, no versatile tool that would allow analyzing economical feasible alternatives of anhydrous alcohol production, bearing in mind environmental issues, is available.

In addition, the explicit consideration of environmental issues offers an alternative route to explore trade-offs. Life Cycle Analysis (LCA) is used in this work. Unlike most environmental approaches to design that focus on reducing the generation of residuals, but without taking into account the impact associated to related processes, LCA allows considering environmental impact as an integral part of the problem. The software used to evaluate the environmental impact is SimaPro®.

Alternative schemes that meet the same specifications of purity and recovery in effluent streams, are examined in this paper. Of particular relevance is the assessment of costs and environmental impact of these processes. A baseline case is defined in our case-study, and structural and parametric changes are made. Different scenarios with respect to economic and environmental issues taking into account the specifications of the final product. Near-optimal scenarios are formulated by defining an operating window for some of environmentally friendly technologies available. The impact of different operating policies for the production of anhydrous alcohol on the environment is examined in detail.

The membrane area, the number of plates in the distillation units, the heat exchange area, and the pressure manipulators are considered as structural variables. Among the most relevant operational variables considered one can mention reflux ratio, pressure ratio, as well as compositions and flow rates of streams.

# (712a) Examining Correlations of Electrostatic Discharge Sensitivity and Electrical Conductivity of Composite Energetic Materials

**Eric Collins**, Texas Tech University, Lubbock, TX, Chelsea Weir, Mechanical Engineering, Texas Tech University, Lubbock, TX and Michelle Pantoya, Mechanical Engineering Department, Texas Tech University, Lubbock, TX

Composite energetic material response to electrical stimuli was investigated and a correlation between the composite's electrical conductivity and ignition sensitivity was examined. The composites consisted of micrometer particle aluminum combined with another metal, metal oxide, or fluoropolymer. Of the nine tested mixtures, aluminum and copper oxide was the only mixture to ignite by electrostatic discharge (ESD) with minimum ignition energy of 25 mJ and an electrical conductivity of 3.45 nS. In an ESD event, energy is delivered through the spark and through joule heating of the particles. For this reason, thermal ignition was also examined using a hot wire. This study showed a similar trend in minimum ignition energy for ignition triggered by a discharged spark compared with ignition triggered by a thermal hot wire source. However, a discharged spark concentrates energy locally supplying greater power for ignition. This study begins to explore critical parameters influencing electrostatic discharge ignition, specifically electrical resistivity and conductivity of particulate mixtures.

# NANOSCALE ALUMINUM, SILICON AND ALUMINUM/SILICON MIXTURE FORMATION FOR USE IN PYROTECHNIC APPLICATIONS

*Mark Miller[1], Zac Doorenbos[2], Matt Puszynski[2], Lauren Armstrong[3], Deepak Kapoor[3] and Jan Puszynski[1,2]*

*[1]Chemical and Biological Engineering Department, South Dakota School of Mines and Technology, Rapid City, SD 57701*
*[2]Innovative Materials and Processes, LLC, Rapid City, SD 57702*
*[3]Armament Research, Development and Engineering Center, Picatinny Arsenal, NJ 07806*

## ABSTRACT

Over the past two decades, work in the area of energetic, pyrotechnic and explosive materials has focused on the use of nanoscale materials to tune energetic outputs and combustion characteristics [1-3]. This work presents a method for the formation of nanoscale aluminum, silicon, and aluminum/silicon composite particulates. The effect of attrition milling parameters (e.g. mill configuration, ratio of media to material, time, tip speed, and type of solvent) on the specific surface area and a reactive metal content of different fuel materials is presented. The combustion characteristics of the different fuel materials were tested as a part of nanothermite composite materials and compared to traditional nanothermite materials derived from commercially available aluminum and silicon nanopowders.

## INTRODUCTION

It has been shown in the literature that the addition of metal nanoparticles to energetics can increase the burn rates and enhance detonation velocity [4]. The combustion enthalpy (-7.4 kcal/g) and high specific surface area (>20 $m^2/g$) of nano-scale aluminum has made this a material of choice as an additive or fuel for energetic material formulations. Silicon is another nanoscale metal that is of interest as an additive. Silicon has a similar energy density compared to aluminum. Nanoscale silicon led to similar results as an additive in energetic material as nanoscale aluminum, but it is less susceptible to aging and has a higher ignition temperature. These characteristics may provide safer and more reliable energetic composites derived from this fuel.

A major limiting factor for the uses of these nanoscale additives in energetic, pyrotechnic and explosive materials is their high cost. The ability to produce nanoscale aluminum and silicon powders at a significantly lower cost can be accomplished using a wet attrition milling process. This milling process allows for the specific surface area of the nanoscale materials to be tuned in the range between 5-26 $m^2/g$ and 21-106 $m^2/g$ for aluminum and silicon, respectively. This same technology can be used for the formation of composite mixtures of aluminum and silicon with specific surface areas ranging between 35-50 $m^2/g$. The stoichiometric ratio of aluminum and silicon in these fuel composites might be adjusted to the desired application. This paper presents the effect of attrition milling parameters on key physical characteristics of aluminum, silicon and aluminum/silicon nanocomposite particulates.

## EXPERIMENTAL

### *Materials*

The following materials were used for the formation of nanoscale aluminum, silicon and aluminum/silicon mixture:

- 0.4 mm zirconium grinding media (Union Process)
- Aluminum flake (Alfa Aesar: 11 micron)
- Silicon metal powder (AEE: 1-5 micron)
- Decane (Sigma-Aldrich: anhydrous >99%)
- Heptane (Fisher Scientific: reagent grade)
- Toluene (Fisher Scientific: reagent grade)
- Isopropyl alcohol (IPA) (Fisher Scientific: anhydrous)

### *Wet Attrition Milling*

Wet attrition milling was conducted by first weighing out a desired amount of zirconia grinding media. The amount of a material to be milled was calculated based on the desired weight ratio of grinding media to a material (15:1-50:1). The material was added to the bottom of the milling container, followed by the addition of grinding media. The desired amount of solvent to cover all grinding media and the powder was then added to the milling container and thoroughly premixed by hand. The milling container was then assembled and placed onto the attrition mill. The picture of the wet attrition mill model 01-HDDM purchased from the Union Process company and located at South Dakota School of Mines and Technology (SDSM&T) is shown in Figure 1. Cooling water was turned on to allow flow through the cooling jacket of the milling container. The material was then milled at the selected operating conditions (mill speed and time). The rheology of the slurry within the milling container was observed and small amounts of solvent were added as needed.

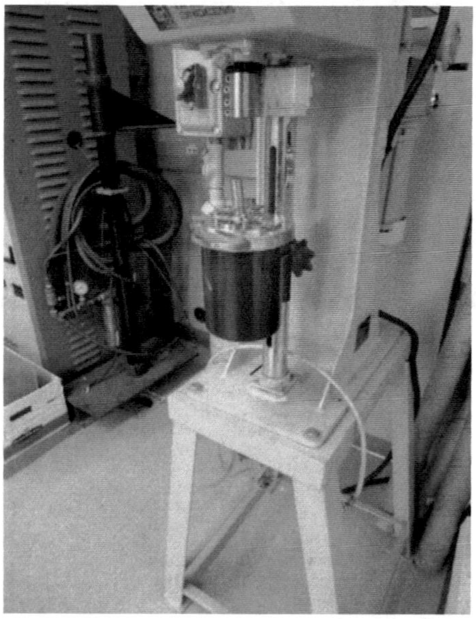

***Figure 1:*** *A photograph showing the wet attrition mill model 01-HDDM purchased from the Union Process company and located at SDSM&T.*

### *Recovery of Wet Milled Material*

In order to recover the milled material, the slurry containing the milling media and material was poured onto a 300 micron (50 mesh) sieve tray with a collection tray for milled powder and the solvent underneath. The grinding media was washed with the fresh solvent until it was clean. Next, a roto-vap was used to remove the solvent. Pior to the exposure of the milled material to ambient air, oxygen passivation process was conducted. The temperature of the material in the roto-vap was reduced to ambient conditions and a small amount of air was bled under vacuum for an extended period of time. This process is dependent on the quantity of the material to be passivated and has to be carefully optimized to prevent its self-heating and possible ignition. The temperature of the passivated nanoparticulate material should be monitored and the intake of oxygen immediately stopped when the recorded temperature rise is more than 5°C. Recovered material from the roto-vap was dried in a vacuum oven for 2 hours at 100°C to insure that all solvent was removed. Next, the material was sieved through 300 micron (50 mesh) sieve tray and collected into an air tight container.

### *Characterization of Milled Material*

Specific surface area (SSA) of the milled materials was determined using BET (model GEMINI III). The reactive content of aluminum nanoflakes was measured using a volumetric reactive aluminum testing setup developed at SDSM&T.

## RESULTS

### *The Effect of Grinding Media Concentration and Attrition Milling Time*

Increasing the weight ratio of grinding media to the milled material has resulted in increased specific surface area (SSA) of milled powders over the same period of time. Figure 2 presents the SSA of silicon milled for 2, 4, and 6 hours of milling in isopropyl alcohol (IPA) at three different grinding media concentrations. The results presented in Figure 3 indicate that the SSA of attrition milled aluminum follows a similar trend as silicon (Figure 2)

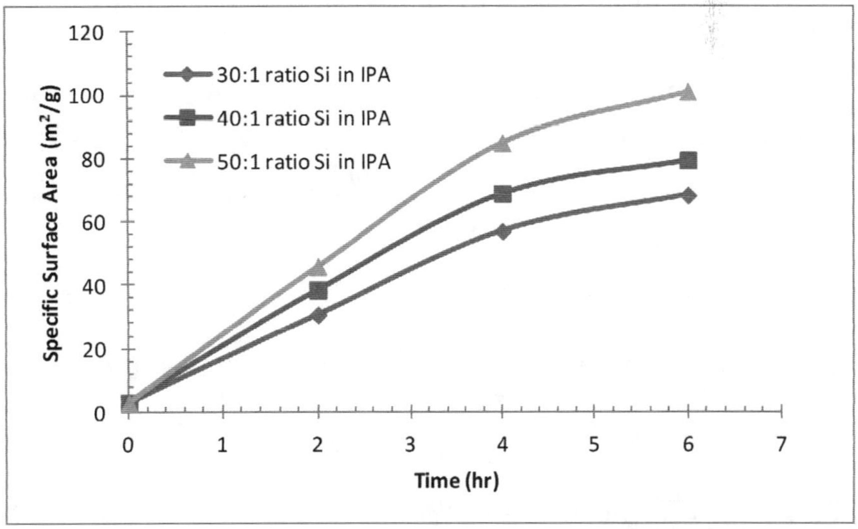


***Figure 2:*** *The effect of grinding media concentration on SSA of attrition milled silicon in IPA as a function of time.*

with increased milling time. When decane was used as a solvent, a slight decrease in SSA was observed compared to IPA solvent. This could be attributed to a partial reactivity of aluminum with this alcohol.

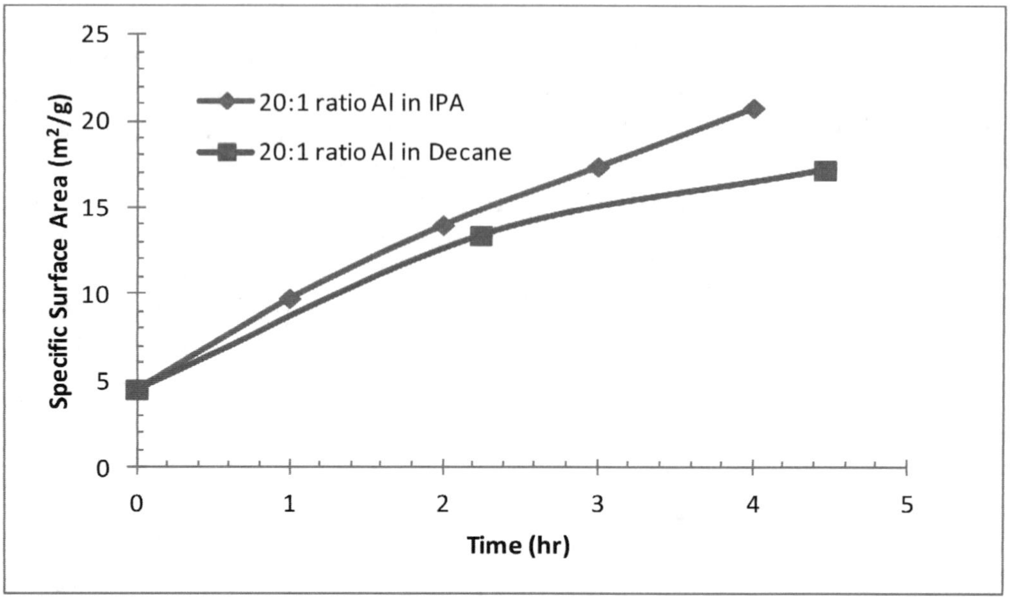


***Figure 3:*** *The effect of attrition milling time on SSA of aluminum for two different solvents (decane and IPA) with a constant grinding media to milled powder ratio.*

### *Effect of Solvents*

Chemical properties of a solvent used in an attritor have a significant effect on powder dispersion and rheology and therefore on milling characteristics. This is especially evident for the milling of silicon, as indicated in Table 1.

***Table 1:*** *SSA results for milled nanoscale silicon using three different solvents with milling parameters.*

| Experiment # | Material | Grinding media to material (g:g) | Time (hrs) | Solvent | SSA ($m^2/g$) |
|---|---|---|---|---|---|
| 1 | AEE 1-5 micron Si | 30:1 | 4 | Heptane | 22.4 |
| 2 | AEE 1-5 micron Si | 30:1 | 4 | Toluene | 33.1 |
| 3 | AEE 1-5 micron Si | 30:1 | 4 | IPA | 43.6 |

The effect of a solvent type was not as significant for the formation of the nanoscale aluminum as shown in Table 2. The reactive content of aluminum nanoflakes was in the range of 84-90 wt% for powders with SSA between 15-17 $m^2/g$.

***Table 2:*** *SSA results for milled nanoscale flake aluminum using three different solvent with constant milling parameters.*

| Experiment # | Material | Grinding media to material (g:g) | Time (hrs) | Solvent | SSA ($m^2/g$) |
|---|---|---|---|---|---|
| 1 | Flake Al | 20:1 | 3.5 | Heptane | 17.6 |
| 2 | Flake Al | 20:1 | 3.5 | Decane | 16.6 |
| 3 | Flake Al | 20:1 | 3.5 | IPA | 15.4 |

### *Morphology of Attrition Milled Aluminum, Silicon and Al/Si Nanocomposites*

SEM images of milled aluminum, silicon, and aluminum/silicon composites were distinctly different from each other. A SEM photograph of the milled aluminum with a SSA of 26 $m^2/g$ formed flakes, which is shown in Figure 4.

***Figure 4:*** *SEM image of milled aluminum flake with SSA of 26 $m^2/g$.*

***Figure 5:*** *SEM image of milled silicon powder with SSA of 79.4 $m^2/g$.*

***Figure 6:*** *SEM image of the milled Al/Si composite with SSA of 48.5 $m^2/g$.*

The milled nanoflake aluminum looked very similar to the starting micron-size aluminum, but had a much higher SSA of 26 $m^2/g$ compared to the micron-size aluminum having SSA of 4.5 $m^2/g$. The milled silicon formed agglomerates consisting of very fine nanoparticles. SEM image of the milled silicon with SSA of 79.4 $m^2/g$ is shown in Figure 5. The milled aluminum/silicon composite appeared to be a combination of both previously described SEM image morphologies. SEM image of the milled aluminum/silicon composite with a SSA of 48.5 $m^2/g$ is shown in Figure 6.

### ***Dynamic Combustion Pressures of Nanothermites Derived from Attrition Milled Aluminum and Silicon/Aluminum Nanoparticulates***

The different milled materials were tested in nanothermite composite materials in a 2.5 mL steel vessel after being flushed with argon. The nanothermite composite materials were then compared with nanothermite materials derived from commercial fuel nanopowders using 80 nm spherical aluminum from NovaCentrix. The dynamic pressure curves are shown in Figure 7. It should be noted that attrition milled silicon and commercially available nanoscale silicon did not ignite under specific ignition conditions set in this study [5].

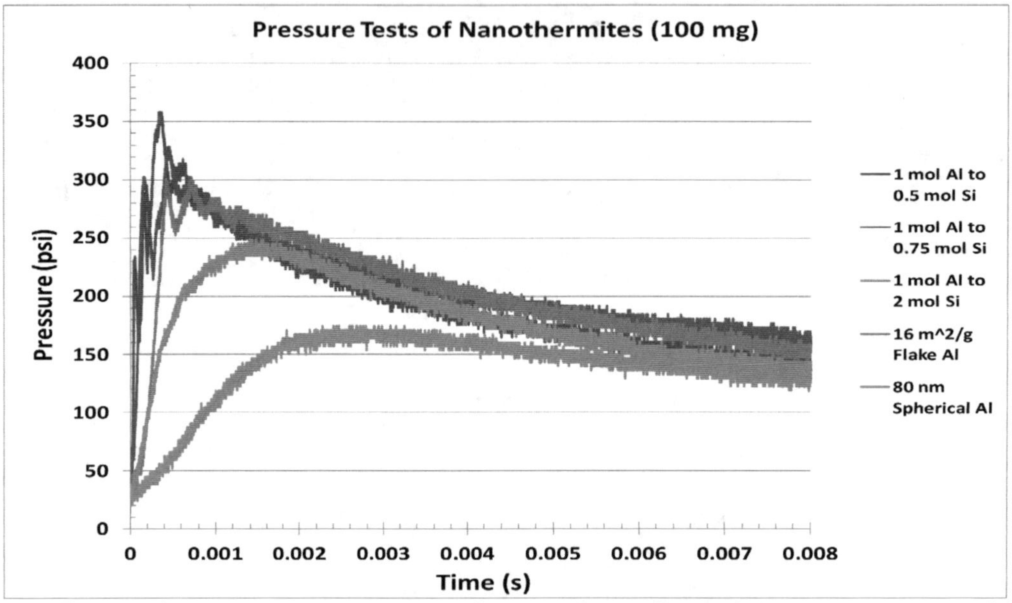


***Figure 7:*** *Dynamic pressure curves of nanothermite mixtures as an oxidizer and nanoscale fuels consisting of bismuth trioxide under current study normalized to 100 mg. Tests were performed in a 2.5 mL steel vessel after being flushed with argon [5].*

## CONCLUSIONS

It was demonstrated in this study that the use of an attrition milling technique can lead to the formation of nanoscale aluminum with a comparable SSA and reactive Al content to 80 nm spherical aluminum (NovaCentrix). The ratio of grinding media to material has a significant effect on the time needed to form fuel materials with a higher SSA. As the ratio of grinding media to milled material increases, shorter milling times are needed to obtain the nanoparticulates with the same SSA. Solvent properties have shown a large effect on silicon attrition milling. The effect of a solvent type on aluminum was much less noticeable, but the reason of using decane compared to IPA or heptane is its much lower vapor pressure. Decane is a much less volatile solvent allowing the viscosity during the milling to be more consistent. Also by using a hydrocarbon instead of an alcohol prevents the possible reaction of aluminum or silicon with the OH groups. Attrition milling processing allows for a predictable way of producing and tuning nanoscale materials for their use in pyrotechnic applications.

## ACKNOWLEDGEMENTS

The authors acknowledge the support from the ARDEC at Picatinny Arsenal, NJ (Contract No. W15QKN-06-D-0006) and the Army Research Laboratory at Aberdeen Proving Grounds (Contract No. W911NF-08-2-0022).

## REFERENCES

1. Son S.F. Yetter R.A. and Yang Y., J. Propulsion and Power, 23(4), 643-644, (2007)
2. Puszynski J.A., Fulian C.J., and Swiatkiewicz J.J., J. Propulsion and Power, 23(4), 698-706, (2007).
3. Piercey D.G., and Klapotke T.M., Central European Journal of Energetic Materials, 7(2), 115-129, (2010)
4. Brousseau P., Dorsett H.E., and Cliff M.W.; 12th International Detonation Symposium, Aug. 11-16, 2002.
5. Doorenbos Z., Walters I., Kapoor D., Balas-Hummers W., and Puszynski J.A., Journal of Energetic Materials and Chemical Propulsion, (submitted 2012)

# (712c) On Mechanism of Thermal Initiation in Fully-Dense 2Al.3CuO Nanocomposite Powder

**Rayon Williams**, Department of Chemical, Biological and Pharmaceutical Engineering, New Jersey Institute of Technology, Newark, NJ, Mirko Schoenitz, NJIT, Newark, NJ, Alexandre Ermoline, Chemical, Pharmaceutical, and Biological Engineering, New Jersey Institute of Technology, Newark, NJ and Edward L. Dreizin, Otto H. York Department of Chemical, Biological, and Pharmaceutical Engineering, New Jersey Institute of Technology, Newark, NJ

Several composite 2Al·3CuO thermite powders with CuO inclusions of different dimensions embedded in Al matrix were prepared by Arrested Reactive Milling. Different milling times were used to adjust the dimensions of CuO inclusions for different powder samples. Thermal initiation in these powders was studied using thermo-analytical techniques (DSC/TGA), ignition on an electrically heated filament, and ignition by electro-static discharge. In ignition experiments, pressure generated by igniting powders was monitored in addition to the produced optical signatures. Each experimental technique involves a different range of heating rates, varied from $10^{-2}$ to $10^{7}$ K/s. The results were compared to prediction of a recently developed model describing kinetics of heterogeneous Al/CuO reaction. The model was expanded to account for oxygen release by the CuO, which is partially reduced by subsolidus reaction and decomposes upon further heating. The effect of the oxygen release on thermal diffusivity of the composite powder was also considered while interpreting the ignition experiments. This paper will present experimental data, comparisons with the model predictions and discuss implications of the present measurements for development of a comprehensive model describing thermal initiation in Al-CuO nanocomposite materials for a broad range of heating rates.

# (712d) Particle Design of HMX Using Supercritical Anti-Solvent Processes

**Bumjoon Seo**, School of Chemical and Biological Engineering and Institute of Chemical Processes, Seoul National University, Seoul, South Korea, Youn-Woo Lee, School of Chemical & Biological Engineering, Seoul National University, Seoul, South Korea, Byung-Chul Lee, Department of Chemical Engineering and Nano-Bio Technology, Hannam University, Daejeon, South Korea and Hyoun-Soo Kim, High Explosive Team, Agency for Defense Development, Daejon, South Korea

HMX (Cyclotrimethylenetrinitramine) is a widely used explosive for military and civil use. Like all the other explosives, the performance and stability of HMX depend on its particle properties, especially size and morphology. Firstly, among the four polymorphs of HMX, β–HMX is desired because it shows greatest performance and stability due to its highest density and stability. On the other hand, the effect of particle size on the performance and stability of HMX is still controversial. However, preparation of nano-energetic materials itself is of great interest because it is a state of the art technology. Supercritical anti-solvent process can be used for the recrystallization of HMX because nano- or microparticles production is possible using this process. Two widely used supercritical anti-solvent processes, ASES (Aerosol Solvent Extraction System) process and GAS (Gas Anti-Solvent) process, were used in this work. It was found that the morphology and crystal phase of HMX is tunable by using different processes and solvents. β-HMX was obtained by GAS process when acetone, dimethyl sulfoxide , cyclohexanone, and γ-butyrolactone were used as the solvents. Also, it was found that additives can be used to change the crystal phase of HMX. Crystal phase changed from α to γ when PVA (polyvinyl alcohol) was co-precipitated using ASES process where dimethyl sulfoxide is used as the solvent. Under all the experimental conditions for GAS process, size reduction from the original mean particle size of 100μm to nearly 10μm was observed. When ASES process was used, although β-HMX was not obtained, nano- or microcrystals of other forms were obtained. The effects of operating parameters such as $CO_2$ addition rate, temperature, initial solution concentration, and stirring rate on the particle size were investigated for the case of recrystallization into β-HMX using GAS process and acetone as the solvent.

# (712e) Nanostructured Aluminum-Fluorocarbon Particles for Energetic Applications

Travis R. Sippel[1], Steven F. Son[2] and Lori J. Groven[1], (1)Mechanical Engineering, Purdue University, West Lafayette, IN, (2)School of Mechanical Engineering, Purdue University, West Lafayette, IN

Metal-fluorocarbon energetics have been considered in a number of energetic applications. Aluminum-fluorocarbon mixtures are of particular interest due to the high exothermicity of aluminum fluorination. Like many heterogenous energetics, producing materials having desirable reaction rates and ignitability often requires the reduction of diffusion distances, which can be accomplished by mechanical activation (MA). In this ongoing work, we report the modification of aluminum by inclusion of fluorocarbons by MA of micrometer aluminum and fluorocarbon mixtures. The morphological characteristics of composites made using high and low energy ball milling will be discussed. In addition, we characterize ignition and combustion of the composite particles using differential scanning calorimetry, oxygen calorimetry, simple flame tests, camera flash ignition, and other high heating rate experiments. Thermal characterization shows MA Al-fluorocarbon composites exhibit drastically modified heating and ignition behavior that are similar to those of nanoparticle physical mixtures for some MA composites, ignition via camera flash is possible.

# (712f) Combustion of Nanoscale Silicon with Polytetrafluoroethylene

Brandon C. Terry[1], Lori J. Groven[2], Paul Redner[3], Youdong Lin[4], Alexander Mukasyan[5], Deepak Kapoor[6] and Steven F. Son[2], (1)School of Aeronautics and Astronautics, Purdue University, West Lafayette, IN, (2)School of Mechanical Engineering, Purdue University, West Lafayette, IN, (3)Advanced Materials Branch, US Army, RDECOM-ARDEC, Picatinny Arsenal, NJ, (4)Chemical and Biomolecular Engineering, University of Notre Dame, Notre Dame, IN, (5)Chemical Engineering, University of Notre Dame, Notre Dame, IN, (6)Materials Division, US Army RDECOM-ARDEC, Picatinny, NJ

Nanoscale silicon (nSi) based composite energetic materials have the capability to improve fuzes, initiation systems, and high radiant intensity applications. There are also advantages, including higher reaction temperatures and a thinner passivation layer, of using silicon as a replacement for nano-aluminum in many reactive compositions. While various oxidizers have been observed to show promising combustion behavior with nSi, the high heat of fluorination for Si/polytetrafluoroethylene (PTFE) combustion suggests that this system is of particular interest.

In this study, four nSi powders with varied specific surface area, oxygen content, and morphology were investigated and their combustion parameters compared. Specifically, three commercially available nSi powders and highly porous nano-silicon powder synthesized by a combustion synthesis method were studied. Reactive mixtures consisted of nSi, nano-featured PTFE, and FC-2175. Combustion experiments in ambient air were conducted to compare flame temperatures, spectral emissions, and deflagration rates. Visual and infrared spectrometers were used to analyze spectral emissions and determine grey body combustion temperatures. It was demonstrated that all investigated mixtures combust at 1 atm and depending on the nSi, the maximum combustion temperature varies from 2000 to 2600 K and burning rate varies from 0.6 to 1.7 mm/s.

# Optimal Design of Integrated Electric-Power an Algal Biofuel Generation Systems

*C. G. Gutiérrez-Arriaga*[1*], *M. Serna-González*[1], *J.M. Ponce-Ortega*[1], *and Mahmoud M. El-Halwagi*[2,3]

[1]*Facultad de Ingeniería Química, Universidad Michoacana de San Nicolás de Hidalgo, Morelia, Michoacán, 58060, México.*

[2]*Chemical Engineering Department, Texas A&M University, College Station Texas, 77843, USA.*

[3]*Adjunct Faculty at the Chemical and Materials Engineering Department, Faculty of Engineering, King Abdulaziz University, Jeddah, 21589. P. O. Box 80204, Saudi Arabia.*

**Abstract**

This paper presents a multi-objective optimization methodology to determine simultaneously, the optimal design of steam power plants, considering simultaneously economic and environmental criteria, in addition to capturing the $CO_2$ from the combustion of different fossil fuels (coal, oil and natural gas) and biofuels (biogas, biomass, softwood and hardwood) selected in the boiler for the production of microalgae and the subsequent biomass conversion to biofuel (biodiesel), also part of this algal biomass produced is used as biofuel in the boiler. This system allows the sustainable reduction of the huge $CO_2$ emissions in the power plants. The methodology considers the optimal selection of the type of fuel to be used because this affects drastically and simultaneously the economic and environmental functions. The methodology is based on a search procedure through genetic algorithms due to the large number of non-convex terms present and to avoid being trapped in local solutions. The environmental objective function is based on the eco-indicator 99 constituted by harmful materials for the health and the ecosystem to measure the global environmental impact of these systems, following the approach of life cycle analysis (LCA), while the economic objective function considers the maximization of gross annual profit, which includes the profit from the electricity produced, tax credit obtained by the reduction of GHG emissions, the revenue generated by the production of biodiesel, and consider the total annual cost including the fuel costs, turbine, boiler, condenser, pumps, feed-water heaters and deaerator, as well as the costs to operate such equipment (fuel, electricity, cooling water, etc.). To show the applicability of the proposed methodology is considered a case study in Mexico. The results are shown through Pareto curves that help to identify a set of optimal solutions that show the relationship between economic and environmental objectives. The problem can be easily solved using the proposed methodology in a relatively small computation time.

**Keywords**: Life Cycle Analysis; Sustainable Energy; Solar Energy; Biofuels; Greenhouse Gas Emissions; Multi-Objective Optimization; Power Plant.

## Introduction.

At present, the increase in the concentration of $CO_2$ in the atmosphere, as well as other greenhouse gases (i.e. CO, $CH_4$, $N_2O$ and $NO_X$) have caused the so-called global climate change. The overall energy consumption and the associated $CO_2$ emissions is increasing. Due to this problem, the Kyoto Protocol (1998) was signed to reduce the greenhouse gases (GHG) emissions. Furthermore, according to the Intergovernmental Panel on Climate Change (IPCC 2007), the higher $CO_2$ emission sources are power plants based on fossil fuels, where mainly coal and oil are used. Therefore, there is needed to develop alternative methods to obtain the required energy at the minimum costs and without the environmental deterioration.

One of the technologies available to combat the problem associated to the GHG emissions is the $CO_2$ microalgae biofixation. A microalgae biofixation system is suitable for the reduction of carbon dioxide emissions in flue gases, and this is presented as a natural and environmentally friendly alternative, where the biomass produced can be used as source for biofuels production (Chisti, 1980; Nagle and Lemke, 1990; Sawayama et al., 1995). Therefore, this paper presents a new methodology that simultaneously considers environmental and economic aspects, and provides the optimal design of power plant systems. Furthermore, this methodology also allows the optimal selection of the different energy sources such as fossil fuels (coal, oil and natural gas) and biofuels (biogas, biomass, softwood and hardwood and algal biomass) in the boiler, where a microalgae systems is used to sequestrate the $CO_2$ emissions and integrated into a biofuels production system obtaining biodiesel and biomass used as heat source. The environmental impact is measured through the eco-indicator-99 accounting for the life cycle, which considers the indicators for emissions, materials harmful to health and the ecosystem (see Azapagic and Clift 1999). To void get trapped prematurely, the proposed approach is based on a combination of genetic algorithms and a nonlinear programming model, where the simultaneous minimization for the economic and environmental impact is considered.

**Problem Statement.**

The problem is to determine the operational parameters of a steam power plant as shown in Figure 1, where the model must determine the optimal operational conditions for the steam power plant such that the maximum profit is obtained accounting for the simultaneous integration of the CO2 emissions captured in the microalgae system which is optimized simultaneously. Next section presents the main components of the considered model.

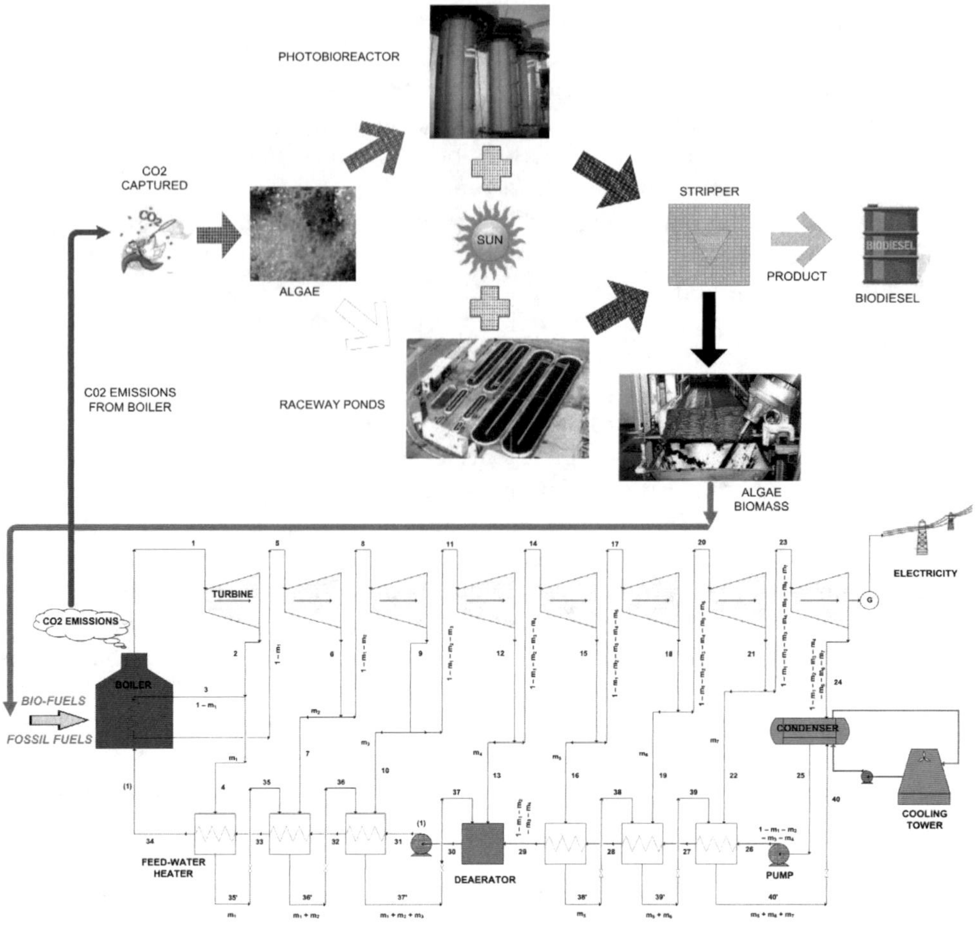


**Fig. 1**. Proposed scheme for the $CO_2$ capture in steam power plants.

## Methodology

### Objective Function

The problem consists of a multi-objective problem, in which the objective function considers two contradict objectives, the first $f_1$ is to maximize the annual gross profit (*GROSS PROFIT*) and the second $f_2$ is the minimization of the environmental impact (*EI*), as follows:

$$Objective\,Function = \{\max f_1;\ \min Ef_2\}$$

### Economic Objective Function

The economic objective function is the maximization of gross annual profit for the design of the steam power system integrated to the microalgae system as follows:

$$Max\,f_1 = \left(PROFIT + TAX\,CREDIT + ALGAL_{Profit}\right) - \left(TAC + ALGAL_{Cost}\right) \quad (1)$$

This objective function considers the profit from the electricity produced in the generator Rankine cycle which results in a economic benefit (*PROFIT*), and tax subsidies (*TAX CREDIT*) obtained by the reduction of GHG emissions by using biofuels and algal biomass as an additional fuel in the boiler, and the revenue generated by the production of biodiesel from algae is also considered ($ALGAL_{Profit}$); minus the total annual cost (*TAC*), which considers the cost of capital for equipment (boiler, turbine, condenser, pump, boiler feedwater and deaerator), the cost of fossil fuel or biofuel used in the boiler and the cost energy used in the Rankine cycle in the pump, moreover, the cost of production of algae ($ALGAL_{Cost}$) accounting for the cost of nutrients, the cost of the system used for the growth of algae (Photobioreactor or Raceway Ponds) and water used in these units.

**Environmental Objective Function**

The environmental objective function is to minimize the overall environmental impact generated by the production of electricity; this is done through the eco-indicator-99, which is based on the methodology of the life cycle analysis. This environmental objective functions accounts for the eco-indicator for the fuels used in the boiler, the construction of the boiler, the construction and operation of the condenser, the cooling tower cooling tower, pumps, and other units, as gollows:

$$
\begin{aligned}
Min\, f_2 = &\left(Ec_1 P_{pump}\right)+\left(Ec_2 P_{fan}\right)+\left(Ec_3 w_P\right)+\left(Ec_4 we_{boiler}\right)+\left(Ec_5 we_{condenser}\right)\\
&+\sum_{t\in T} Ec_t we_t + \sum_{t\in T} Ec_h we_h \qquad (2)\\
&+\left[\sum_{f\in F} Ec_f^{Fossil} F_f^{Fossil} + \sum_{bf\in BF} Ec_{bf}^{Biofuel} F_{bf}^{Biofuel} + \sum_{bm\in BM} Ec_{bm}^{Biofuel} F_{bm}^{Biomass}\right]
\end{aligned}
$$

where $Ec_1$ is the eco-indicator of the energy used in the pump, $Ec_2$ is the eco-indicator of the energy used in the fan of the cooling tower, $Ec_3$ is the eco-indicator of the energy used in the pump of the Rankine cycle, $Ec_4$ and $Ec_5$ are the eco-indicators for the production of stainless steel required by the boiler and the condenser, respectively, $Ec_t$ and $Ec_h$ are the eco-indicators for the production of stainless steel required by the turbine *t*, and heater *h*, respectively, $we_{condenser}$ is the amount of steel in the condenser, $we_{boiler}$ is the steel used for the boiler, $w_P$ is the power needed to operate the feedwater pumps, $P_{fan}$ is the power of the cooling tower fan, $P_{pump}$ is the pumping power of the cooling tower, $Ec_f^{Fossil}, Ec_{bf}^{Biofuel}$ and $Ec_{bm}^{Biomass}$ are the eco-indicators for fossil fuel *f*, biofuel *bf* and biomass *bm*, respectively. Also, $F_f^{Fossil}, F_{bf}^{Biofuel}$ and $F_{bm}^{Biomass}$ are the flowrates of fossil fuel *f*, biofuel *bf*, and biomass *bm*, respectively.

**Variables to be Optimized**

To generate an efficient optimization method, genetic algorithms (GA) together with a nonlinear programming mode as it is shown in Figure 2 are used as a search method.

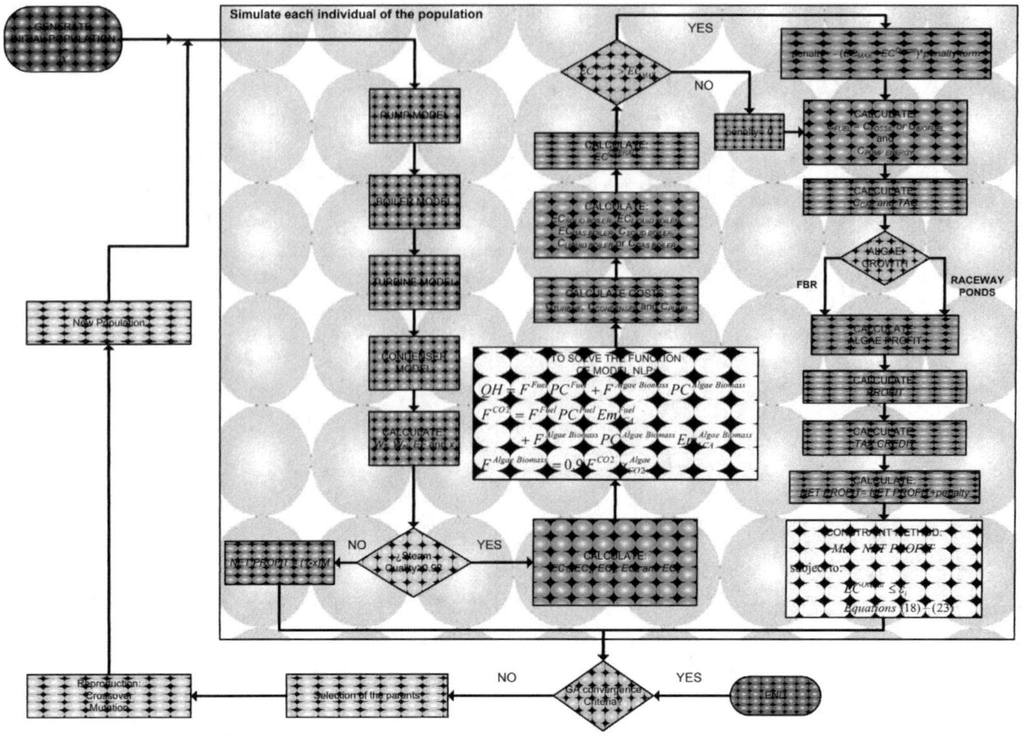

**Fig. 2**. Solution algorithm, using genetic algorithms and NLP.

The modules used in the simulation of steam power plant are solved sequentially for each component of the system (pump, boiler, turbine and condenser). To solve these modules, the proposed method uses several subroutines to obtain the thermodynamic properties of water and steam (i.e. enthalpy, entropy, volume) through the IAPWS formulation (Wagner et al. 2000), this formulation is the most accepted worldwide for calculations thermodynamic properties in the industry for steam power plants.

**Results and Discussion.**

A case study is used to show the applicability of the proposed methodology. Figure 3 shows the results obtained for the case study. Fist Figure 3a shows the Pareto curve for the profit versus eco-indicator for a photo-bioreactor, whereas Figure 3b shows the Pareto curve for Raceway ponds integrated to the steam power plant. These figures are implemented for several values of tax credits for the $CO_2$ reductions. It is noteworthy that in significant economic benefits can be observed in the integrated solutions and at the same time the associated environmental impact is not considerable.

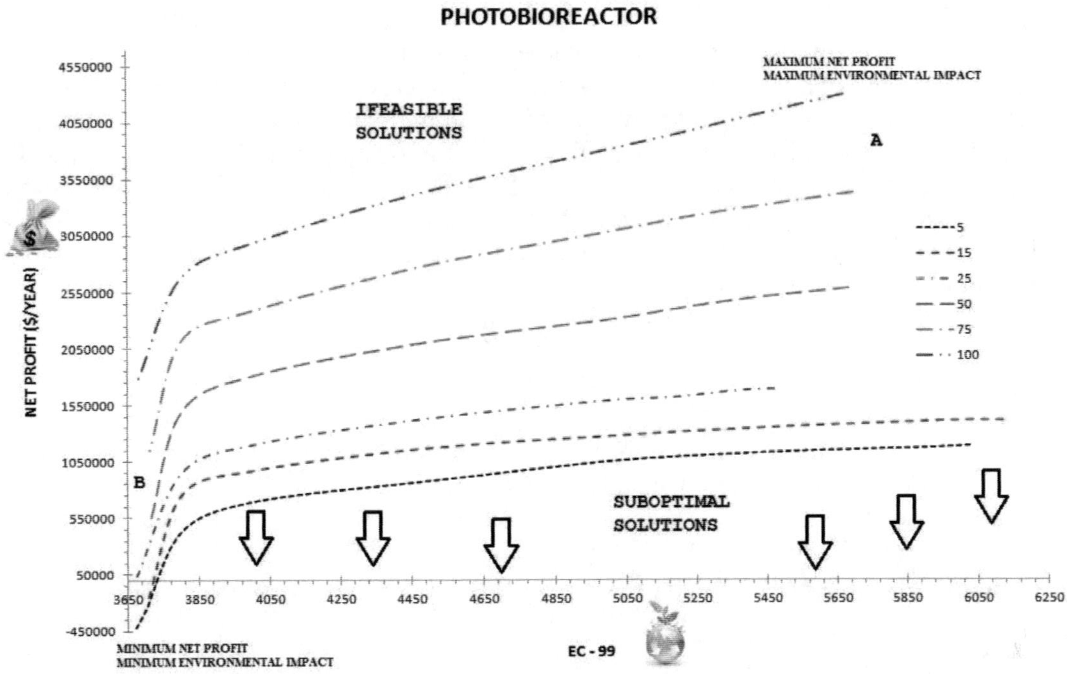


**Figure 3a.** Pareto curve for a *TAX* credit of \$5/ton $CO_2$ eq to \$100/ton $CO_2$ eq for the Photobioreactor.

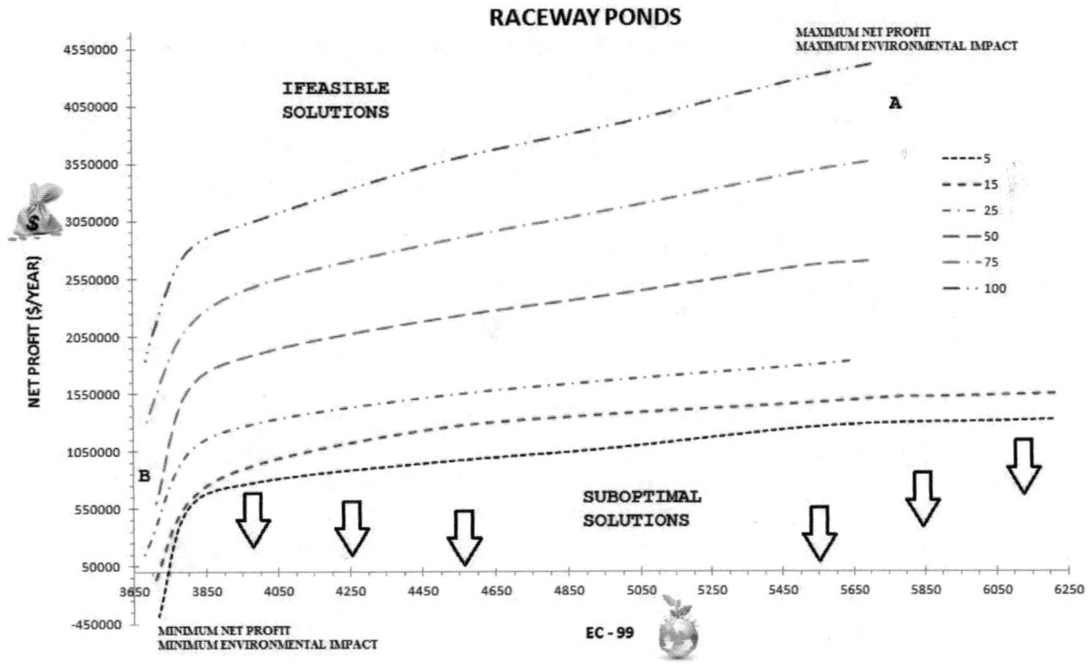


**Figure 3b.** Pareto curve for a *TAX* credit of \$5/ton $CO_2$ eq to \$100/ton $CO_2$ eq for the Raceway ponds.

**Conclusions.**

This paper proposes integrated approach for the design of a steam power system integrated with a microalgae system, where both systems are optimized simultaneously to maximize the net profit and at the same time to reduce the associated environmental impact. To avoid get trapped prematurely in a suboptimal solution, a solution approach based on GA and NLP is implemented. The thermodynamic correlations as well as the design equations are based on rigorous models. Results show that it is possible to get good economic solution and at the same time to reduce the associated environmental problems through the proposed methodology. Also, the results can be represented in Pareto curves where the decision maker can select the solution that best compromise between the contradict objectives shows.

## References


Azapagic A, Clift R (1999) Application of life cycle assessment to process optimization. Comput Chem Eng 23:1509-1526.

Benemann JR (1997) CO2 mitigation with microalgae systems. Energy Convers Manage 38:S475-S479.

Bruno JC, Fernandez F, Castells F, Grossmann IE (1998) A rigorous MINLP model for the optimal synthesis and operation of utility plants. Chem Eng Res Des 76:246-258.

Chisti Y (1980) An unusual hydrocarbon. J Ramsay Soc 27-28:24-26.

Chisti Y (2008) Biodiesel from microalgae beats bioethanol. Trends Biotechnol 26(3):126-131.

Goldberg DE (1989) Genetic algorithms in search, optimization, and machine learning. Addison-Wesley, Reading, MA.

IPCC (2007) Intergovernmental Panel on Climate Change. IPCC fourth assessment report: Climate Change.

Liska AJ, Yang HS, Bremer V, Walters DT, Erickson G, Klopfenstein T, Kenney D, Tracy P, Koelsch R, Cassman KG (2009) BESS: Biofuel energy systems simulator; Life cycle energy and emissions analysis model for corn-ethanol biofuel. Vers. 2008.3. 1, www.bess.unl.edu.

Molina Grima E, Acién Fernández FG, García Camacho F, Chisti Y (1999) Photobioreactors: light regime, mass transfer, and scale up. Progress Ind Microbiology 35:231-247.

Nagle N, Lemke P (1990) Production of methyl-ester fuel from microalgae. Appl Biochem Biotechnol 24-25:355-361.

Protocolo de Kyoto de la Convención Marco de las Naciones Unidas sobre el Cambio Climático. Naciones Unidas, 1998. Disponible en: www.unfcc.int.

Sawayama S, Inoue S, Dote Y, Yokoyama S-Y (1995) CO2 fixation and oil production through microalgae. Energy Convers Manag 36:729-731.

Wagner W, Cooper JR, Dittmann A, Kijima J, Krestzschmar HJ, Kruse A, Mares R, Oguchi K, Sato H, Stocker I (2000) The IAPWS industrial formulation 1997 for the thermodynamic properties of water and steam. J Eng Gas Turb Power 122:150.

Wang M, Wu Y, Elgowainy A (2007) Operating manual for GREET: Version 1.7. Center for Transportation Research, Energy Systems Division, Argonne National Laboratory, Iowa.

# (716b) Degradation of of Rhodamine B Dye Over Mn-Doped Titania Nanotubes

Manohar Palla[1], Rakesh Sadu[2], R. Asapu[2], George W. Spencer[2], Jewel Andrew Gomes[3], Zhanhu Guo[3] and **Daniel H. Chen**[3], (1)Dan F. Smith Dept. of Chemical Engineering, Lamar University, Beaumont, TX, (2)Chemical Engineering, Lamar University, Beaumont, TX, (3)Dan F. Smith Department of Chemical Engineering, Lamar University, Beaumont, TX

Mn-doped titania nanotubes were fabricated by solution combustion synthesis followed by the hydrothermal treatment in the removal of Rhodamine B, a dye. Combustion synthesis has emerged as a facile and economically viable technique for the preparation of advanced ceramics, catalysts and nanomaterials. Stoichiometric amounts of transition metal nitrates and fuel (Glycine) are mixed in a solution for the preparation of nanomaterials by using the single step solution combustion method at 350° C. The $Mn_xTi_{(1-x)}O_{(2-\delta)}$ (x ranges from 0.05-0.25) nanomaterials are made in the anatase phase by the solution combustion synthesis. The nanotubes are then fabricated using the hydrothermal method using 10 N NaOH at temperature of 130°C for 24 hours.

The prepared Mn-doped titania nanotubes were characterized using JEOL 2100 field emission Gun Transmission Electron Microscope (JEM 2100F TEM) and HITACHI S-3400 N Scanning Electron Microscope (SEM) attached with Energy Dispersive X-ray Spectrometry (EDX). X-ray diffraction (XRD) analysis was performed on a Bruker D8 Focus diffractometer equipped with a Sol-X detector using a copper radiation source, and specific surface area analysis was done using the Brunauer, Emmett, and Teller (BET) method based on $N_2$ adsorption at 77 K with a NOVA 1000 series analyzer. The thus prepared Mn-doped titania nanotubes contain both monoclinic and anatase phases as evidenced by the XRD and EDX results. The surface area for the nanotubes decreases with increased doping of Mn. Further, TEM analysis shows the formation of nanotubes, which are open ended and are approximately 100-120 nm length with inside and outside diameter estimated to be 6 nm and 12 nm.

The UV-Vis spectrum for the degradation of 4 ppmw Rhodamine B (RhB) over the 0.1 g (0.067 wt %) $Mn_{0.15}Ti_{0.85}O_{(2-\delta)}$ nanotubes shows the successive blue shift from the absorption peak at 554 nm (Rhodamine B) to 498 nm (Rhodamine), indicating the stepwise N'de-ethylation from dark pink to light green. All the experiments were conducted in dark without external energy supply. Degradation of Rhodamine B follows a second order kinetics as shown by the linear plots between $1/C_a$ vs time. 15% Mn is the optimum doping with the hioghest degradation rate constant. The activity of $Mn_{0.15}Ti_{0.85}O_{(2-\delta)}$ increases with temperature with an activation energy of 40777 kJ/kmol for the degradation of RhB.

# (716c) Bio-Inspired Green Approach to Manufacture Fibers

**Kadhiravan Shanmuganathan**, Robert K. Sankhagowit, Prashanth Iyer and Christopher J. Ellison, Department of Chemical Engineering, The University of Texas at Austin, Austin, TX

Recent advancements in fiber production techniques have facilitated new applications for fibrous materials in diverse fields such as optoelectronics, regenerative medicine, piezoelectrics, ceramic materials, etc. Current polymer fiber manufacturing methods almost always use solvents or heat to lower polymer viscosity for processing. A greener approach to making fibers remains a challenge. However, spiders and silkworms have developed benign ways of making silk fibers with high strength and toughness. Their approach of chemically linking small functional units into long chain molecules and solid fibrillar structures while simultaneously extruding the fibers is fundamentally different from current synthetic fiber manufacturing methods, where extrusion of pre-formed long chain polymers is facilitated with organic solvents or heat. Drawing inspiration from nature, a method will be described which uses light to trigger a thiol-ene chemical interaction to rapidly transform small reactive liquid mixtures into solid thread-like structures as they are forced out of a capillary at high speeds. Besides being manufactured without using solvents/volatile components or heat, these fibers are mechanically robust and have excellent chemical and thermal stability.

# (716d) Reducing Water Usage in Bioprocessing

**Demetri Petrides**, INTELLIGEN, INC., Scotch Plains, NJ and Charles Siletti, INTELLIGEN, INC., Mount Laurel, NJ

Water is the primary material used in bioprocessing. While water is not the most expensive raw material the availability of large amounts of purified water and subsequent waste treatment capacity may limit site selection opportunities. This presentation uses bioprocess simulation to explore several strategies. Single-use systems, for example, provide significant reductions in water usage, but there is a practical scale limit. Platform selection can have a significant impact on water usage. For example microbial product expression may be less water-intensive than cell culture. For the majority of products, water reduction can be accomplished through process improvements. Titer optimization, can reduce upstream water usage by 40%. Single-use technologies allow for considerable water savings by eliminating the need for equipment and line cleaning. Single use systems have practical size limitations. Buffer and media preparation strategies like in-line dilution reduce vessel size and resulting water consumption. Alternative downstream technologies can also provide significant reductions in water usage.

# (716e) Green Approach to Produce Bio-Jet Fuel From Microalgae

**Marian Elmoraghy** and Ihab Farag, Chemical Engineering, University of New Hampshire, Durham, NH

**Green Approach to Produce Bio-jet Fuel from Microalgae**

Marian Elmoraghy, Ihab H. Farag*

Chemical Engineering Department, University of New Hampshire, Durham, NH

* Correspondence, ihab.farag@unh.edu

**Introduction**

Renewable jet fuel for the aviation industry, also termed bio-jet fuels could reduce flight-related greenhouse-gas emissions by 60 to 80 percent compared to fossil fuel based jet fuel. The green bio-jet fuel is made by blending microalgae bio-fuels with conventional petroleum-derived jet fuel to provide the necessary specification properties. The major advantage of using microalgae oil for biodiesel is the high oil production capacity by microalgae, as they could produce up to 58,700 L oil per hectare.

Sustainable production of bio-jet fuel requires minimizing the energy requirements and reducing fresh water usage while simultaneously lowering the production cost of bio-jet fuel. These are challenges to the biodiesel production from microalgae. Considerable efforts have been made to develop efficient and cost-effective photobioreactors for microalgae growth. Yet, the high cost of installing and operating artificial light sources in conventional photobioreactors with artificial illumination systems remains a major problem.

**Objectives**

The goal of this research is to develop an economical process to produce bio-jet fuel while minimizing the energy requirements, reducing water usage and meeting the jet fuel specifications. To accomplish this goal the following objectives were defined. 1- Investigate minimizing the energy requirements by replacing fluorescent lights with light emitting diodes (LEDs); 2- Investigate the use of municipal waste water in growing microalgae to reduce fresh water usage; 3- Study the one-step production of biodiesel using in situ algal biomass transesterification process to reduce production time and cost, and 4-investigate biodiesel blending with jet-fuel in order to obtain bio-jet fuel that has specific properties.

**Microalgae Growth Using LEDs:**

LEDs are lighter than the fluorescent lights and small enough to fit into virtually any photobioreactor (PBR). Other advantages of LEDs include a longer life-expectancy, lower heat generation, and a greater tolerance for switching on and off. LEDs use less electric energy than fluorescent light to produce the same light intensity hence their use lowers the energy requirement for algae growth, and make the process greener by eliminating the CO2 emitted in generating the excess electric power needed for the fluorescent light.

**Production of bio-jet fuel:**

There are three routes to produce bio-jet fuel from microalgae. Figure 1 shows these three routes. The first route involves using microalgae oil to produce Bio-SPK (Bio derived synthetic paraffinic Kerosene) by cracking and hydro-Processing. This can be used for kerosene-type fuels include jet A, jet A-1, JP-5 and JP-8. FT-SPK (Fischer–Tropsch Synthetic Paraffinic Kerosene) is the second route which involves pyrolysis of solid biomass to produce pyrolysis oil or gasification to produce a syngas which is then possessed into FT SPK. The third route involves algae growth, harvesting, oil extraction and transesterification or (in situ process) to produce biodiesel. The biodiesel will be blended with Jet fuel to produce bio-jet fuel.

**Approach:**

One of the purposes of this research is to improve energy efficiency of the microalgae cultivation in photobioreactor by comparing LEDSs with fluorescent light sources, and the use of wastewater versus reverse osmosis (RO) water. Chlorella Vulgaris microalgae were grown in wastewater and in RO water. Two LED color panels were used for this comparison: a red panel and a red-blue Panel. The measured light intensities were maintained at a value of 2000 Lux. The absorptivity and the cell counts measurements were recorded while the algae were growing using the spectrophotometer and the microscope. Then, the algae were harvested after the maximum growth was reached. The biomass freeze dried algae were determined after centrifugation and freeze drying processes. Research is still progressing towards production of biodiesel through the in situ process and production of bio-jet fuel.

**Discussion:**

The highest algae growth (3.2 g freeze dried algae/L over a growth period of 18 days = 178 mg. freeze dried algae/L-day) was obtained using red-blue LEDs and RO water. The lowest algae growth (1.21 g of freeze dried algae/L over a growth period of 18 days = 67 mg freeze dried algae/L-day) was obtained using fluorescent light and waste water. Algae growth (3.2 and 1.63 gm. freeze dried algae/L over a growth period of 18 days = 178 and 91 mg. freeze dried algae/L-day) were obtained using RO water and waste water respectively with red-blue LEDs. Also, Algae growth (1.95 and 1.25 gm. freeze dried algae/L over a growth period of 18 days = 108 and 69 mg. freeze dried algae/L-day) were obtained using RO water and waste water respectively with red LEDs. The results indicate that red-blue LEDs are more effective than red LEDs using either RO water or waste water. Since the power requirements of the fluorescent light and the LEDs are 68 and 45 Watts respectively, the total saving energy by LEDs is about 34%. Light capture efficiencies of LEDs and fluorescent light, based on the heating value of the algae produced compared to the incident light energy over the growth period, are 0.72% and 0.48% respectively.

*Figure 1 Production of Bio-jet fuel through three different routes.*

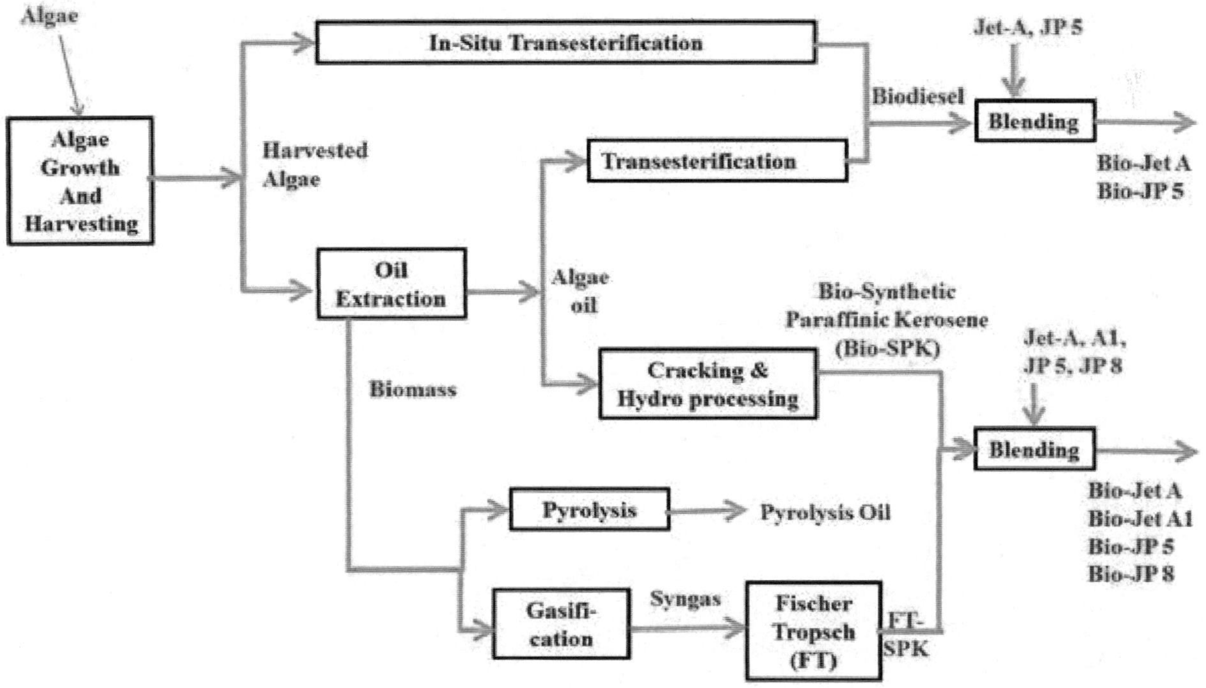

# (716f) Impact of Floc Size On Ethanol Tolerance of the Self-Flocculating Yeast SPSC01 and Underlying Mechanism

Chuang Xue and Feng-wu Bai, School of Bioscience and Biotechnology, Dalian University of Technology, Dalian, China

High cost is the biggest challenge in fuel ethanol production, in which energy consumption comprises about 30%, the second largest only after feedstock consumption, and thus saving energy consumption is one of the most effective strategies to make fuel ethanol more economically competitive. Very high gravity (VHG) fermentation can significantly increase ethanol titer in the fermentation broth, which not only saves energy consumption for the downstream distillation but also for the waste distillage treatment due to its significant reduction. However, high ethanol concentration severely inhibits yeast cells, and stuck fermentation frequently occurs, with more sugars unfermented and ethanol yield compromised correspondingly. Therefore, exploration of the mechanism underlying the ethanol tolerance of yeast cells is one of the prerequisites for developing effective VHG ethanol fermentation process.

The impact of the size of the yeast flocs on the ethanol fermentation was studied by controlling their size at 100, 200 and 300 mm, which was characterized by their average chord monitored by the focused beam reflectance measurement. The experimental results illustrated that the yeast flocs with the size of 300 mm exhibited maximum ethanol production, ethanol tolerance and glucose consumption, with 110.0 g $l^{-1}$ ethanol produced, 58.0% viability remained for the yeast cells after 20% (v/v) ethanol shock for 5 h and glucose uptake rate of 286.69 C-mmol $l^{-1}$ $h^{-1}$. Moreover, the retention of the self-flocculating yeast was improved for the yeast flocs with relatively large size and higher biomass concentration was achieved within the bioreactor, contributing to more glucose consumed and carbon flux to ethanol production. The decrease in the fluxes to the growth of the yeast cells such as DNA, RNA, protein and lipid biosynthesis as well as the increase of ATP consumption for the maintenance indicated that improved ethanol production resulted from the increased yeast floc size raised more energy burden on the yeast cells, and glucose uptake of the yeast flcos was compromised under VHG conditions, making the improvement of glucose uptake rate the key to improve VHG ethanol fermentation.

# SUBSTANTIAL INVOLVEMENT BY DOE TO ADDRESS SCALE-UP CHALLENGES FOR BIOREFINERIES

*Gene Petersen and Glenn Doyle, US Department of Energy, Golden, CO; Art Wiselogel, CNJV, Golden, CO; Carol Babb, SAIC, Denver, CO; and Andras Marton, Independent Project Analysis, Inc., Ashburn, VA.*

## Extended Abstract

Since 2003, the Department of Energy (Department) has been substantially involved in cost-sharing the development of over thirty integrated biorefinery projects to process biomass feedstocks into transportation fuels and biobased products. These projects range in scale from developing engineering designs for pilot operations, to constructing integrated pilot- and demonstration-scale biorefineries, as well as commercial biorefinery demonstrations. Under direction of public law (Energy Policy Act of 2005), the Department was authorized to "carry out a program to demonstrate the commercial application of integrated biorefineries[1]." Under that mandate and subsequent law (e.g., the American Recovery and Reinvestment Act of 2009), the Department's Biomass Program conducted three separate competitive Funding Opportunity Announcements (FOAs) from 2006-2009 to solicit applications from the community to develop integrated biorefineries at these various scales. This presentation provides a brief overview of the careful selection process employed to meet the mandates of public law and describes how the Department utilizes project management best practices to address the challenges of scale-up and completion of these projects. An oft voiced criticism is that the federal government tries to pick technology winners. In fact, the Department's overall objective is to foster development of a robust, commercially-competitive biorefinery industry by selecting a broad suite of technology types that appear technically mature enough to be successful, not to advance any particular technology. A further objective is to reduce the risk associated with designing, constructing and operating first-of-a-kind biorefinery demonstrations such that the return on the investment can become attractive to private investment. In this vein, the overall benefit of the industry-federal partnerships will be discussed along with the practices and processes used to help both recipients of federal funds and the Department achieve successful development of Integrated Biorefineries.

Among the key best practices the Department has utilized to assist in the scale up of these new technologies and to manage the competitive process and oversight of these projects are the following:

### *Competitive Process and Selection*

- Utilized competitive FOAs that identified the Department's minimum feedstock throughput and biofuels/bioproduct production threshold objectives and outlined the evaluation criteria for selecting integrated biorefinery projects. This allowed the industry the flexibility to propose projects intended to meet the criteria. In all three FOAs, the Department did not specify any particular technology or approach to meeting the objectives especially since the FOAs acknowledged the different levels

[1] Energy Policy Act of 2005, §932

of technology maturity by accepting applications for facilities ranging from pilot to demonstration scale.

- Conducted comprehensive merit reviews that compared responsive proposals against the published evaluation criteria using non-conflicted experts to determine the competitive range (i.e., technically meritorious applications) to be recommended to the Selection Official.
- Independent Project Analysis, Inc. prepared risk assessments on responsive proposals to compare them to its industry project management best practice benchmark database to determine each proposal's risk profile and state of readiness to begin construction.
- Program policy factors as outlined in each FOA were then used to assess the balance of the Biomass Program's integrated biorefinery investment portfolio and, along with the merit review results, recommendations for selection were submitted to the Selection Official.
- Based on the above factors and availability of Federal funds, the Selection Official made selections and notified the Contracting Officer to negotiate financial assistance awards.

***Post-selection Award and Project Management Practices of DOE***

- Upon selection, DOE chose to execute Federal financial assistance agreements instead of contracts to allow for greater flexibility in adjusting to technological, financial and project management challenges.
- The Biomass Program adapted and implemented relevant sections of Department of Energy Order 413.3B, *Program and Project Management for the Acquisition of Capital Assets* (DOE Order 413.3B) to the management of integrated biorefineries, which are process and capital intensive projects. Figure 1 compares the guidelines in DOE Order 413.3B with the development of a typical capital construction project and DOE's substantial involvement.
- DOE employed the industrial consulting firm, Independent Project Analysis, Inc. to comprehensively compare each selected proposal against its project management best practices database to assess the project cost and schedule risks and to provide the Department with an estimate of the amount of contingency needed to account for the identified risks.
- DOE's Golden Field Office contracted with RW Beck (now SAIC) to be the Department's Independent Engineer (IE) to assist in review of initial applications, and then evaluate and monitor progress on the selected projects. Also, the IE performs independent performance test analysis to assess actual biorefinery operations as compared to the target performance criteria. The IE's support to the Department is comparable to the role it would fulfill for a private financial institution involved in the project.
- Best project management practices were invoked, where applicable, to strengthen the Department's and the recipient's controls over the execution of the project. This includes the use of:
    - Separate budget periods to control the flow of federal cost share funds
        - Budget period 1: Engineering, Front-end loading design 2 and 3, National Environmental Policy Act analysis, permitting, secure intellectual property, secure site location and infrastructure needs,

        - secure feedstock agreements, identify and secure off-take agreements, and secure non-Federal cost share commitments
        - Budget period 2: Final design, construction, commissioning, startup
        - Budget period 3: Operations
    - DOE Critical Decision(CD) points
        - CD-2: Approve the performance baseline
        - CD-3: Approve start of construction
        - CD-4: Approve start of operations
    - Project management plans that invoke a Work Breakdown Structure framework and identify risks and barriers to be overcome
    - Resource-loaded Gantt charts and critical path schedules
    - Milestones
    - Go/No go decision points
    - Stage-gates to address technological maturity challenges
    - Earned value management
    - Responsible, Accountable, Consulted, Informed (RACI) charts
    - Risk Mitigation Plans and Risk registers
    - Contingency
    - Continual review of lessons learned to facilitate improving the process of federal and industry co-development of these biorefineries.

***Examples of Good Practices and Lessons Learned***

- The foremost challenge experienced by integrated biorefineries is successfully and continuously feeding solid biomass to the throat of the reactor. Adaptation of "commercial" technology to these first-of-a-kind technologies has been plagued by numerous challenges, resulting in changes in design, materials of construction, operating parameters, etc.
- The value of integrated pilot plant tests for scale up purposes cannot be overemphasized: For these first-of-a-kind demonstration- and commercial-scale IBRs, integrated pilot plant tests have proven to be invaluable in validating the design and establishing a credible performance baseline that controls scope creep, re-engineering, and minimizes change control actions during construction, all of which increase schedule delays and capital costs. Without the benefit of fully integrated pilot plant data to inform the demonstration or commercial designs, the risks to IBR cost and schedule are magnified. This reinforces key guidance provided to DOE by IPA, Inc.
- A major lesson learned is not to "assume" that a commercially available feed system will successfully integrate with a conversion technology. Similarly, unless "commercial" technology has been successfully integrated with new process technology at an appropriate scale before being commercialized, it cannot be assumed that it will work at the larger scales.
- The more new technology process steps, the higher the risk to successfully executing the project. A corollary to this would be, the more new technology process steps, the greater the need for fully integrated pilot plant tests of durations sufficient to validate and "lock" the demonstration- or commercial-scale design.
- Convening the appropriate team with the right skill and resource mix to successfully execute major capital projects depends on the stage of technical and organizational maturity: Oftentimes, due to the perceived need to protect intellectual property rights

and/or to keep costs at a minimum, R&D/entrepreneurial entities try to take on the challenges of scaling-up bench- or pilot-scale projects without having the necessary skills mix and resources to succeed. Delays in project schedules and cost increases often reflect the inability to quickly ramp up on the learning curve to execute major capital projects. R&D and entrepreneurial organizations need to educate themselves on the risks associated with scaling and plan to secure the appropriate skills and experience mix, either internally or through contracts, to improve their chances of success.

- Key R&D and/or pilot plant testing not complete: The urgency to be "first to market" sometimes results in R&D and/or integrated pilot plant testing not being performed or completed in order to inform a comprehensive ±10% design upon which to base EPC bids that are competitive and less likely to experience significant change actions.
- Adopting formal project management best practices improves the chances of success: Once the decision is made to build a demonstration or commercial biorefinery, the early adoption of formal project management practices and best practices will significantly strengthen the chances for a successful outcome. Assuring alignment with business objectives, establishing a well-defined performance baseline (scope, schedule, cost and risk), managing to this baseline, and controlling scope creep will greatly enhance the execution of the project.
- Pre-award meetings to clarify expectations for negotiating awards and meeting federal requirements were very valuable and provided companies with better roadmaps for proceeding with their project.
- Regular teleconferences (weekly or fortnightly) and face-to-face meetings were valuable tools to facilitate negotiations and monitor progress.
- Utilizing separate budget periods aligned with Critical Decisions (e.g., Budget Period 2 aligned with Critical Decision 3 – Approve Start of Construction) approach provided the requisite time needed to work out issues prior to actual start of construction. In one case, where a Budget Period 1 approach (see above for description of the activities typically involved in BP1) was not used, the project did not achieve successful operation.
- Annual comprehensive project reviews that involved DOE and the recipients allowed DOE to be better informed about actual progress and allowed DOE to provide some useful lessons from the collective experiences of other biorefinery efforts.

***Benefits of Industry-Federal Partnerships***

- Due to the reluctance of the private financing industry to invest in the "first" plant, the Department's overall objective in cost sharing in pilot-, demonstration- and commercial-scale integrated biorefineries is to "buy down" some of that inherent risk, advance the industry closer to being competitive with fossil-based fuels, and help the industry attract private investment for the 2$^{nd}$ through Nth plants.
- Through these financial assistance agreements, the Department and industry have demonstrated that the Federal government can play an effective role in seeding development of technologies that can help the nation move closer to energy independence while creating domestic employment opportunities.

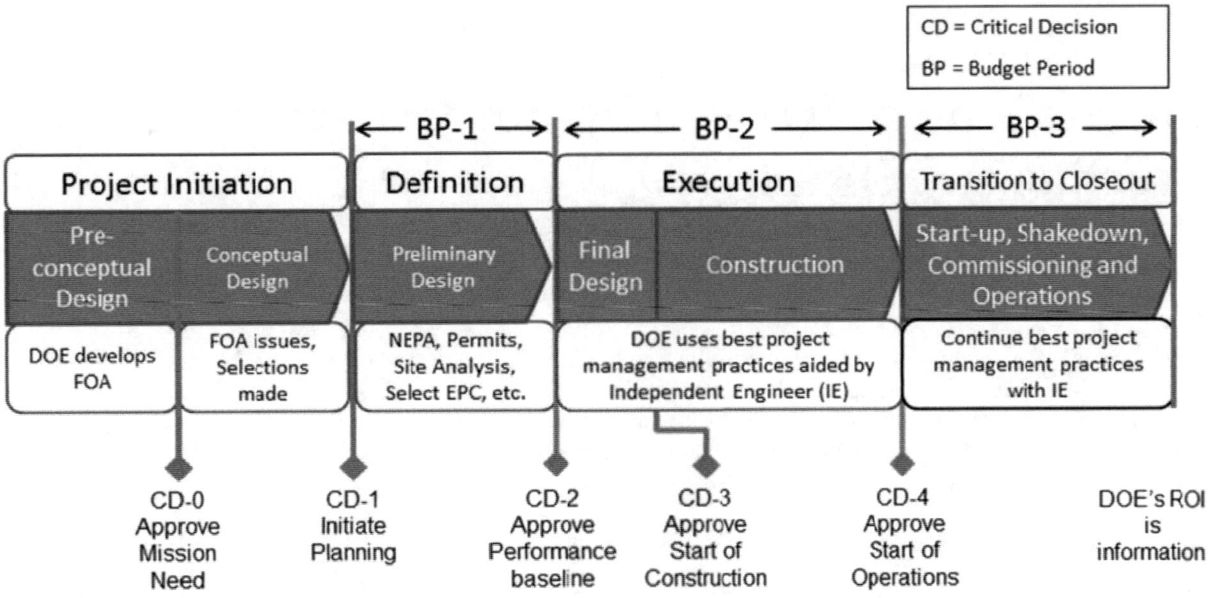


Figure 1 – Integration of Project Management for Biorefinery Development

# (780b) Scale up of IH2 Process for Converting Biomass to Gasoline and Diesel

Terry Marker, Martin Linck, Michael Roberts, Pedro Ortiz-Toral, Jim Wangerow and Larry Felix, Gas Technology Institute, Des Plaines, IL

The $IH^2$ process is a new technology which directly converts biomass such as wood, cornstover, and lemna, into high quality liquid hydrocarbon transportation fuels using hydropyrolysis. Initial work was done on a small bench scale semi-continuous unit, which processes 0.45kg/hr of biomass. Technoeconomic and LCA studies of a commercial process based on these bench tests is very favorable, showing that gasoline and diesel can be produced from biomass at less than $2.00/gallon and have greater than 90% greenhouse gas reduction compared to fossel fuels.

Now the $IH^2$ process has been scaled up to a continuous unit which processes 50 kg/day of biomass. The new pilot plant unit helped us define key design parameters and also determine catalyst life and stability. GTI will discuss controlling process variables and show continuous data from the 50kg/day pilot plant. The 50kg/day tests lay the ground work for our design of a 50 t/day demo $IH^2$ unit which should lead directly to 2000 t/day commercial scale process in 2015-2016.

# (780c) Piloting a Multiproduct Biorefinery

**James J. Foster**, Archer Daniels Midland Company, Decatur, IL and Michael Erpelding, Research, ADM, Decatur, IL

For some biorefineries, a single product is the goal of the project. Sometimes it is ethanol or an oil or perhaps a chemical compound, but it is focused on one product that will make or break the project. While it is good to be focused, it can lead to tunnel vision that can prematurely crash a project.

We believe a better method is run the biorefinery as having a number of co-products which allow a number of different outlets for the conversion of biomass. In our current biomass project, partially funded by the Department of Energy, we can produce ethanol, butyl acrylate, fiber and lignin. Butyl acrylate can be used as a drop in replacement for current petroleum based acrylates, fibers can be used for further conversions, and the lignin can be used as fuel source for power or steam generation while the fibers can be used as a feedstock for further conversation or sold as animal feed.

As part of the presentation we will list some of the challenges with the start-up of the pilot plant and the concerns of building a full scale plant including the technical issues and regulatory issues.

# (780d) Scale-up of Ethanol Production From Sweet Sorghum Stems Using Advanced Solid-State Fermentation Technology

**Shi-zhong Li**, Institute of Nuclear and New Energy Technology, Tsinghua University, Beijing, China and Lei Zhang, Institute of New Energy Technology, Tsinghua University, Beijing, China

Compared with conventional submerged liquid fermentation (SLF), solid-state fermentation (SSF) can considerably reduce the process costs of substrate pretreatment and water usage, avoid the pollution problems resulted from the disposal of liquid residues. However, due to the inherent barriers of heat and mass transfer, SSF has not yet proved to be feasible in large-scale ethanol production. Accordingly, Tsinghua University had developed and operated a process of advanced solid-state fermentation (ASSF) at 127 $m^3$ scale using fermentable sugars from sweet sorghum stems in Inner Mongolia, China, in 2010. The field trial achieved high ethanol productivity that were comparable to conventional SLF, indicating that mass ethanol production in SSF with lower costs and less pollution problems has the great potential for its industrial application. In this study, to further optimize the process of ASSF, an improved rotary-drum fermenter at 555 $m^3$, the largest solid-state fermenter in the world so far, was constructed and deployed recently in Inner Mongolia, China, which was operated as continuous way for high process efficiency. During the process, the solid substrate and seed culture continue to be fed into the reactor and the fermented residues was conveyed immediately into distiller, making the whole process keep running without interval that improve the process efficiency to a great extent. Consequently, the continuous process achieved ~91% of theoretical ethanol yield that was parallel to 127 $m^3$ scale in the batch fermentation. The retention time of solid substrate (fermentation time) was only less than 30 h. In the meantime, tech-economic analysis (TEA) from National Renewable Energy Laboratory (NREL) indicated that 16 tons of sweet sorghum stems (with 13.5 % sugar content and 70% moisture) can be used to produce 1 ton ethanol; the fermented residues can be used to feed 1 cattle; the fuel ethanol production cost is \$615.4/ton (49cent/litter) and the sweet sorghum stem price is at \$30/t. In this study, the results of continuous fermentation and distillation at 555 $m^3$ scale are agree with the data from bench scale and pilot plant well. The TEA further demonstrates that ASSF is a cost-effective process that is highly feasible to be applied at industrial and commercial scale for fuel ethanol production, and eventually is able to improve the biofuel industry.

# (780e) Evaluation of Bio-Oil Corrosion Characteristics

**Patrick Johnston**, Center for Sustainable Environmental Technologies, Iowa State University, Ames, IA and Robert C. Brown, Mechanical Engineering, Iowa State University, Ames, IA

**Evaluation of Bio-oil Corrosion Characteristics**

Patrick A. Johnston[1] and Robert C. Brown[1,2]

[1] *Center for Sustainable Environmental Technologies;* [2] *Department of Mechanical Engineering*

*Iowa State University*

*Ames, IA 50011*

Petroleum crude oils contain naphthenic acids that are one of the main causes of corrosion in the petroleum industry during storage, transport, and refining. "Naphthenic Acid Corrosion" NAC has been related to three parameters: Total Acid number (TAN), temperature, and flow rate [1, 2, 3, 4]. Bio-oils do not contain naphthenic acids but do however contain other organic acids such as acetic, formic, glycolic and propanoic acid that may have similar consequences to metal surfaces. A comprehensive corrosion study needs to be completed to determine the actual effects of the organic acids in bio-oil on select metal surfaces. As a result of this study, acid number and corrosion potential of pyrolysis bio-oils can be correlated and ranked to produce a fraction that is non-problematic with typical petroleum refining equipment.

The proposed project will evaluate and compare corrosion and acid number (AN) on fast pyrolysis bio-oil stage fractions produced at Iowa State University (ISU). The corrosion characteristics will be evaluated using a vacuum distillation apparatus that will simulate a distillation tower in a petroleum refinery. Metal specimens will be subjected to both the liquid and vapor phase of the bio-oil. Method D664 "Standard Test Method for Acid Number of Petroleum Products by Potentiometric Titration" will be used to determine acidy of the bio-oil samples. Ion Chromatography (IC) and High Performance Liquid Chromatography (HPLC) will be used to quantify organic acid concentration. The corrosiveness will be ranked gravimetrically according to mass loss during the study. Selected metal specimens will also be examined by Scanning Electron Microscopy (SEM) equipped with energy-dispersive spectroscopy (EDS) to determine morphological and composition changes on the metal surfaces.

**References:**

1. A. Jayaraman, H. Singh, and Y. Lefebvre, Rev. Inst. Fr.Pet.**41**, 265 (1986).

2. C. M. Cooper, Hydrocarbon Process.**51**, 75 (1972).

3. E. Babaian-Kibala, H. L. Craig, G. L. Rusk, et al., Mater.Perform. **32**, 50 (1993).

4. E.B. Zeinalov, V.M Abbasov, and L.I. Alieva, Petro. Chem.**49**, 185 (2009).

# AUTHOR INDEX

# AUTHOR INDEX

# AUTHOR INDEX

# AUTHOR INDEX

# AUTHOR INDEX

# AUTHOR INDEX